Application of Biology to Cultural Heritage

Application of Biology to Cultural Heritage

Editors

Maria Filomena Macedo
António Portugal
Ana Catarina Pinheiro
Ana Zélia Miller

MDPI • Basel • Beijing • Wuhan • Barcelona • Belgrade • Manchester • Tokyo • Cluj • Tianjin

MDPI

Editors

Maria Filomena Macedo
Universidade NOVA de Lisboa
Portugal

António Portugal
University of Coimbra
Portugal

Ana Catarina Pinheiro
Évora University
Portugal

Ana Zélia Miller
Évora University
Portugal

Editorial Office
MDPI
St. Alban-Anlage 66
4052 Basel, Switzerland

This is a reprint of articles from the Special Issue published online in the open access journal *Applied Sciences* (ISSN 2076-3417) (available at: https://www.mdpi.com/journal/applsci/special_issues/biology_cultural_heritage).

For citation purposes, cite each article independently as indicated on the article page online and as indicated below:

LastName, A.A.; LastName, B.B.; LastName, C.C. Article Title. *Journal Name* **Year**, *Volume Number*, Page Range.

ISBN 978-3-0365-3315-5 (Hbk)
ISBN 978-3-0365-3316-2 (PDF)

Cover image courtesy of Maria Filomena Macedo

Contents

About the Editors

Maria Filomena Macedo (Associate Professor). Filomena Macedo is a Professor of "Biology in Conservation" and "Preventive Conservation" at the Department of Conservation and Restoration of the Faculty of Sciences and Technology, NOVA University of Lisbon. M.F. Macedo was been actively involved in several national and international projects since 1992 and was the leader of the Portuguese team in two Luso-Spanish Bilateral Projects. From 2016 to 2019, she was also the Principal Investigator (PI) of the project "CleanART", supported by the Foundation for Science and Technology (Fundação para a Ciência e Tecnologia), Ministério da Ciência, Tecnologia e Ensino Superior. She has supervised or co-supervised a total of 38 postgraduate theses: 6 PhD theses, 21 master theses (with 2 ongoing) and 11 bachelor theses (before Bologna) in the areas of Preventive Conservation and Biodeterioration of Cultural Heritage. M. F. Macedo is also the author of more than 110 publications, including 56 articles published in peer-reviewed international papers listed in JCR.

António Portugal (Assistant Professor). António Portugal has a PhD in Molecular Biology, an MSc in Ecology and a degree in Biology. He is an Assistant Professor of the Department of Life Sciences, University of Coimbra, where he teaches Genetics, Genetic Resources, Applied Ecology and Bioremediation within several degrees and master programmes. He is a researcher of the Centre for Functional Ecology, University of Coimbra, where he is the head of the Mycology and Biodeterioration Lab. He has been the PI of several research projects dealing with Biodeterioration of Cultural Heritage. He is also Director of FitoLab—Laboratory for Phytopathology, Instituto Pedro Nunes. This laboratory is devoted to the detection and research of plant pests and diseases, with the aim of improving plant health in agriculture and forestry. FitoLab is recognized by the Portuguese Authorities for the detection of several quarantine organisms in the fields of microbiology, virology, mycology and nematology. The required methods and EPPO standards to work with quarantine organisms are implemented in FitoLab.

Ana Catarina Pinheiro (Researcher). Ana Catarina Pinheiro is a pharmacist and a Conservator-Restorer with a PhD in Conservation Sciences. She is currently working in the HERCULES Laboratory, Évora University, where her main lines of interest include occupational exposure, sustainability and the biodeterioration of organic supports, especially medieval parchment and tawed leather. She has, throughout her career, participated in several projects—from stone to paper conservation—and has assumed teaching roles in Preventive Conservation and Pharmacology. She is engaged in international efforts to import sustainability into Cultural Heritage and leads the Sustainability in Conservation (Ki Culture) Portuguese branch. Ana Catarina Pinheiro has published 19 papers in specialized peer-reviewed journals and is the author/co-author of seven book chapters.

Ana Zélia Miller (Researcher). Ana Zélia Miller is a geomicrobiologist with a PhD in Conservation Sciences (2010) from the New University of Lisbon. During her PhD thesis, she pioneered the implementation of stone bioreceptivity experiments using multi-species phototrophic cultures for simulating competition and/or synergy among microorganisms under laboratory conditions. Currently, she is a Ramón y Cajal researcher at the Instituto of Natural Resources and Agrobiology of Seville from the Spanish Research Council (IRNAS-CSIC), where she leads the Geomicrobiology and Biogeochemistry research group, focused on the scientific underpinnings for microbe–mineral interactions in complex environmental systems, including volcanic caves and stone cultural heritage. She has implemented novel approaches to blending classical microbiology, molecular biology, metagenomics, mineralogy and geochemistry in a holistic effort to understand microbial diversity, biosignatures and minerogenesis in volcanic caves from Canary (Spain), Selvagens (Portugal), Easter (Chile) and Galapagos (Ecuador) Islands. Ana Zélia Miller is Principal Investigator of 7 research projects, co-author of 75 papers in international peer-reviewed journals and has more than 100 communications in national and international congress meetings.

applied sciences

Editorial

Application of Biology to Cultural Heritage

Maria Filomena Macedo [1,2,*], Ana Zélia Miller [3], Ana Catarina Pinheiro [3] and António Portugal [4,5]

1 Department of Conservation and Restoration, Faculty of Science and Technology, Universidade NOVA de Lisboa, 1099-085 Lisboa, Portugal
2 VICARTE, Research Unit Vidro e Cerâmica para As Artes, Faculdade de Ciências e Tecnologia, Campus Caparica, Universidade NOVA de Lisboa, 2829-516 Caparica, Portugal
3 HERCULES Laboratory, University of Évora, 7000-809 Évora, Portugal; anamiller@uevora.pt (A.Z.M.); acmsp@uevora.pt (A.C.P.)
4 Centre for Functional Ecology, Department of Life Sciences, University of Coimbra, 3000-456 Coimbra, Portugal; aportuga@bot.uc.pt
5 Fitolab-Laboratory for Phytopathology, Instituto Pedro Nunes, 3030-199 Coimbra, Portugal
* Correspondence: mfmd@fct.unl.pt

Citation: Macedo, M.F.; Miller, A.Z.; Pinheiro, A.C.; Portugal, A. Application of Biology to Cultural Heritage. *Appl. Sci.* **2022**, *12*, 841. https://doi.org/10.3390/app12020841

Received: 27 December 2021
Accepted: 11 January 2022
Published: 14 January 2022

This Special Issue of the *Applied Sciences*, entitled "Application of Biology to Cultural Heritage" aimed to cover all the latest outstanding progress of biological and biochemical methods developed and applied to cultural heritage. As you can see cultural heritage biodiversity and biodeterioration has received much research attention in recent years. This Special Issue intended to provide a comprehensive examination of the science of biology in various fields and areas, and its practical application for the preservation of cultural heritage. Full research articles and reviews on all aspects of biological causes, modes of action, biocidal treatment, protection, and prevention of cultural heritage, are here presented as well as long term studies on the biodeterioration of cultural heritage sites and monuments. Analyses and testing of macro- and microorganisms affecting the preservation of cultural heritage were also addressed.

The knowledge that has arisen from the published papers about the studies on new techniques and new products applied to the cultural heritage area may now be translated into new conservation and restoration treatments in similar objects, sites and supports. This was our main goal and was achieved in a mission that we carried out with great pleasure!

This issue addresses researchers from both academia and industry, working in microbiology and biotechnology. We hope that you enjoy and find it useful in your current and future work, research projects and careers.

In this Special Issue eleven excellent papers (including review and full research articles) were published and all went through a hard and demanding reviewing work to assure maximum quality. In the following paragraphs, a summary of these 11 papers, with their most relevant contributions, is presented.

In the paper by Klügl and Di Pietro [1] an atlas of micromorphological degradation of archaeological birch bark, dating from the Neolithic to the Middle Age, was developed in order to describe the decay of archaeological birch bark. The characteristics of contemporary non-degraded mature birch phellem was first investigated and compared with the characteristics of contemporary decayed in-nature phellem. The morphology of 13 samples from ice-logged, waterlogged and in-cave recovered archaeological birch barks was then investigated, using light and electron transmission microscopy and, the authors made, for the first time, a comprehensive description of microscopic degradation features of birch bark and correlated those with macroscopic visible features. These results can be very useful for conservators working with archaeological birch bark.

The next three studies are related to stone biodeterioration and possible treatments.

Bartoli et al. [2] present a very interesting paper on the efficiency of biocides on stone cultural heritage. The authors describe and investigate the biocide effects of two different biocides encapsulated in two different silica nanosystems. The in vitro experiment was performed using *Chlorococcum* sp., a green algae commonly found in biodeteriorated stone. The two biocides selected were: Zosteric sodium salt (ZS), a green biocide, and the commercial biocide, 2-mercaptobenzothiazole (MBT), widely used in the treatment of cultural heritage. The analyzed systems were the following: silica nanocapsules (NC) and silica nanoparticles (MNP) not loaded with biocides, two nanosystems loaded with ZS and MBT, and free biocides. The results confirmed the interest of the silica nanosystems coupled with biocides as a potential protective coating for stone monuments. Nanoparticles, synthesized for the formulation of a multifunctional coating, proved better than nanocontainers as being promising tools in the conservation of stone cultural heritage.

The paper by Santo et al. [3] describes a multidisciplinary study on the state of conservation of white marbles from the Florence Cathedral and the microbial community involved in their deterioration. This remarkable study is focused on the widespread dark discoloration of marble analyzed in two differently exposed sites of the Cathedral: NW and SE. The authors used chemical and petrographic analyses, in situ and ex situ microscopy, and cultivation and identification of microorganisms. According to the obtained data the darkening is mainly due to the growth of black fungi and dark cyanobacteria at both the study sites. The biodiversity of the lithobiontic community seems to be similar in the investigated areas, especially that of the photoautotrophic components. On the contrary, the pattern of microbial colonization on/inside marble seems to be different at the NW and SE sites depending on the distinct climatic conditions of the two study areas, in particular on solar radiation exposure, which influences marble's bioreceptivity. This work presents the first report on the lithobiontic community inhabiting the Florence Cathedral marbles. These results are very important for future interventions to control microbial growth.

In the Roman wall of Lugo (NW Spain), declared a World Heritage site by UNESCO in 2000, three separate treatments were applied, in the past 20 years. In this interesting work, Prieto et al. [4] used a combined laboratory and field research to examine the possible alterations caused by herbicide treatments applied to the construction materials of the Roman Wall (schist and some granite, bound with mortar). Their results showed that herbicidal treatments should focus on environmentally friendly technologies with low risk to human health and should not damage the construction materials. In this respect, *Origanum vulgare* L. and *Thymus vulgaris* L. essential oils seem to be the appropriate, as they did not provoke any mineralogical alterations or changes in the salt content and original colour.

Nanhai No. 1, a Chinese Song Dynasty shipwreck, was discovered in 1987 near the Chuanshan archipelago in the South China Sea, Guangdong Province. In 2007, Nanhai No. 1 was salvaged and displayed in "the Crystal Palace" of the Guangdong Maritime Silk Road Museum. Due to the high humidity and high temperature in the storage environment of Nanhai No. 1, and the fact that wooden cultural relics are natural organic matter, this provides suitable conditions for the growth and reproduction of microorganisms. Microorganisms are very harmful to wooden relics. Some bacteria and fungi can destroy wood by degrading cellulose, hemicellulose and lignin. In this fascinating paper by Huang et al. [5], the authors report the inhibitory effect of cinnamaldehyde on the main destructive microorganisms of Nanhai No. 1 Shipwreck. The results showed the inhibitory effects of cinnamaldehyde, a natural biocide that proved to be effective as a control agent for many of the main destructive microorganisms identified in the Nanhai No. 1 Shipwreck. This is an important step towards the development of safer products – both for the environment and the safety of the professionals responsible for the application of such products.

In this newsworthy work from Fuentes, Carballeira and Prieto [6] the granite biodiversity is analysed. The microorganisms' diversity present in historical rural granite buildings was identified for the first time. This granite has been used throughout history in Galicia (NW Spain), forming the basis of much of the region's architecture. For the identification of the granite biofilms the authors used a combination of culture-dependent and next generation sequencing (NGS) techniques. For this work three granite-built churches located in rural areas of Galicia, in the northwestern part of the Iberian Peninsula, were selected. Since the orientation of buildings is one of the factors that most strongly influences the development and composition of subaerial biofilms, as it determines the amount of light and the availability of water on the substrate, the role of exposure was also considered by comparing the biodiversity on north and west walls of the three churches. The authors also compared their results with previous work, made on biodiversity on historical granite buildings in urban settings. The results proved that orientation is an important factor regarding both the diversity and abundance of microorganisms on the walls, with environmental factors associated with the north orientation factors favouring a higher diversity.

This excellent paper, from Dyda et al. [7], also addresses the stone biodiversity topic. In this case the biodiversity of different biocenoses on the historical sandstone of the Northern Pergola in the Museum of King John III's Palace at Wilanow (Poland) are analysed. The authors aimed at describing seasonal changes of microbial community composition in situ in different biocenoses: winter and summer, of the same historical sandstone. Another objective of this work was to describe the seasonal changes of microbial community composition in situ in relation to biodeterioration patterns or with anthropogenic factors, e.g., air pollution. The microbial biodiversity was analyzed by the application of Illumina-based next-generation sequencing methods. The metabarcoding analysis on fungi, bacteria and lichenized fungi are presented and compared between winter and summer. The authors found that the diversity of organisms in the biofilm ensures its stability throughout the year despite the differences recorded between winter and summer. On the other hand, analyses of bacterial and fungal communities' biodiversity in biocenoses from historic sandstone Pergolas revealed that the observed changes were mostly associated with the biodeterioration patterns.

This Special Issue also presents four very excellent reviews.

In this very interesting work by Macedo, Vilarigues and Coutinho [8], the authors review the 21st century literature (2000 to 2021) regarding the biodiversity and biodeterioration of glass-based historical building materials, particularly stained glass and glazed tiles. These authors study the provenance of the stained glass and glazed tiles with biological colonization, presented in the 21st century literature. They also compile a list of fungi, bacteria and phototrophic microorganisms reported on glass-based historical building materials and make a critical overview regarding the organism's biodiversity and the colonized materials. They also compare the biodeterioration patterns from case studies with laboratory-based colonization experiments, showing that many deterioration patterns and corrosion products are similar. Finally, the authors make suggestions regarding further studies that are urgent to do in order to fully understand the biodiversity and biodeterioration of stained glass and glazed tiles.

In addition, in this special issue, there is an excellent review by Ortega-Morales and Gaylarde [9] dealing with bioconservation of historic stone buildings. In this review, the authors seek to provide updated information on the innovative bioconservation treatments that have been or are being developed. According to the authors there is an urgent need to replace traditional conservation treatments by more environmentally acceptable, biologically-based, measures, including bioconsolidation and biocleaning. Bioconsolidation can use whole bacterial cells or cell biomolecules while biocleaning can employ microorganisms or their extracted enzymes to remove inorganic and organic surface deposits such as sulphate crusts, animal glues, biofilms and felt tip marker graffiti. The great advantage of biorestoration based on microbial cells or their products, compared with traditional chemical, physical and mechanical methods, is that biological treatments are not destructive of the underlying substrate, simply removing unwanted overlying materials, in the case of biocleaning, or producing new stone, in the case of biocalcification.

The interesting review paper by Trovão and Portugal [10] attempts to provide an overview of biodeteriorative profiles obtained by amended plate essays, while also providing a summary of currently known fungal species putative biodeteriorative abilities. Both the essays available and the results concerning distinct fungal species biodeteriorative profiles obtained by amended plate essays, remain scattered and in need of a deep summarization. In this way, the authors also provide a series of checklists that can be helpful to microbiologists, restorers and conservation workers when attempting to safeguard cultural heritage materials worldwide from biodeterioration. This review highlighted that, so far, isolates from more than 200 fungal species have been showed to exhibit biodeteriorative abilities when studied by specific plate essays. This review can be very interesting also for the broad target public from researchers to end users.

Finally, an excellent review by Passaretti et al. [11] on metal heritage cleaning by biologically derived gel formulations, is presented. The authors review biologically derived gel formulations already proposed for artistic and historical metal as reliable tools for cleaning. An overview on the common practices for cleaning metallic surfaces in cultural heritage is presented. The potentialities of gel-alternatives and in particular those with a biological origin are discussed. In this review water-gels (i.e., hydrogels) and solvent-gels (i.e., organogels) together with particular attention to bio-solvents, are discussed in detail. Nevertheless, most of metals conservators rarely use this bio-gels. This could be due to the relatively high price, the unawareness of either risk related to traditional methods or existence of reliable alternatives, and the effort required to change long-established habits in favour of unprecedented methodologies.

Author Contributions: Conceptualization, M.F.M., A.Z.M., A.C.P. and A.P.; methodology, M.F.M., A.Z.M., A.C.P. and A.P.; writing—original draft preparation, M.F.M., A.Z.M., A.C.P. and A.P.; writing—review and editing, M.F.M. and A.P.; visualization, M.F.M. and A.P.; supervision, M.F.M. and A.P.; funding acquisition, M.F.M., A.Z.M., A.C.P. and A.P. All authors have read and agreed to the published version of the manuscript.

Funding: This work was supported by National Funds through FCT-Portuguese Foundation for Science and Technology under the contracts CEEC:CEECIND/01147/2017 (AZ Miller) and EECIND/02598/2017 (A.C. Pinheiro). This work was also funded by FCT/MCTES through national funds (PIDDAC) under the project of the R&D Unit Centre for Functional Ecology—Science for People and the Planet (CFE) UIDB/04004/2020, and through VICARTE RESEARCH UNIT (UIDB/00729/2020).

Acknowledgments: We would like to thank all the authors and peer reviewers for their valuable contributions to this special issue. This issue would not be possible without their valuable and professional work. The MDPI management and staff are also to be congratulated for their untiring editorial support for the success of this project.

Conflicts of Interest: The authors declare no conflict of interest.

References

1. Klügl, J.; Di Pietro, G. Atlas of Micromorphological Degradation of Archaeological Birch Bark. *Appl. Sci.* **2021**, *11*, 8721. [CrossRef]
2. Bartoli, F.; Zuena, M.; Sodo, A.; Caneva, G. The Efficiency of Biocidal Silica Nanosystems for the Conservation of Stone Monuments: Comparative In Vitro Tests against Epilithic Green Algae. *Appl. Sci.* **2021**, *11*, 6804. [CrossRef]
3. Santo, A.; Cuzman, O.; Petrocchi, D.; Pinna, D.; Salvatici, T.; Perito, B. Black on White: Microbial Growth Darkens the External Marble of Florence Cathedral. *Appl. Sci.* **2021**, *11*, 6163. [CrossRef]
4. Prieto, B.; Sanmartín, P.; Cancelo-González, J.; Torres, L.; Silva, B. Impact of Herbicide Treatments on the Construction Materials in the Roman Wall of Lugo, Spain (UNESCO World Heritage Site). *Appl. Sci.* **2021**, *11*, 5276. [CrossRef]
5. Huang, X.; Han, Y.; Du, J.; Guo, P.; Wang, Y.; Ma, K.; Li, N.; Zhang, Z.; Li, Y.; Pan, J. Inhibitory Effect of Cinnamaldehyde on Main Destructive Microorganisms of Nanhai No. 1 Shipwreck. *Appl. Sci.* **2021**, *11*, 5262. [CrossRef]
6. Fuentes, E.; Carballeira, R.; Prieto, B. Role of Exposure on the Microbial Consortiums on Historical Rural Granite Buildings. *Appl. Sci.* **2021**, *11*, 3786. [CrossRef]
7. Dyda, M.; Laudy, A.; Decewicz, P.; Romaniuk, K.; Ciezkowska, M.; Szajewska, A.; Solecka, D.; Dziewit, L.; Drewniak, L.; Skłodowska, A. Diversity of Biodeteriorative Bacterial and Fungal Consortia in Winter and Summer on Historical Sandstone of the Northern Pergola, Museum of King John III's Palace at Wilanow, Poland. *Appl. Sci.* **2021**, *11*, 620. [CrossRef]
8. Macedo, M.F.; Vilarigues, M.G.; Coutinho, M.L. Biodeterioration of Glass-Based Historical Building Materials: An Overview of the Heritage Literature from the 21st Century. *Appl. Sci.* **2021**, *11*, 9552. [CrossRef]
9. Ortega-Morales, B.; Gaylarde, C. Bioconservation of Historic Stone Buildings—An Updated Review. *Appl. Sci.* **2021**, *11*, 5695. [CrossRef]

10. Trovão, J.; Portugal, A. Current Knowledge on the Fungal Degradation Abilities Profiled through Biodeteriorative Plate Essays. *Appl. Sci.* **2021**, *11*, 4196. [CrossRef]
11. Passaretti, A.; Cuvillier, L.; Sciutto, G.; Guilminot, E.; Joseph, E. Biologically Derived Gels for the Cleaning of Historical and Artistic Metal Heritage. *Appl. Sci.* **2021**, *11*, 3405. [CrossRef]

applied sciences

MDPI

Article

Atlas of Micromorphological Degradation of Archaeological Birch Bark

Johanna Klügl [1,2,*] and Giovanna Di Pietro [2]

1 Archaeological Service of the Canton Bern, Brünnenstrasse 66, 3018 Bern, Switzerland
2 Bern Academy of the Arts, Institute Materiality in Art and Culture, Fellerstrasse 11, 3027 Bern, Switzerland; giovanna.dipietro@bfh.ch
* Correspondence: johanna.kluegl@be.ch

Abstract: In this paper we present an atlas of micromorphological degradation of archaeological birch bark for the first time. We analysed the morphology of 13 samples extracted from ice-logged, waterlogged and cave-retrieved objects dated from the Neolithic to the Middle Age by means of light microscopy (LM) and transmission electron microscopy (TEM). We then compared their morphology to that of a contemporary sample, both intact and decayed. In all samples, 13 morphological characteristics that can be associated with fungal, bacterial, chemical, mechanical and light degradation are defined and described, and example LM and TEM images are provided. This novel atlas provides conservator-restorers a much-needed tool to relate the macroscopic appearance to the microscopic structure of birch bark objects. The most important macroscopic features allowing estimation of the state of preservation at the cell level are colour changes, loss of pliability, presence of delamination and increased brittleness. Colour change and delamination can be connected to microscopic features, and microscopic analysis can trace whether they were caused by biotic, chemical or physical decay. However, increased brittleness cannot be connected to a specific microscopic feature.

Keywords: phellem; birch bark; decay; fungi; bacteria; light microscopy; transmission electron microscopy; archaeology; ice patch; waterlogged

Citation: Klügl, J.; Di Pietro, G. Atlas of Micromorphological Degradation of Archaeological Birch Bark. *Appl. Sci.* **2021**, *11*, 8721. https://doi.org/10.3390/app11188721

Academic Editor: Maria Filomena Macedo

Received: 5 August 2021
Accepted: 14 September 2021
Published: 18 September 2021

1. Introduction

Birch bark is a waterproof material that has been widely used since Mesolithic times because it is widely available and easily harvested and processed. In recent decades, some key archaeological objects made of birch bark have been discovered, among them a bow case from the Schnidejoch pass in the Swiss alps. Conservation strategies for birch bark objects often mimic those for wooden objects, but data on birch bark decay pathways are needed if appropriate conservation methods are to be developed. It is a well-known fact that archaeological wood suffers mainly from biodeterioration. When it dries in an uncontrolled way, the biotic decay of the cell walls results in cell shrinkage and collapse, macroscopically visible as severe volume reduction, warping, cracking and even disintegration. In order to develop appropriate drying procedures, it is therefore relevant to ask if birch bark suffers from similar biotic decay. To answer this question, we first review the current knowledge on the morphology, chemical composition and degradation of birch bark. Then, we present novel light microscopy and electron transmission microscopy results describing the macroscopic appearance and micromorphology of ice-logged, waterlogged and cave-retrieved samples dated from the Neolithic to the Middle Age and compare them with the morphology of intact, oxygen-exposed, decayed-in-nature contemporary samples.

Although degradation is always object-dependent, our aim is to present an atlas of possible micromorphological degradation patterns of archaeological objects made of birch bark in order to help conservator-restorers relate the macroscopic condition to the microscopic structure. This is a novel and much-needed tool that will help in assessing the state of conservation of birch bark objects, the first step in making a decision about active and passive conservation measures.

1.1. Morphology of Birch Bark

The portion of the bark of the birch tree that is used to manufacture objects is the outer smooth white layer, called the phellem in botany. The function of the birch phellem is to prevent transpiration and to provide thermal [1] and parasite [2] protection. It consists of two cell types produced by the phellogen, arranged in alternating and compact radial layers, providing the characteristic laminated structure of the bark. One cell type has thin walls and a broad lumen filled with a white substance, betulin, while the second cell type has thick walls and a small lumen filled with a red-brown material composed of phenolic compounds [3–5] (Figure 1). The ratio of thin- to thick-walled cell layers depends on the *Betula* species and the specific tree history [3].

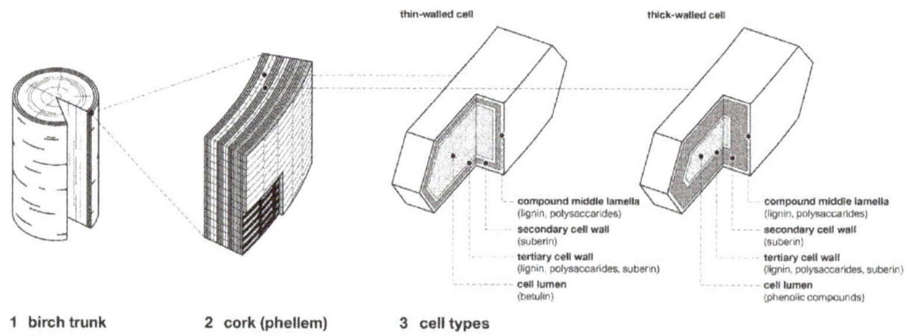

Figure 1. Schematic representation of the morphology and chemical composition of birch bark cells: 1. location of bark in trunk; 2. alternating layers of thick-walled and thin-walled phellem cells; 3. structure and chemical composition of thin- and thick-walled cells.

The cell walls are composed of a primary, secondary and tertiary layer, and the individual cells are joined together by the middle lamella, often called the compound middle lamella, as it can hardly be distinguished from the primary wall [6]. The phellem cells are arranged in such a way that the radial middle lamella creates a line. The thin-walled cells are heavily compressed in the outer portion of the bark but start off with a square shape. The compression and folding of the thin-walled cells can be explained by the formation of the phellem. To adjust to the growing size of the phellem, the bark cells are pushed outwards: the cells stretch in a tangential direction, while the thin-walled cells are compressed [7]. The folding of radial cell walls has also been observed in phellems of other trees, e.g., *Querus suber* [8], *Pseudotsuga menziesii* [9] and *Prunus serrula* [10].

The secondary walls of both types of cells are composed of suberin, a biopolymer that makes the cork impermeable. Gas and water exchanges occur through characteristic openings called lenticels [11,12]. These are formed by fanned-out bands of continuous, heavily suberized, thick-walled cell layers and interrupted by void spaces filled with irregular cells. The intercellular spaces between the thick-walled cells visible in the lenticels' radial sections are absent in the normal phellem [12,13].

The suberization of the cell walls takes place after their generation from the phellogen [14]. In this phase, the channels connecting the cells, so-called plasmodesmata, become closed and the phellem cells die. Plasmodesmata with a diameter of 50 nm have been found mainly in tangential suberized cell walls of *Betula* sp. phellem [12]. Dead cells are less nutritious and have better resistance to microbial decay [15].

1.2. Chemical Composition of Birch Bark

The chemical composition of the outer bark of *Betula* sp. has been investigated several times [16–18]. The most recent analyses [19] confirm that it is composed of suberin (36.2%), lignin (14.3%) and polysaccharides (10.3%) located in the cell walls, and extractives

(32.2%) located in the cell lumen, specifically betulin in the thin-walled cells and phenolic compounds in the thick-walled cells [20]. Kiyoto and Sugiyama [21] recently elucidated the chemical composition of the different layers of the cell wall. The middle lamella and the primary cell wall are composed of lignin, cellulose and polysaccharides, the secondary cell wall by suberin, and the tertiary cell wall is subdivided into a middle layer with the same composition as the primary cell wall and a suberized inner layer (Figure 1). This composition was found in both cell types, and thin-walled cells have been reported to be less suberized than thick-walled cells [12]. Electron micrographs of suberized cell walls reveal a lamellar structure arising from the polyaliphatic and polyphenolic parts of suberin connected by glycerol via ester bonds. The light areas consist of aliphatic suberin, and the dark lamellae are rich in polyphenolic substances [22].

1.3. Degradation and Preservation Condition of Archaeological Birch Bark

Degradation of archaeological birch bark artefacts causes macroscopic colour fading [23], discoloration [24], delamination [25,26], deformation–distortion [27] and increased water absorption [28], but the connection between the burial conditions of archaeological bark and the dominant degradation pathways has never been investigated in detail. Different authors have shown that when Mesolithic and Neolithic archaeological birch bark artefacts are recovered from wetlands [29], caves [30,31] or waterlogged environments [32], they can be brittle and lose substance. This increased brittleness can have chemical and/or biological origins. Concerning the chemical origins, gas chromatography/mass spectrometry (GC/MS) examinations of waterlogged Neolithic alpine artefacts [25] detected depolymerization and oxidation, attributed to higher hydrolysis of susceptible groups, such as epoxy linkages.

Concerning the biological origins, investigations of artefacts recovered from the wetland site Star Carr dating from the ninth millennium BC revealed fungal hyphae within the bark structure, and pyrite and gypsum crystals on the surface [29]. The fragility of the artefacts was attributed to both the mineral deposits and the microbial attack. Gypsum crystals within the bark structure were also found in birch bark torches from the Kosackenberg caves dated 2300–1800 BC and Iron Age [30], in a quiver from Xinjiang (Northwest China) dated 618–907 AD [33] and in Bronze Age barks from salt mines [34]. Specific investigations of the biotic decay of birch bark could not be found in the literature.

Tree phellems that consist of suberized cells, such as birch cork, are understood to be very resistant to biological decay [35,36] because they are constructed to protect the plant/tree from infection by pathogen microorganisms [37]. Studies performed in the 1970s on the degradation of contemporary barks under natural burial conditions and in the laboratory reported that the phellem was the last component affected by fungi [38]. The decay was reported to start after 3 years of exposure, and microorganisms grew within the thick secondary tangential suberized cell walls. The growth of hyphae mechanically damages the cells by creating oval cavities, and this might lead to weight loss. Early studies on conifer barks showed that the weight loss caused by fungi in bark was five times smaller than that in wood, and that despite a weight loss of up to 10%, the physical properties may remain unchanged [39]. Several authors have shown that enzymes produced by fungi can degrade suberin [40–44], but the detailed pathways are still largely unknown, because suberin is strongly cross-linked and an unchanged extraction was not possible until new extraction techniques were developed [45,46]. Martins et al. [47] revealed that cutinasen, lipases and long-chain alcohol-modifying enzymes released by *Aspergillus nidulans* fungus depolymerises suberin. The long-chain fatty acids are broken down via peroxisomal β-oxidation. A very recent publication found that auxin-regulated GDSL lipases could degrade suberin [48]. Studies of enzymes from cold Arctic [49] and Antarctic [50,51] environments show that enzymes are active at temperatures close to 0 °C. Some enzymes have limited function, but activity at 0–15 °C is common for many of them [52].

Regarding archaeological wood, it is known that when it is recovered from low-oxygen sediments or under water, bacteria are the main cause of slow degradation [53–55], and when it is exposed to air, fungi lead to a very fast metabolism controlled by temperature and humidity [56,57]. The absence of water and oxygen and low temperatures inhibit biotic decomposition, and changes due to chemical and physical processes may become evident [58]. However, very extreme conditions are required for microbial degradation to be completely suppressed [59]. Studies of wood biodegradation from Arctic regions have shown that predominantly soft rot is found [60,61], while brown rot plays a minor role and white rot could not be unequivocally identified [62]. It is an open question whether similar patterns of biotic decay are present in archaeological birch bark.

2. Materials and Methods

The morphology of birch bark has been examined visually, with light microscopy (LM) and transmission electron microscopy (TEM). In order to describe the decay of archaeological birch bark, the characteristics of contemporary non-degraded mature birch phellem was first investigated and compared with the characteristics of contemporary decayed in-nature phellem. The morphology of 13 samples from ice-logged, waterlogged and in-cave recovered archaeological birch barks was then investigated (Table 1).

2.1. Samples

The reference bark was a mature birch phellem collected from a felled *Betula pendula* tree about 30 years old in the district of Vex (Valais, Switzerland) in November 2014 (Figure 2). The naturally decayed bark was harvested from a fallen tree about 15 years old in Borgne Valley, Valais (Figure 3). The trunk was weathered in nature, laying on a sandy subsoil with stones. The fragment was chosen for its pronounced biological degradation, visible macroscopically as green, grey and brown-black staining, deposits and cracks. Such degradation was present especially at the lenticels and the inner bark (Figure 3). The ice-logged samples were taken from the Neolithic bow case from the Schnidejoch ice patch, Switzerland (Figure 4), and from three birch bark fragments retrieved from the Lendbreen ice patch, Norway (Figure 5). The Schnidejoch bow case has been described in a number of publications [63–65]. This object is dated to around 2800 BC and consists of three parts, a lid and an upper and lower body, that were found under different circumstances and in different conditions. Each part is made of 2 to 3 layers of superimposed outer birch bark strips sewn together with lime bast. The lid was subjected to a harmful ethanol treatment that extracted most of the alcohol-soluble components from the bark. It is now stored dry at ambient temperature. The upper and lower body are both untreated and stored damp and frozen at −26 °C. The three Lendbreen birch bark fragments were found in a retreating ice patch located at the northern slope of Lomseggen ridge, Oppland (Norway). The ice patch extends from 1690 to 1920 m a.s.l. and the finds are dated between 300 and 1500 AD, with an accumulation of finds around 1000 AD [66]. The waterlogged sample was extracted from a Neolithic birch bark retrieved in Lake Moossee in Switzerland (521 m a.s.l) during an excavation in 2011 (Figure 6). The layer from which the bark was retrieved dates between 4500 and 3800 BC. Since excavation, this bark, which was originally not harvested but detached naturally from a tree, as the phloem is still attached to the bark, is stored in distilled water at 4 °C. The last fragment (Figure 7) was recovered from a cave named Kosackenberg, in the Kyffhäuser mountains west of Bad Frankenhausen in Germany and excavated in the 1950s. It is dated to the Bronze Age (23001800 BC) and, after unburial, has been preserved untreated in a sealed container.

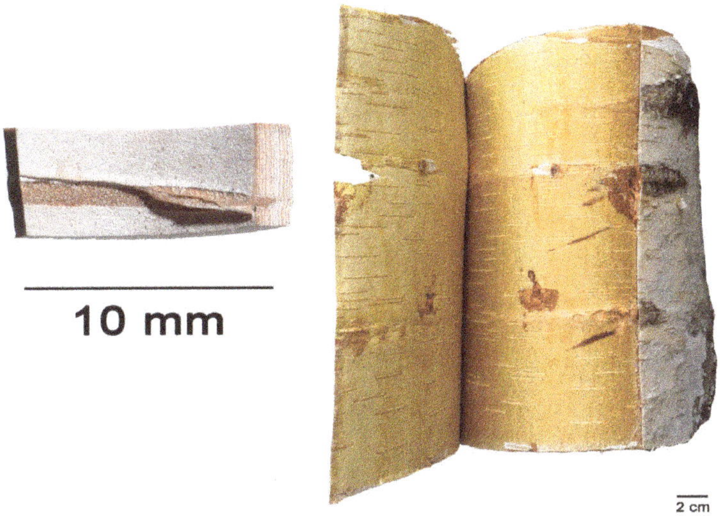

Figure 2. Reference sample (**left**) was extracted from contemporary outer birch bark harvested from a mature tree with strong and pliable phellem (**right**). Sample has white powdery surface due to betulin content and slight delamination is present at the lenticel. Thick- and thin-walled cell bands are recognizable macroscopically as red and white bands, respectively, along the radial section.

Figure 3. Birch bark fragment harvested from fallen trunk degraded in nature. Outer surface, especially when in contact with the ground, shows green, grey and brown discoloration and vertical cracks. Lenticels have cracks and overlays and are areas no longer clearly defined. On the inner side, remains of brittle and brown-black stained phloem are present and black circular deposits indicate areas of increased degradation. Location of analysed sample 0 is indicated with red rectangle.

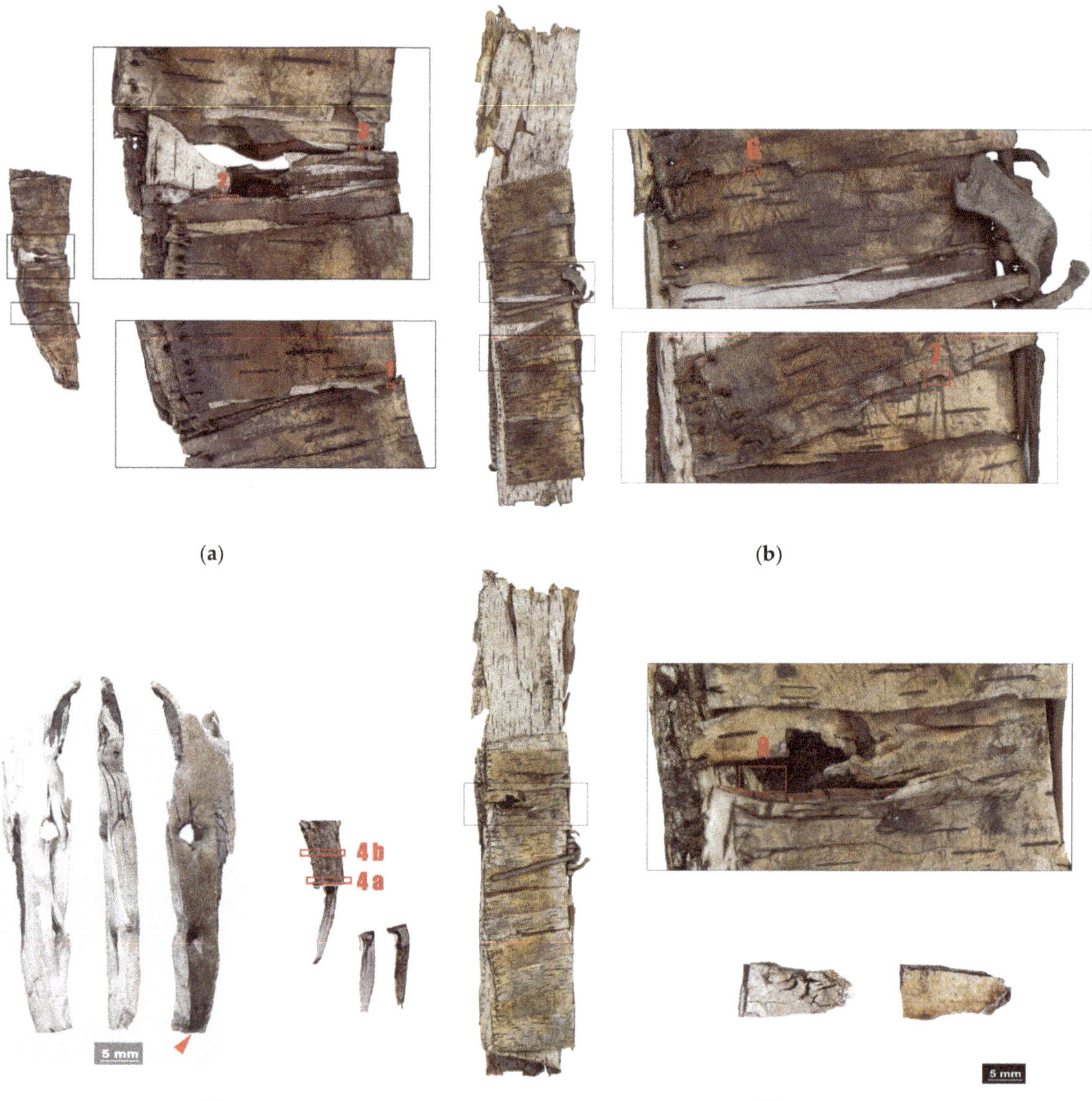

Figure 4. (**a**) Lower and (**b**) upper part of Schnidejoch bow case body, ground-facing side; (**d**) upper part, upward facing. Samples were taken from outer bark strips (1, 6, 7), middle bark strips (2) and inner lining (3, 4, 8). Outer and middle bark strips of bow case body have outer bark surface oriented towards inside, and inner lining has outer bark surface oriented towards outside. Sample 4 was extracted from a fragment probably from (**c**) lid inner lining. Sample 5 was a loose fragment found close to the lower part (not shown).

Figure 5. Left: Lendbreen fragments; **right**: analysed samples. Sample 9 was taken from a weathered fragment; lenticels were fully eroded but sample was not brittle. Surface and cambium side were grey and faded. Cut surface was dark brown with a few reddish-brown areas. Sample 10 was taken from a rolled birch bark fragment, stiff but not brittle, with delaminated areas and large cracks. Cambium side (outer side in image) was compact and with a light, mainly grey-brown and white colour. Some holes from insect damage were present and normally elevated lenticels were recessed. On cut surface, outer layers were grey-brown, and only inner layers still exhibited original red-brown colour. Sample 11 was taken from a long narrow strip rolled on one end. Fragment was very strong, hard to cut and still pliable. Surface was weathered, coloured matte whitish-grey; red and brown colours were missing. Cut surface revealed unaltered colours in inner layers and grey outer layers.

Table 1. Analysed samples and their location, dating and burial conditions.

Sample	Object	Location	Date	Burial Conditions		
				Environment	T, RH	O_2
Ref	Intact phellem	Valais, CH	Contemporary	tree		Ambient
0	Rotten phellem	Valais, CH	Contemporary	In-nature rotten tree		Ambient
1	Bow case, lower body	Schnidejoch	≈2800 BC (2905–2622 cal BC)	Ice patch	0 °C, 100% (*)	Low
2	Bow case, lower body	Schnidejoch	≈2800 BC (2905–2622 cal BC)	Ice patch	0 °C, 100%	Low
3	Bow case, lower body	Schnidejoch	≈2800 BC (2905–2622 cal BC)	Ice patch	0 °C, 100%	Low
4	Bow case, probably from lid	Schnidejoch	≈2800 BC (2881–2579 cal BC)	Ice patch	0 °C, 100%	Low
5	Bow case, lower body	Schnidejoch	≈2800 BC (2905–2622 cal BC)	Ice patch	0 °C, 100%	Low
6	Bow case, upper body	Schnidejoch	≈2800 BC (3011–2668 cal BC)	Ice patch	0 °C, 100%	Low
7	Bow case, upper body	Schnidejoch	≈2800 BC (3011–2668 cal BC)	Ice patch	0 °C, 100%	Low
8	Bow case, upper body	Schnidejoch	≈2800 BC (3011–2668 cal BC)	Ice patch	0 °C, 100%	Low
9	Fragment	Lendbreen	≈500 AD (430–540 cal AD)	Ice patch	−3 °C (**)	Low
10	Fragment	Lendbreen	≈900 AD (891–981 cal AD)	Ice patch	−3 °C	Low
11	Fragment	Lendbreen	≈1450 AD (1430–1470 cal AD)	Ice patch	−3 °C	Low
12	Fragment	Lake Moossee	4500–3800 BC	Waterlogged	≈4–16 °C (†)	Anoxic
13	Fragment	Kosackenberg	2300–1800 BC	Cave	4–7 °C, ≈95% (‡)	Ambient

* [67], ** [68], † [69], ‡ [30].

Figure 6. Left: birch bark fragments from Lake Moosee (521 m a.s.l) were recovered during an excavation in 2011 from a waterlogged sediment. Small, waterlogged fragments floating on water retain phloem, indicating that they were not harvested from the tree but detached naturally. During cutting, many fractures occurred, and small pieces broke off. **Right**: sample 12 does not have original white colour apart from a fresh fracture with splintered surface. Inner bark has dark-brown colour and is very brittle, and lenticels are not well defined and have frayed edges.

Figure 7. Sample 13 was extracted from a birch bark fragment recovered in the Kosackenberg cave. Inner and outer surfaces have reddish-brown overlay with white blooming. Normally alternating white and reddish-brown layers are not present, as reddish-brown colouring of thick-walled cells is completely lost. Lenticels have also lost their normally dark colour and are hardly distinguishable from normal phellem. Fragment is very light and extremely fragile; handling leads to fragmentation and loss.

2.2. Description of Embedding and Sectioning

Samples 1–3, 6–8 and 12 were wet, while the others were already air-dried. They were all documented and cut with a razor blade to a rectangular size of about 7 mm × 3 mm. The cutting provided information about the brittleness and risk of delamination. First attempts showed that non-embedded sectioning did not allow for a thorough study of deterioration due to the separation of the layers during cutting. Further, the lumen content of the thin-walled cells obscured the cell wall study and had to be removed [25]. After test staining with safranine aniline blue, toluidine blue, a polychromatic stain commonly used to highlight fungal structures in plant cells walls [70] or to examine erosion bacteria degradation in waterlogged wood [71], was used.

Thin sections were obtained by immersing the samples in 70, 80 and 96% ethanol (Alcosuisse, Switzerland) and twice in 100% ethanol (Merck, Darmstadt, Germany) for at least 3 h each at room temperature. The samples were then immersed twice in 100% acetone (Merck, Darmstadt, Germany) for at least 2.5 h at room temperature and in increasing concentrations of Epon/acetone solution (1:2, 1:1, 2:1). Finally, they were embedded in 100% Epon (Fluka, Switzerland) and left to harden at 60 °C for 5 days.

Thin sections of 1 μm were produced for LM with a UC6 ultramicrotome (Leica Microsystems, Vienna, Austria) and put on glass slides with 1 mL of distilled water. The glass slides were placed on a hot plate with a temperature of 55 °C to stretch the sections. After evaporating the water, they were stained with an aqueous solution of 0.5% toluidine blue O (w/v) (Merck, Darmstadt, Germany) for 3 min at 55 °C. They were then washed in water, dried on the hot plate and mounted with Entellan® (Merck, Darmstadt, Germany).

Ultrathin sections (75 nm) were cut from the same embedded samples for electron microscopy (TEM). The sections, mounted on 200 mesh copper grids, were stained with Uranyless (Electron Microscopy Sciences, Hatfield, PA, USA) and lead citrate (Leica Microsystems, Vienna, Austria) with an ultrastainer (Leica Microsystems, Vienna, Austria).

2.3. *LM and TEM*

The permanent slides were investigated with an Olympus BH-2 light microscope using transmitted and polarized light at different magnifications (04×, 10×, 40×, 60×). Photographs were taken with a Jenoptik ProgRes SpeedXT digital camera. If needed, Helicon Focus 6 stacking was used to produce a fully focused image.

TEM ultrathin sections were examined with a transmission electron microscope (CM12, Philips, Eindhoven, the Netherlands) equipped with a digital camera (Morada, Soft Imaging System, Münster, Germany) and image analysis software (iTEM). ImageJ was used to quantify dimensions in the sections.

3. Results

The micromorphology of cell structure and degradation was investigated on transverse and radial sections. To facilitate the comparison among the results, mostly radial sections are shown in this paper.

We first present a careful description of the micromorphology of the reference material and the contemporary bark decayed in nature, followed by an analysis of the 13 archaeological samples. The morphological features were systematically described, and 13 degradation characteristic patterns were identified and named with abbreviations to facilitate their localisation on the LM and TEM images (Table 2). For each feature, example LM and TEM images are presented. Macroscopic images of the samples are shown in the previous section. The occurrence of degradation features in the samples is summarized in a table.

Table 2. Occurrence of degradation features in samples. x indicates the feature was present in the analysed sample. Fractures, unfolding of thin-walled cells and hyphae are also present in the reference material, a piece of contemporary intact birch bark.

	Fra	Un	No-bi	In-st	Swe	Ir	Det	Rem	Re-th	Hy	Vo	Me	Acc
Ref	x	x								x			
0	x	x	x	x-fu	x-fu	x	x-fu-wl	x-fu	x-fu	x	x-fu	x-fu	
1		x	x	x-fu	x-fu		x-fu-cl	x-fu	x-fu			x-fu	
2	x	x	x	x-ov	x-fu		x-fu-cl	x-fu		x			
3	x	x	x		x-ov	x	x-ov			x (rarely)			
4	x	x, x-st			x-ov, x-fu	x	x-ov-wl, x-fu-cl	x-fu					
5		x					x-ov-wl, x-fu-cl	x-ov					
6	x	x	x	x-fu	x-fu		x-wl-fu	x-fu	x-fu	x	x-fu	x-fu	
7	x	x	x	x-fu	x-fu		x-fu-cl, x-fu-wl	x-fu		x	x	x	
8	x	x	x	x-ov	x-ov	x	x-ov	x-ov					
9	x	x						x-ov, x-fu		x (rarely)			
10	x	x		x-ov	x-fu, x-ov			x-ov, x-fu		x	x-fu		
11	x	x			x-fu			x-fu		x	x-fu		
12	x	x	x	x-ov	x-ov		x-fu?-wl, x-fu?-cl	x-fu?		x (rarely)	x-ba	x-ba	
13	x	x	x	x-ov	x-fu		x-fu-wl	x-fu		x (rarely)			x

The following features may be associated with fungi or bacteria, or overall in the sample. To distinguish among these cases, the suffix fu (fungi), ba (bacteria) or ov (overall) is added to the abbreviation:

- Fra: Fracture of radial cell walls;
- Un: Unfolding of compressed thin-walled cell; if it is in conjunction with stretching of the thick-walled cells, it is marked as Un-st;
- No-bi: Absence of birefringence of secondary wall of thick-walled cell;
- In-st: Intense toluidine blue staining;
- Swe: Swelling and/or disaggregation of the secondary cell walls;
- Ir: Increased irregularity of cell shape and size leading to irregular cell arrangement;
- Det: Detachment, either of cells from each other (Det-cl) or secondary cell walls from the compound middle lamella (Det-wl) or the tertiary wall;
- Rem: Removal of phenolic compounds in the lumen of thick-walled cells or the compound middle lamella or in the secondary cell walls;
- Re-th: Reduced cell wall thickness;
- Hy: Hyphea in thin-walled cell lumen;
- Vo: Circular/oval voids in the corners of secondary thick-walled cell walls with a diameter of about 0.6 μm, or smaller voids in the secondary cell walls with diameter of about 1.2 μm aligned along or in the compound middle lamella; the former are

caused by mechanical penetration in the wall by hyphae, the latter by the penetration of bacteria;

- Me: Complete dissolution of cell structure;
- Acc: Accumulation of foreign substance in the cell lumen of thin-walled cells.

3.1. Reference Material: Contemporary Intact Birch Bark

The bark morphology of the reference sample was examined with light microscopy (Figure 2). The radial thin sections revealed a compactly arranged and regular cell structure consisting of an equal number of thin-walled and thick-walled cell bands, with the former being more hydrophilic than the latter (Figure 8). The thin-walled cells were heavily compressed in the outer portion of the bark but originally had a square shape, visible in the centre of the radial thin section. The shape of the lumen varied and was generally lenticular. The compound middle lamella and tertiary walls had a deeper blue colour than the secondary walls, an indication of their greater affinity for water [12]. The barely coloured thick-walled cell walls, on the other hand, showed strong birefringence. Suberin had a crystalline structure showing birefringence [72].

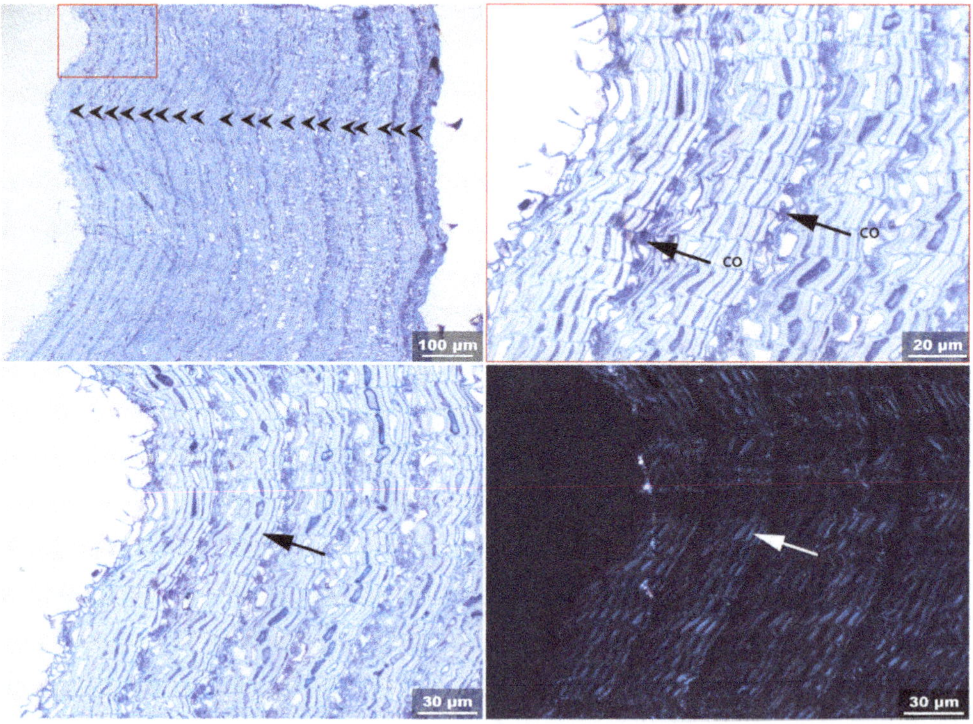

Figure 8. Reference sample. LM images of radial thin sections (**left**, outer surface; **right**, inner side) reveal thick-walled and thin-walled cell bands (arrowheads, **top left**). Tips of arrows indicate boundaries between thin-walled and thick-walled cell bands. Compressed thin-walled cells are marked with "co" (**top right**). Water affinity of thin-walled cell walls, compound middle lamella and tertiary walls can be appreciated from the intensity of blue staining. Thick-walled cell walls are hydrophobic (light blue) and show strong birefringence (**bottom**, **left** and **right**). Sample shows overall strong birefringence of thick-walled cells. While thin-walled cells usually have only very weak birefringence, the birefringence of thin-walled cell layer at boundary with thick-walled cells is just as strong as that of thick-walled cells.

The TEM investigation of the reference sample shows the arrangement of cells and the layered structure of cell walls (Figure 9). In the thick-walled cell bands, the radial compound middle lamella was about 0.24 ± 0.03 μm thick with a net-like structure in the centre and dark rims. The tangential compound middle lamella was 0.08 ± 0.01 μm thick and electron dense, an indication of higher lignification [73,74]. Interruptions in the tangential compound middle lamella were present. At the transition between the thick- and thin-walled cell bands, the radial compound middle lamella built a thick corner with ends susceptible to fracture. The thick-walled cells were filled with electron-dense compounds with an irregular shape. In the thin-walled cell bands, it was not possible to distinguish between radial and tangential compound middle lamella, as the cells were compressed. The compound middle lamella had a homogeneous structure and was about 0.08 ± 0.02 μm thick. The secondary wall is suberized, and therefore had a lamellar structure with very fine dark lines, arranged around dense globular particles. The tertiary wall is the boundary layer at the lumen and was very dense. The thin-walled cells were partly filled with a very dark multiform compound.

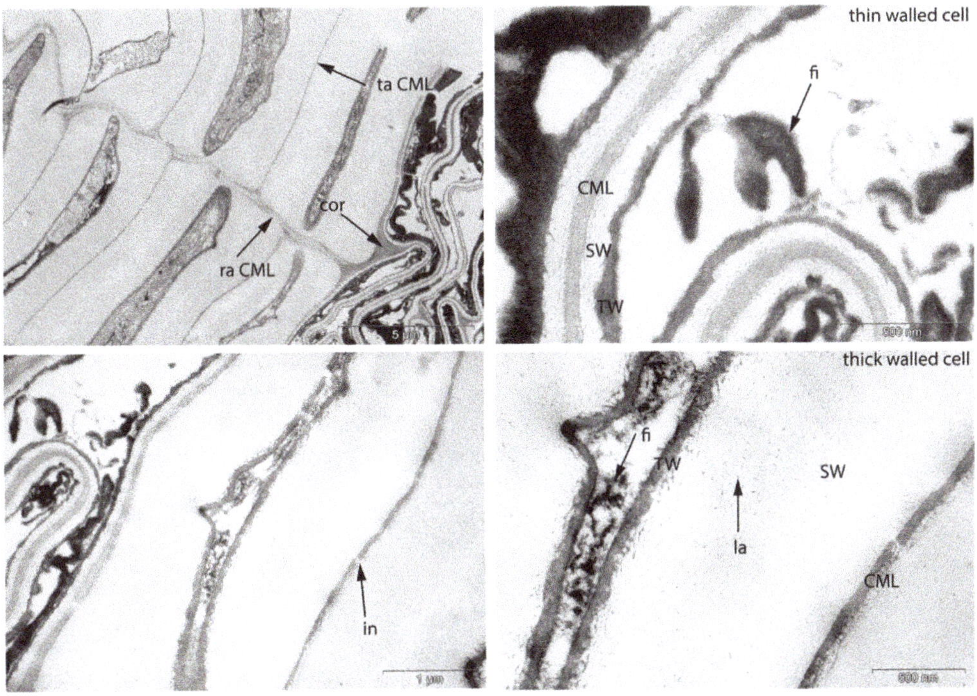

Figure 9. Reference sample. TEM images of thick-walled and thin-walled cells. The compound middle lamella (CML) is thicker in radial direction (ra) than tangential direction (ta) in thick-walled cell bands (**top left**). In tangential direction, interruptions of composite middle lamella are present (in, **bottom left**). There is a clearly pronounced corner (cor) at the transition between thick- and thin-walled bands. Details of thin-walled cell (**upper right**) and thick-walled cell (**bottom right**) are presented. SW, secondary wall; TW, tertiary wall; la, lamella of secondary suberized cell wall; fi, filling material.

Close to the lenticel, the reference sample was delaminated (Figure 2, left). LM images show that separation of the layers occurred at the boundary between thin- and thick-walled cell bands. Fracture occurred in the radial thin-walled cell walls (Figure 10, left). The consequence was a tangential separation (delamination) of the layers and the unfolding of thin-walled cells. The dark blue and purple balls and bubbles and the pinkish cloudy structures visible on the outermost layers of the toluidine-stained radial thin sections indicate the presence of hyphae. TEM images of the same area show fungal structures in the lumen of the cells and a hypha penetrating from one cell to another.

The breakthrough has thin dark coloured edges, an indication that the cell wall was altered by enzymatic degradation.

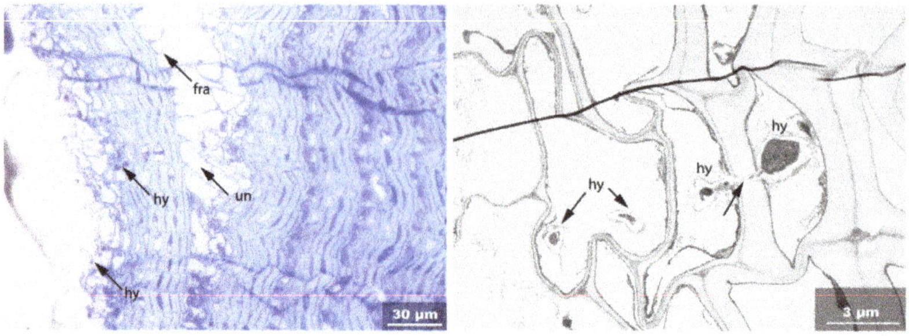

Figure 10. Reference sample. LM images (**left**): delamination due to breaking of boundary radial cell walls (fra) accompanied by unfolding of compressed thin-walled cells (un). TEM images (**right**): hyphae within thin-walled cell lumen and penetrating all wall layers.

Figure 11 shows a lenticel. Lenticels are fanned out layers with a convex shape, more pronounced towards the phloem. Proceeding tangentially from the normal phloem to the lenticel, the thick-walled cell bands form a continuous structure. The transition between normal phloem and lenticel involves about 4–5 cells where the cell lumen of the thick-walled cells changes from a square to ellipsoid shape, while the walls of the thin-walled cells become heavily fractured. The radial cohesion of the thick-walled cells has numerous failures. The remains of the fractured thin-walled cells are strongly stained, an indication of their highly hydrophilic nature. The thick-walled cells display no birefringence in the centre of the lenticels, while this increase approaching the transition to the normal phellem.

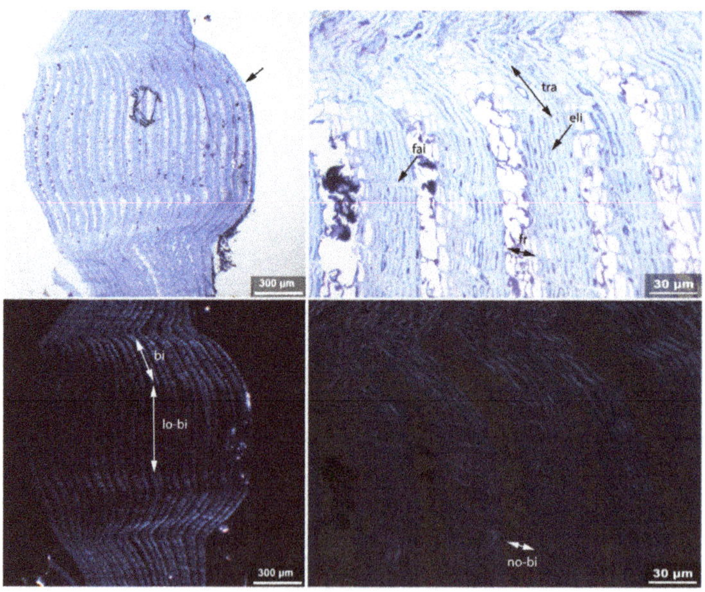

Figure 11. Reference sample. Lenticel with fanned out thick-walled bands (**top left**). Detailed image of cell structure (**top right**) shows ellipsoid lumen (eli) of thick-walled cells and fractures (fr) of thin-walled cells. Birefringence of thick-walled cells is low at lenticel centre (lo-bi, **bottom left**) and rises at the transition to normal phellem (bi), while it is absent in thin-walled fractured cells (no-bi, **bottom right**).

TEM images (Figure 12) show that within the tangential thick-walled cell walls of the lenticel, plasmodesmata about 25 to 30 nm in diameter are present. They are unbranched and have a tubular shape, and they connect two cells. They are filled with electron-dense material, and occasionally a deposit is visible on the inner surface of the cell. In the lenticel, small intercellular spaces can be found, while they are absent in the normal phellem.

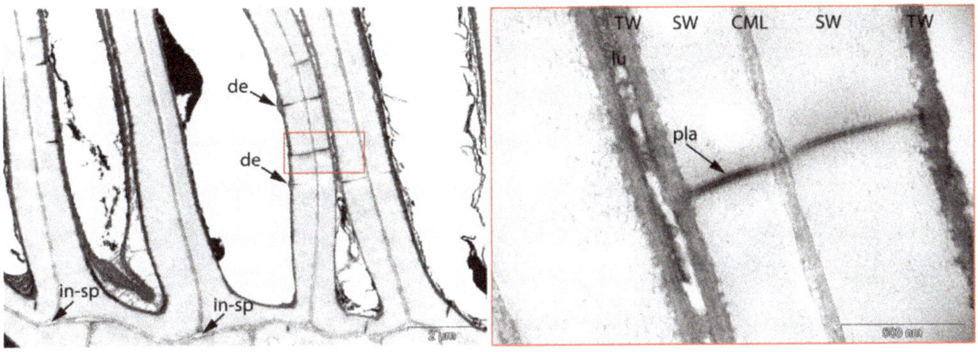

Figure 12. Reference sample. TEM images of thick-walled cell of lenticel with plasmodesmata (pla) in tangential wall and intercellular spaces (in-sp) in radial cell corners. Tubular plasmodesmata are filled and occasionally have a deposit (de) towards the cell lumen.

3.2. Sample 0: Contemporary Birch Bark Degraded in Nature

This sample presents all identified degradation features except for decay by bacteria and the accumulation of foreign substances, which was present only in sample 13, retrieved from a cave environment.

In the LM pictures, it is possible to identify degraded areas characterized by unfolding of the thin-walled compressed cells and an irregular arrangement of the cell structure. Birefringence is strongly present in the well-preserved areas, whereas it is completely missing in the degraded areas. All layers of thick-walled cells are visible in the well-preserved areas (Figure 13). The toluidine staining is stronger on the suberized secondary cell walls and weaker on the compound middle lamella. The phenolic compounds in the cell lumen are absent.

The penetration of the secondary wall by hyphae caused circular voids located at the cell corners. Further, swelling and loosening of the wall is visible, leading to complete secondary wall disintegration. Finally, only a dark blue unstructured substance remains. Hyphae are visible in the lumen of the cells, leading to decay in parts of the tertiary walls (Figure 14, bottom left and top right).

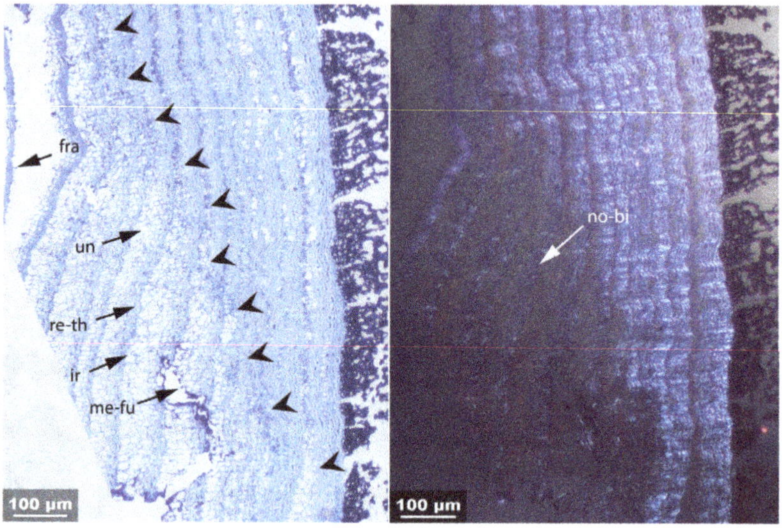

Figure 13. Sample 0. LM overview under transmitted light (**left**) and polarised light (**right**). Arrows on the left mark transitions between differently preserved areas. Degraded areas are characterized by unfolding of thin-walled cells (un), increased irregularity of cell arrangement (ir) and reduction of cell wall substance, especially suberized secondary wall of thick-walled cells (re-th) up to complete dissolution of cell structure (me-fu). Fracture of thin radial walls and layer separation are also visible (fra). In degraded area, thick-walled cells do not show birefringence (no-bi).

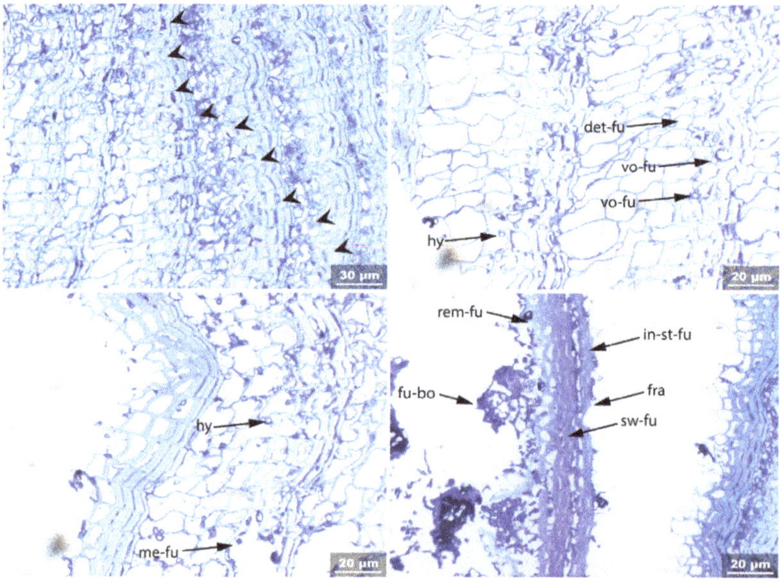

Figure 14. Sample 0. LM overview of sample with clear boundary between well-preserved area and fungal degraded area (arrowheads). Secondary walls of thick-walled cells are penetrated mainly in the corners by hyphae (vo-fu), and hyphae are present in thin-walled cell lumen (hy). These adhere to the tertiary cell wall and sometimes cause it to detach (det-fu) (top right). At bottom left, complete dissolution of cell structure is visible (me-fu). At bottom right, fungal structures can be seen (fu-bo). These cause a decrease in lignin-containing middle lamella (rem-fu) and phenolic lumen filling of thick-walled cells. At the same time, darker staining (in-st-fu) and swelling of secondary cell wall occur (sw-fu). Fractures due to breaking of radial thin-walled cells (fra) are also visible.

The TEM pictures show that the secondary cell walls, which are infested with hyphae in the corners, causing oval voids, are swollen, their structure is loosening and their shape is irregular (Figure 15). Advanced degradation is indicated by disintegration of the wall; in some cases the cell wall seems to be completely digested, while in others an electron-dense granular residue is left. Similar compounds have been identified in sclereid of beech bark in the area around the hyphae, and interpreted as remnants of not completely degraded lignin [75]. Connected to the infestation with hyphae, the phenolic compounds in the lumen, compound middle lamella and secondary walls were removed (Figure 16). This led to separation of the cells along the compound middle lamella.

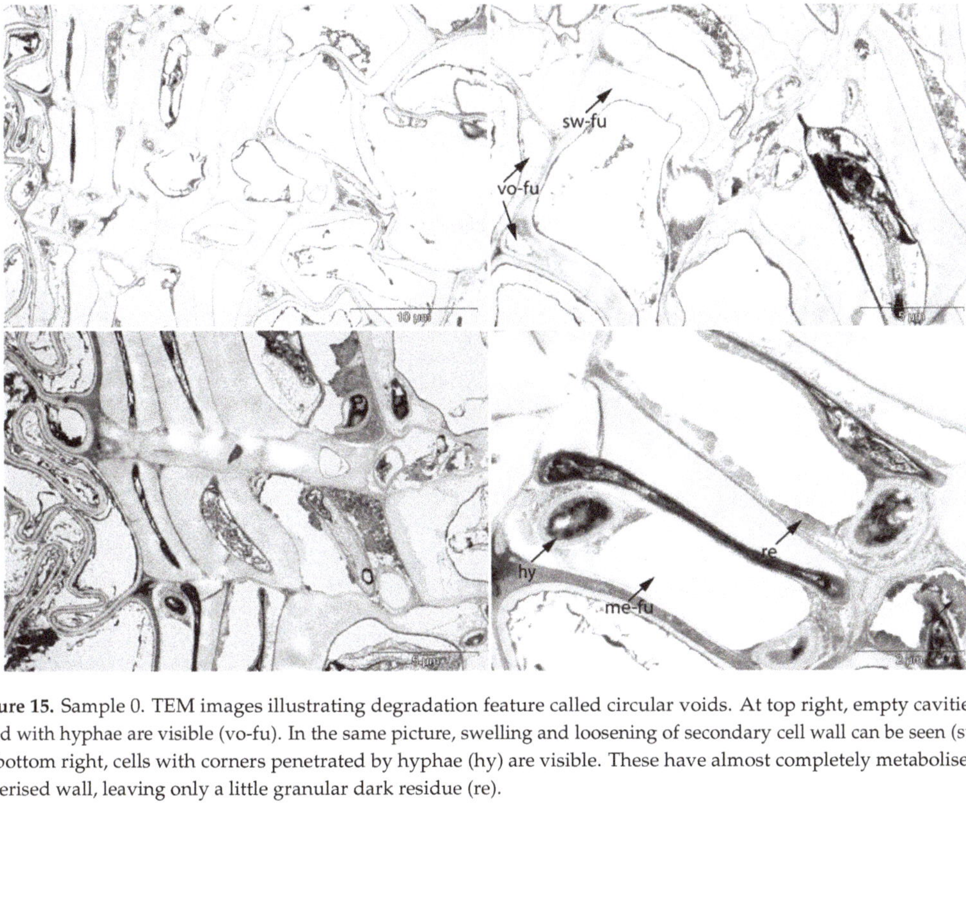

Figure 15. Sample 0. TEM images illustrating degradation feature called circular voids. At top right, empty cavities not filled with hyphae are visible (vo-fu). In the same picture, swelling and loosening of secondary cell wall can be seen (sw-fu). At bottom right, cells with corners penetrated by hyphae (hy) are visible. These have almost completely metabolised the suberised wall, leaving only a little granular dark residue (re).

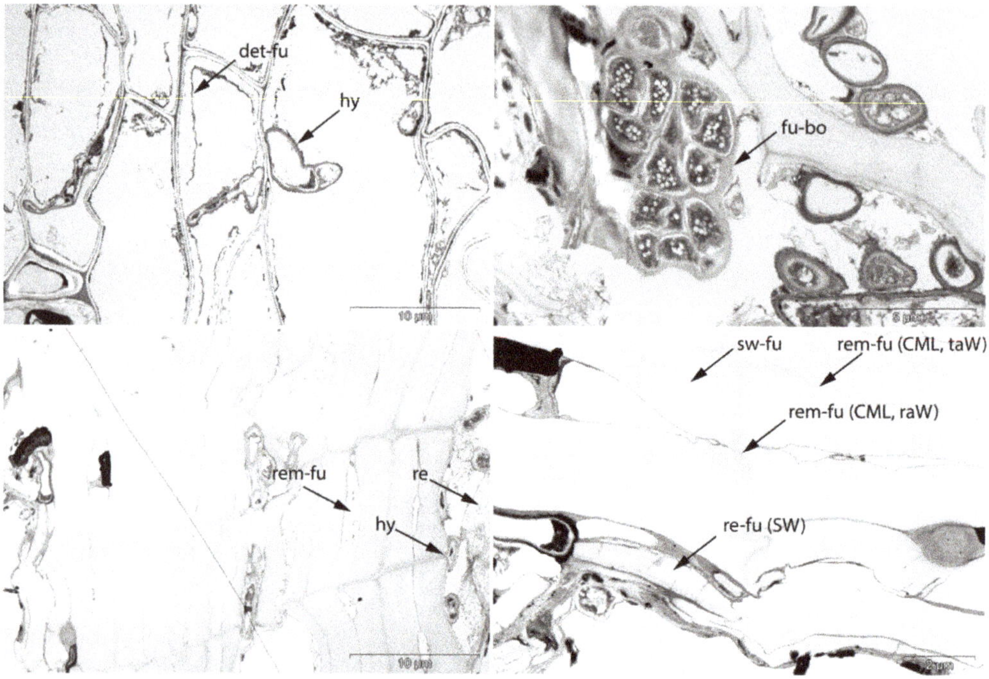

Figure 16. Sample 0. TEM images. **Top left**: hyphae (hy) in thin-walled cells and partial detachment of tertiary cell wall layer (det-fu). **Top right**: fungal body (fu-bo) and hyphae in thick-walled cell bands caused removal of lignin from middle lamella (rem-fu, CML) (**bottom right**) and phenolic content from cell lumen (rem-fu) (**bottom left**). Suberised cell wall swelled and lost narrow lamellae, an indication of suberisation (sw-fu). Wide light and dark bands appear, and in an advanced stage of degradation, the thickness of the wall is reduced (re-fu, SW).

3.3. Archaeological Samples

3.3.1. Sample 1

In the archaeological samples, the presence of biotic decay was macroscopically visible by surface discoloration. Swelling of the cell walls, removal of substance and increased staining were commonly found microscopic features of biotic decay. One example is provided by a dark discoloured area on sample 1, extracted from the lower body of the Schnidejoch bow case on the ground-facing side (Figure 4, top left). The cut section of the sample had a brown colour with alternating white betulin-filled bands and showed no delamination or breaking-off pieces due to the cutting process. LM images reveal a very heterogeneous preservation condition. The area marked 1 in Figure 17 is characterized by an intact regular cell structure with compressed thin-walled cells and slight birefringence of the thick-walled cells. The blue coloration of the secondary cell walls is slightly darker compared to the reference material. The area marked 2 is located at the surface of the sample and is characterized by the presence of fungi. The effect of fungi can be observed in both LM and TEM images. They cause swelling and disintegration of the suberized cell walls (Figure 17), which acquire a dark blue colour upon staining and exhibit no birefringence. Further, fungi cause degradation of the compound middle lamella, reducing its thickness (Figure 17). Finally, the cells are detached and the phenolic filling material in the thick-walled cells are degraded (Figure 17).

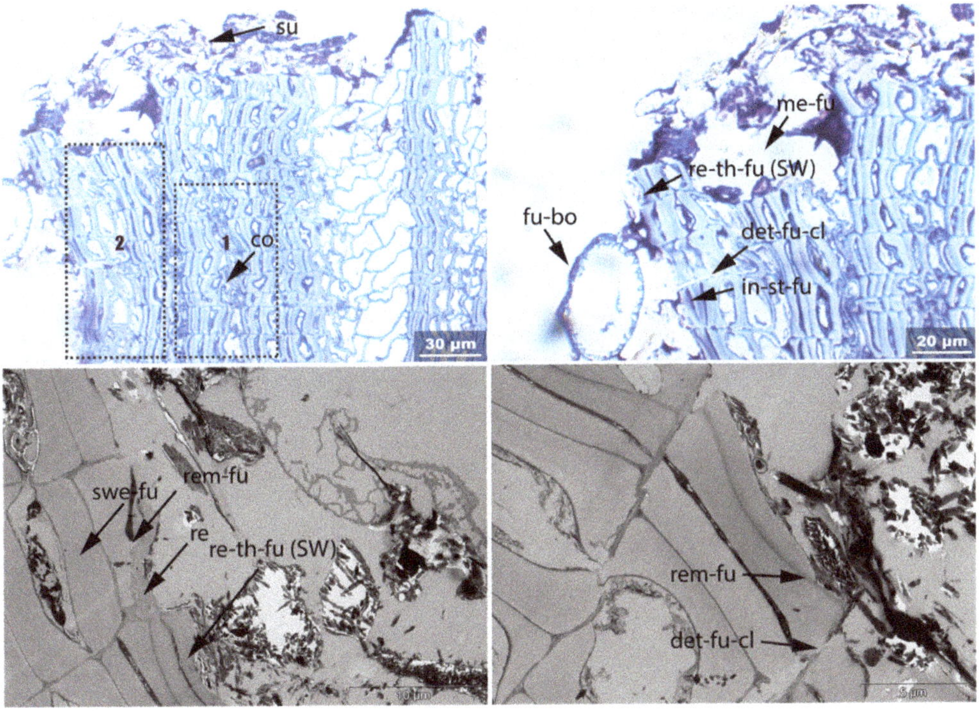

Figure 17. Sample 1. Top left: LM image of thin section. Area 1 has very well-preserved cell structure; area 2 is biodegraded. Superficial deposits are visible (su). In area 1, thin-walled cells are still compressed (co) and secondary cell wall layer of thick-walled cells is still birefringent (not shown). In area 2, clear change is noticeable in secondary cell wall layer: dark blue staining (in-st-fu), swelling (swe-fu) and dissolution of the wall (re-th-fu). Middle lamella (rem-fu) is also degraded (rem-fu), eventually leading to radial detachment of cells (det-fu-cl) (top right). Cavity in upper area (top right) between superficial and cell structure can be seen, in which complete cell structure dissolution occurred (me-fu). Changes to cell walls can be seen in more detail in TEM: thick-walled suberised cell walls are clearly swollen and loosened (swe-fu), after which they lose thickness (re-th-fu, SW)), leaving unstructured residue (re). Phenolic content of cell lumen is also degraded (rem-fu) (bottom left). Alterations of compound middle lamella can be seen at bottom left; wall loses thickness and electron density area is reduced (rem-fu).

3.3.2. Sample 2

Fractures and unfolding of the compressed thin-walled cells are particularly visible in sample 2. Sample 2 (Figure 18, top left) was extracted from the middle bark strips in a dark-brown discoloured area. The cut section showed delamination and breaking-off pieces due to the cutting process. LM images (Figure 18, top right) reveal strong tangential delamination due to the breaking of radial cell walls. The secondary cell walls of the thick-walled cells have a dark blue colour, an indication of higher water absorption. No birefringence of thick- and thin-walled cells was detected, and the thin-walled cells were no longer compressed (bottom left).

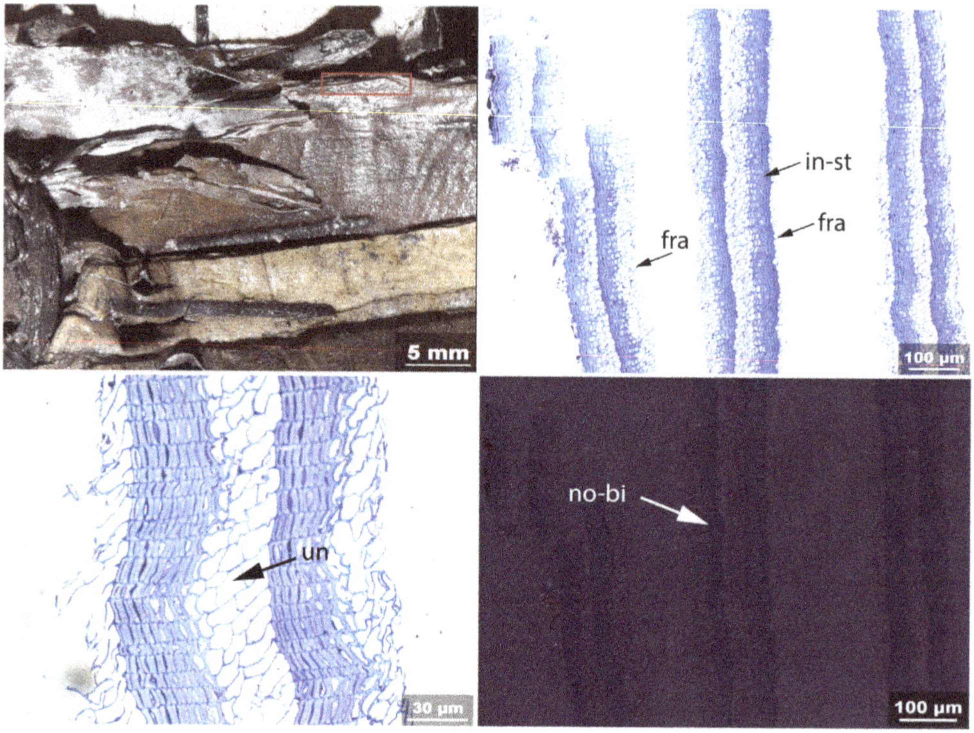

Figure 18. Sample 2. **Top left**: location of sample. **Right**: LM of thin sections. Tangential delamination is visible due to fracture (fra) of radial thin-walled cells. Increased toluidine staining of thick-walled cells is also appreciable (in-st). **Bottom left**: thin-walled cells in normal phellem are overall unfolded (un). **Bottom right**: birefringence is absent in both thick- and thin-walled cells.

3.3.3. Sample 3

Increased irregularity of the cell structure and detachments are particularly visible in sample 3. Sample 3 was extracted from the inner lining and contained a lenticel; no surface discolouration was present. LM images reveal that all cell wall layers were present, despite being swollen and irregular (Figure 19). In some areas the thin-walled cells were still compressed. Many broken radial cell walls at the boundary between thick- and thin-walled cell layers were present, but no complete delamination was detected. Slight fungal degradation was present on the left outer layer, which caused a dissolution of cell walls and a reduction in the middle lamella and phenolic compounds. Thick- and thin-walled cells displayed no birefringence.

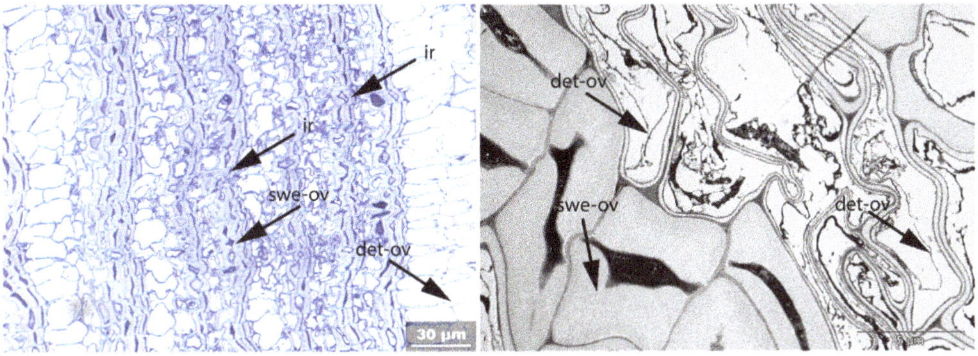

Figure 19. Sample 3. **Left**: LM image. Secondary cell walls of thick-walled cells are swollen (swe-ov). Deformation of cells has also taken place, leading to irregular structure (ir). Parts of tertiary cell wall of thin-walled cells are detached (det). **Right**: TEM image. Swelling of secondary cell wall of thick-walled cells is visible, together with detachment of parts of tertiary cell wall of thin-walled cells. In sample 3, this phenomenon is widespread and not localised.

3.3.4. Sample 4

A further degradation feature is the detachment of the secondary cell wall from the compound middle lamella and the tertiary wall. This is visible in sample 4. Unfolding of thin-walled cells and stretching of thick-walled cells is also present. Sample 4 was taken from the bow case inner lining, but the original position on the object is unknown, possibly the lid. The birch bark is slightly pliable and retains some strength. The surface is compact but shows cracks and delamination in the top half (Figure 4c). The bark is somewhat brittle, especially the lenticel. The outer side of the bark is grey, the inner brown. In the lower area, a black cloudy discoloration is visible that is also present on the edges. Red-brown colours are strongly reduced.

Thin section 4a shows a higher degree of decay than 4b. This sample is characterized by unfolding of thin-walled cells, elongation in the transverse direction of thick-walled cells and increased staining. Along with this, the structure becomes irregular. While in sample 4b the cell structure is regular, the walls of the thin-walled cells are still slightly folded and there is only slight light blue staining of the secondary cell walls, an indication of low water sorption.

The detachment of the secondary cell wall from the compound middle lamella and the tertiary wall affected large areas of thin section 4a and only slightly affected thin section 4b (Figure 20). TEM images (Figure 21, bottom right) show that due to this separation, voids and threadlike structures appear between the wall layers. The very thin, delaminated tertiary cell wall layer is present as a ring within the lumen. The swelling of the secondary cell wall causes a wave deformation of the secondary wall layer. The compound middle lamella is broken, and interruptions are present. The original structure of the suberin lamella has disappeared and broad light and dark bands with the same orientation as the former suberin lamella have formed. The degradation of the compound middle lamella led to a separation of the cells and a degradation of the phenolic content, visible as reduced cell lumen content.

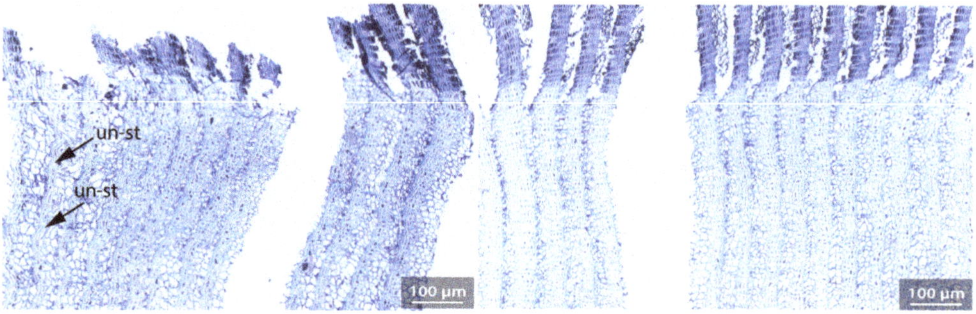

Figure 20. Sample 4. Two thin sections, about 4 mm apart, were made from the embedded sample: thin section 4a (**left**) and thin section 4b (**right**). Thin section 4b reveals a much better-preserved cell structure than 4a: it is regular, walls of thin-walled cells are still slightly folded and there is only slight light blue staining of secondary cell walls, an indication of low water sorption; thin section 4a shows unfolding of thin-walled cells and stretching of thick-walled cells (un-st).

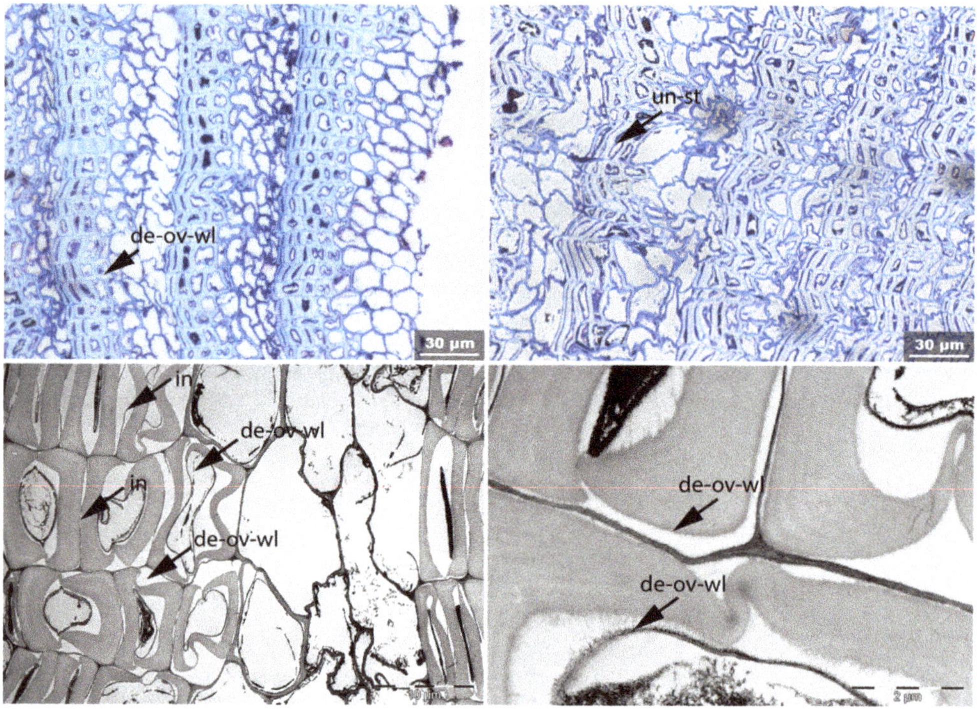

Figure 21. Sample 4a. Detachment of secondary cell wall from compound middle lamella is found throughout (de-ov-wl, **top left**). Elongation of thick-walled cells (un-st) is present (**top right**). TEM (**bottom left**) shows detachments of secondary wall from compound middle lamella and tertiary wall (de-ov-wl). Very thin, delaminated tertiary cell wall layer is present as a ring within the lumen. Compound middle lamella is broken and shows large interruptions (in). Swelling caused wave deformation of secondary wall layer. Separation and threadlike structures appear between secondary and tertiary wall (**bottom right**). Original structure of suberin lamella has disappeared and broad light and dark bands with the same orientation as the former suberin lamella are formed.

3.3.5. Sample 6

Fungal degradation causes circular/oval voids in the secondary wall of thick-walled cells, and an advanced state of decay leads to complete dissolution of the structure (Figure 22). Both are found in sample 6, extracted from the upper part of the bow case body on the ground-facing side. The area where the sample was taken was fully exposed to the soil. All exposed surfaces showed discolouration macroscopically. The LM reveals strong fungal degradation starting from both exposed sides. The strongest degradation is found at the surface layers, and some of the innermost layers are not affected. The circular voids are mainly located in the corners and were caused by the penetration of hyphae. In contrast to the decayed in-nature sample 0, the remains of hyphae in the holes are no longer present in sample 6. In an advanced stage of decay, swelling of the secondary cell walls occurred, followed by their disintegration. The skeleton of the middle lamella was the last to be metabolised. A superficial structure that consists of cell fragments, fungal structures and cloudy, dark blue parts remains on the sample surface.

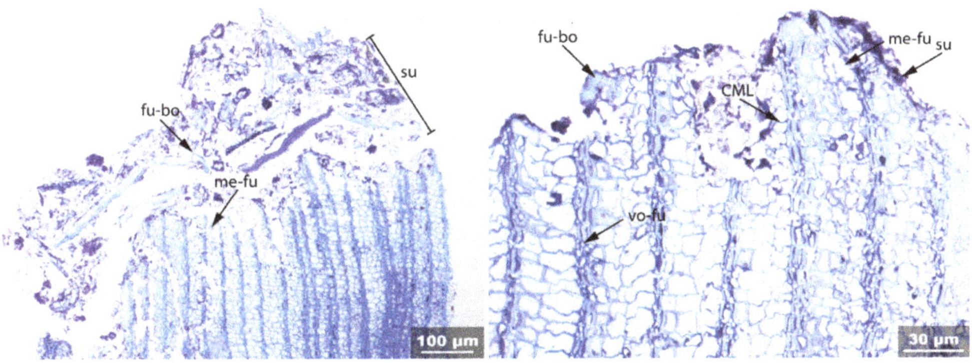

Figure 22. Sample 6 shows a superficial structure (su) consisting of cell fragments, fungal structures (fu-bo) and dark blue cloudy remains (**left**). Empty circular/oval voids in corners (vo-fu) of suberized cell walls of thick-walled cells caused swelling and then dissolution of secondary wall (**right**). At top of sample, complete disintegration of cell structure (me-fu) is visible. Fungal structures cause a loss of cell wall material (secondary cell wall, cell lumen filling material) and in an advanced state complete disintegration of all cell wall layers (me-fu).

3.3.6. Sample 10

A further degradation feature is the removal of phenolic cell wall components from all cell wall layers and cell lumen-filling substances, leaving behind a wall that is bleached but unchanged in thickness. This phenomenon was only found on the surface, up to 0.06 mm deep. An example is sample 10, extracted from a rolled birch bark recovered in permafrost in Lendbreen. Both surfaces were white-grey and lost the red-brown colour, which was retained in the inner layers (Figure 5). The bleaching involved the superficial five cell layers (Figure 23). The compound middle lamella, the secondary and tertiary walls and the cell lumen lost components that bind with the toluidine dye and cause birefringence. The thickness of the walls was unchanged, and no disaggregation of the structure occurred. After a transition of a few cell layers, in which the secondary cell wall was slightly swollen and more strongly stained, the deeper layers showed an unchanged appearance, birefringence was present to a large extent and the thin-walled cells were still compressed. Thick and thin-walled cells were stained with a similar blue tone, slightly darker than that of the reference material.

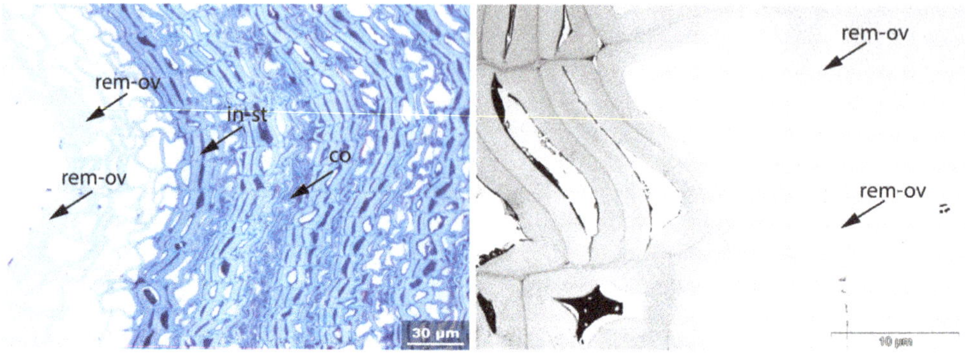

Figure 23. Sample 10. **Left**: LM. Loss of phenolic compounds from cell wall layers and lumen is visible in left outer bands (rem-ov). Deeper layers are unchanged and display birefringence (not shown). Thin-walled cell walls are compressed (co). In transition zone between bleached and unchanged phellem a slight swelling (swe) and increased blue staining (in-st) of secondary cell wall is visible. **Right**: TEM. Overall loss of phenolic components is significant and a change in appearance of the thin lamella structure of secondary cell wall is visible.

3.3.7. Sample 11

In some cases, fungal infestation can be clearly visible by the presence of hyphae within the lumen of thin-walled cells, but the cell structure retains regularity, birefringence and water-repellent properties. An example is sample 11, extracted from a pliable and considerably strong Lendbreen fragment. The colour of the fragment was grey-whitish matte with a weathered surface; red and brown colours were missing superficially but were retained in the inner part of the sample (Figure 5).

LM shows a very regular, compact birefringent structure that reflects a good preservation condition. The walls of thick-walled cells have a pale colour and are therefore still water-repellent (Figure 24, left). Round cavities in the secondary walls and hyphae are found within the lumen of thin-walled cells. The hyphae are arranged along the tertiary wall and appear to metabolise it or parts of it from the lumen (Figure 24, right). At the contact point between the hyphae and wall, the wall thickness is often reduced. It can be assumed that hyphae are able to mechanically break through cell walls, as otherwise they would not have been able to distribute themselves so extensively throughout the cell structure. However, the secondary cell wall and the composite middle lamella are not degraded over a wider area by this type of fungus.

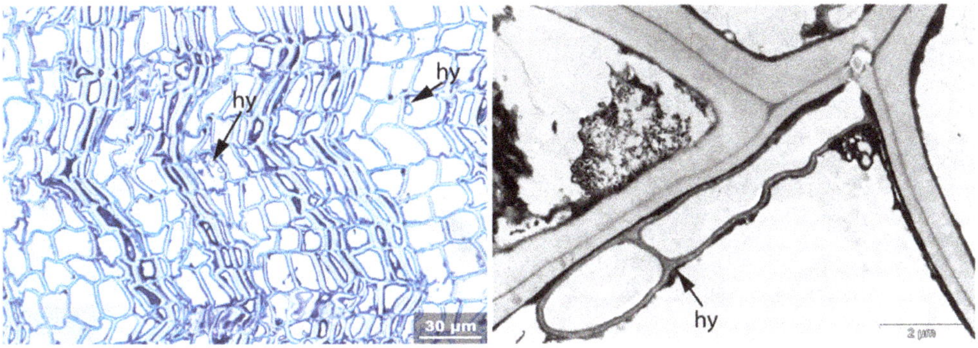

Figure 24. Sample 11. **Left**: LM. Hypha present in lumen of thin-walled cells distributed over entire thin section. Hyphae are aligned along tertiary wall and cause reduced thickness. **Right**: TEM. Division of hyphae into two cells (septum) is present. Defects are visible in a corner of the secondary cell wall and the middle lamella, possibly caused by hyphae.

3.3.8. Sample 12

The waterlogged sample found in anoxic sediment showed biotic degradation mainly by bacteria, and to a much lesser extent by fungi. Bacterial degradation was recognizable as many small voids in the secondary cell walls and complete metabolism of the cell structure. Macroscopically, the strength of this Neolithic sample was considerably reduced; many fractures occurred during cutting. Due to the waterlogged condition, the susceptibility to layer separation could not be determined. The bark can still be identified as birch bark, but the surface is brown.

LM thin sections (Figure 25) reveal a regular cell structure with un-folded and overall swollen thin-walled cells. Fractures in the radial thin cell walls are present in some areas, but complete delamination did not occur. The secondary walls show increased staining and no birefringence (not shown). The phenolic content in the thick-walled cells is still present and very dark in colour, sometimes fractured. The fracture (in a radial direction) sometimes affects not only the lumen filling substance, but also the wall. Detachments of cells along the middle lamella and of cell wall layers within individual cells are present (not shown).

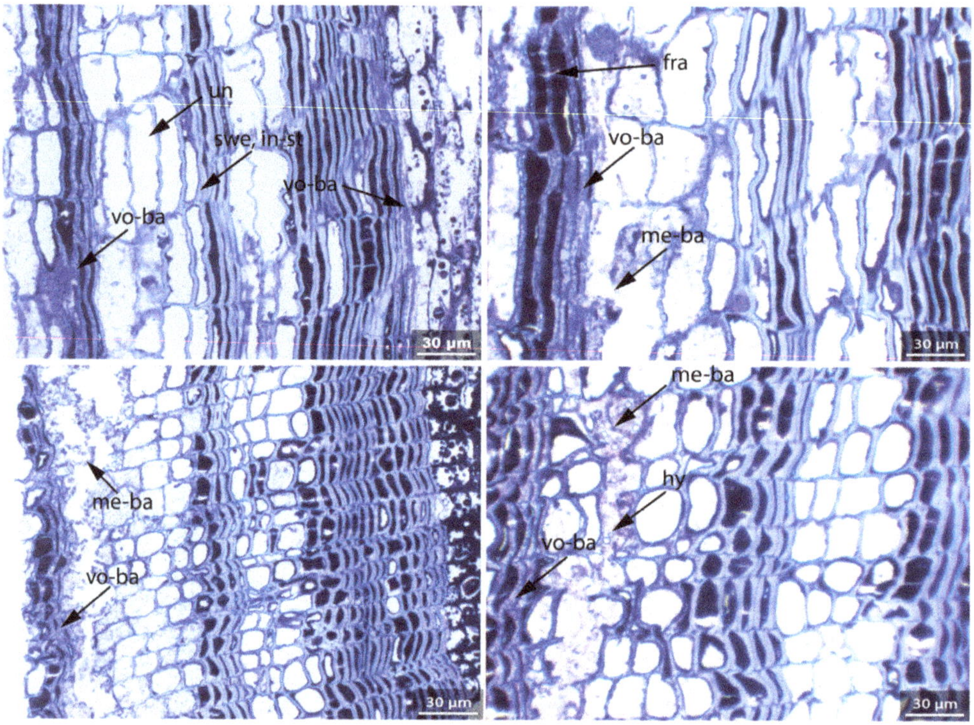

Figure 25. Sample 12. **Top left** and **right**: LM of cross-section. **Bottom left** and **right**: LM of radial thin section. Bacteria created voids (vo-ba) in secondary cell wall and, in advanced stage, metabolised all cell wall layers and phenolic lumen content of thick-walled cells (me-ba). Cloudy purple material was produced that accumulated in the lumen of thin-walled cells. Decay is strongest on sample surface, on left outer cell bands, indicating that bacteria penetrated from there. Few hyphae are found in the sample (hy). Entire sample shows increased staining (in-st), swollen cell walls (swe) and unfolded thin-walled cells (un). Radial fractures running across phenolic content and cell walls are present (fra).

The TEM examination showed that bacteria were mainly located in the secondary cell walls and were aligned along the middle lamella (Figure 26, right) and the tertiary wall, causing voids. Their location indicates that they metabolised carbohydrate-containing components of the compound middle lamella and the tertiary wall. In the advanced stage of degradation, the middle lamella and the phenolic components in the cell lumen of the thick-walled cells are also degraded. Based on the characteristic degradation traces, they can be identified as tunnelling bacteria [76].

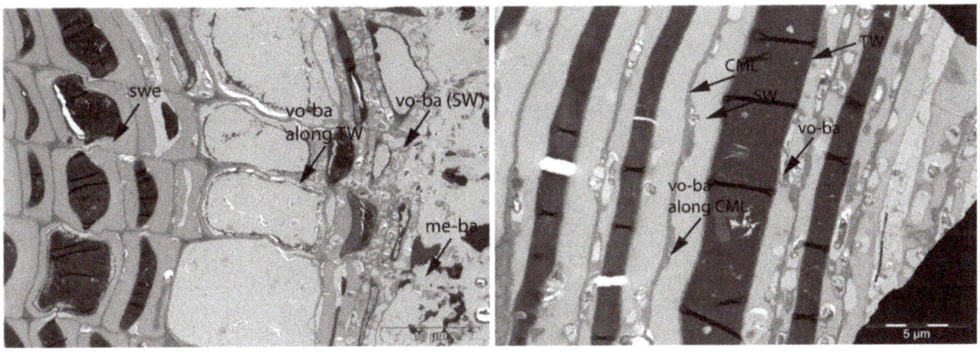

Figure 26. Sample 12. TEM: swelling and loosening of secondary walls of thick-walled cells (swe) and voids created by bacteria in secondary cell wall of thick- and thin-walled cells. Bacteria are aligned along compound middle lamella and tertiary wall, creating tunnels with concentric rings, characteristic of tunnelling bacteria. Based on the appearance, the presence of erosion bacteria is possible. Bacteria were able to digest all cell wall layers and phenolic lumen content (me-ba).

In addition to the biotic degradation, TEM sections show that the structure of the secondary cell wall has changed considerably: the wall is swollen, with broad light and dark stripes, and in some places, it is only preserved as a loose net with many defects. The cohesion of the cell wall layers is also greatly reduced, resulting in cell detachment (not shown).

3.3.9. Sample 13

The final degradation feature identified microscopically in the archaeological samples was the accumulation of foreign substances within the cell lumen. This was found in sample 13, recovered from the Kosackenberg cave. Both inner and outer surfaces of the sample had a reddish-brown overlay, while the internal layers had a light, slightly pink colour and the normally white and reddish-brown alternating bands were no longer perceptible. White blooming was present on both surfaces. The reddish-brown colouring of the thick-walled cells was completely lost. The lenticels lost their normally dark colouring and were hardly distinguishable from normal phellem. The sample was extremely fragile, and handling caused the loss of small fragments. The cohesion of layers was very reduced (Figure 7).

LM reveals a regular structure with several fractures of radial cell walls and complete layer separations (delamination). The thin-walled cells are unfolded, and increased staining of the thick secondary cell walls is visible; the closer to the surface, the darker the shade of blue. Over the entire thin section, a thick, dark-coloured, partially birefringent residue is visible on the surface (Figure 27). Furthermore, purple-blue cloudy material is found in the lumen of the thin-walled cells. It can be assumed that this is gypsum, as this was identified in analyses of birch cork from the same site [30]. As gypsum is poorly soluble in water, it is assumed that the ions were dissolved in water and then precipitated inside the cells and accumulated in the lumen. The phenolic content of the thick-walled cells is strongly reduced. In the cell layers near the surface, sporadic hyphae are found. Furthermore, close to the phloem, the secondary cell walls of thick-walled cells are extremely swollen or absent (not shown).

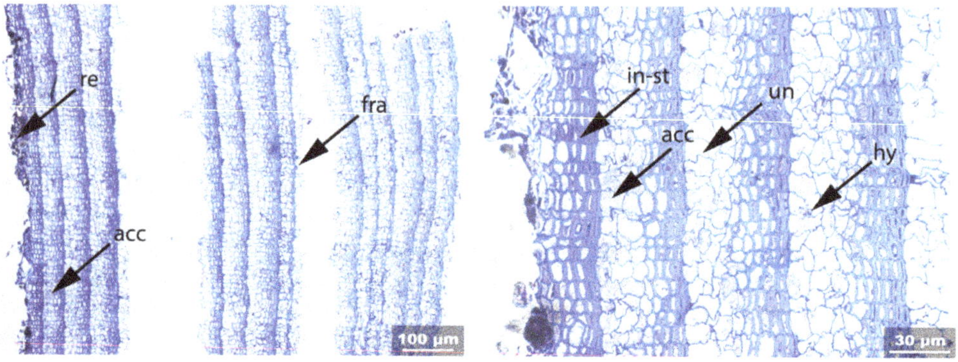

Figure 27. Sample 13. LM shows many fractures of radial thin-walled cells (fra) leading to delamination. Thin-walled cells are unfolded (un); thick-walled secondary cell walls reveal increased staining, especially close to the surface. Thick dark-coloured residue (re) covers the surface. An accumulation of a cloudy substance in the lumen of thin-walled cells (acc) is present; some lumens are completely filled.

4. Discussion

The identification and systematic description of microscopic degradation features and the macroscopic characteristics such as colour, brittleness and pliability of the 15 birch bark samples analysed in this study allowed us to establish a correlation between some macroscopic and microscopic features.

If delamination was macroscopically visible, fractures and detachments of thin-walled cell walls were found. If the sample surface showed discoloration, decay at a microscopic level was found. Bacterial decay was only found in the waterlogged sample, while fungal decay was common. However, if biotic decay is very localised, it might not lead to macroscopic colour changes. If the samples were exposed to sunlight, they lost their red-brown colour and appeared strongly weathered, and removal of the phenolic compounds from the cell walls and lumen could be found in the very outer layers. However, it was not possible to establish a correlation between macroscopic brittleness and microscopic degradation features. For example, the Lendbreen samples are macroscopically strongly weathered but microscopically intact, except for the outer layers. Similarly, the Kosackenberg sample is macroscopically extremely brittle, but microscopically all the cell wall layers have been preserved. Microscopic biotic decay might be present in well-preserved and pliable samples such as our contemporary reference sample. More importantly, the occurrence of microscopic degradation features is very local in birch bark and can vary within the same fragment.

The next step in interpreting the collected microscopic data was to try to establish any causality for the identified degradation features. Clearly, the presence of hyphae as well as voids in the cell walls, and the reduction in cell wall thickness and the complete disintegration of cells are all connected to biotic decay, mostly fungal, except for the waterlogged sample, in which bacteria were found. Biotically produced enzymes can also lead to the swelling of cell walls, removal of phenolic compounds, reduction in wall thickness and detachment of cells from each other or from the compound middle lamella [77], but in this case, these features were restricted to areas close to the hyphae. When the loss of phenolic compounds is present overall but confined to the upper layers, and when macroscopically the surface has lost its brown-red colour, photocatalysis of the phenolic compounds due to sunlight has probably taken place [23,78]. Irregularity of the cell structure and, consequently, the cell arrangement, swelling of the cell walls, increased toluidine blue staining and loss of birefringence can also be found in the absence of signs of biotic decay, and we suggest that they are due to chemical degradation of the cell walls. In particular, loss of birefringence can be connected to a loss of crystallinity of suberin [72,79]. Toluidine blue staining, though, is not selective for suberin, and a thorough investigation of suberin degradation should be performed using different methods.

Accumulation of foreign substances, probably gypsum, was only detected in the fragments recovered from the cave environment and might be due to the deposition of salts dissolved in the water the samples were exposed to. Finally, fracturing and unfolding of thin-walled cell walls and elongation of thick-walled cell walls are mechanical degradation features that might have chemical, biotic or physical causes. In particular, unfolding of thin-walled cells has been detected in combination with every degradation feature and is a reliable microscopic degradation marker. Unfolding has been found in other phellems upon exposure to solvents or heat [9,80]. The age of the samples did not correlate with the state of preservation, at either the macroscopic or microscopic level.

5. Conclusions

Careful examination of sections of 15 samples of birch bark, two contemporary and 13 archaeological, using light and electron transmission microscopy, allowed us to provide, for the first time, a comprehensive description of microscopic degradation features of birch bark and correlate with macroscopic visible features. Further, we could establish causes for a number of these features. Understanding the correlation between macroscopic and microscopic features is essential in order to develop passive or active conservation measures. The most important macroscopic features that allow us to predict the microscopic state of the samples' preservation are colour changes, such as surface discolorations and loss of the red-brown colour, and changes in mechanical properties, such as a loss of pliability, presence of delamination and increased brittleness. Colour changes and delamination may be connected to microscopic features, and a microscopic analysis can trace whether they are caused by biotic, chemical or physical decay. The most relevant microscopic feature that was detected in association with every degradation form was unfolding of thin-walled cells. This is therefore a reliable microscopic degradation marker. However, the increased brittleness could not be connected to a specific microscopic feature and would need to be assessed macroscopically, for example by quantifying the loss of fragments while handling or cutting the samples. Future research should concentrate on understanding possible causes for the increased brittleness of some birch bark archaeological objects, establishing whether there are correlations with the burial environment, and developing remedial measures specific to such objects.

Author Contributions: J.K. and G.D.P. conceived this study. J.K. prepared samples and performed the LM and TEM and analysed the images. J.K. interpreted the degradation and prepared images. G.D.P. was responsible for funding acquisition and project administration. J.K. and G.D.P. wrote the paper. All authors have read and agreed to the published version of the manuscript.

Funding: This research was funded by the Swiss National Science Foundation (SNSF), grant number 159662 (http://p3.snf.ch/Project-159662, accessed 20 July 2021).

Institutional Review Board Statement: Not applicable.

Informed Consent Statement: Not applicable.

Data Availability Statement: The datasets used and/or analysed during the current study are available from the corresponding author on reasonable request.

Acknowledgments: The authors would like to thank Jostein Bergstøl and Jörg Hägele-Masnick for the kind donation of archaeological birch bark samples; Thomas Volkmer, Gaspard Clerc, BFH Biel, Robertas Ursache, and Fritz Schweingruber, WSL Zürich, for testing cutting, embedding and staining methods on birch bark; Paul Safko for the preparation of orthophotos from the bow case used in Figure 4, and Nadim Scherrer for technical support in microscopy. A special thanks to Beat Haenni, University of Bern, for the skilled embedding and sectioning of the samples. The authors also would like to thank Adriano Boschetti, Archaeological Service of the Canton of Bern, and Albert Hafner, University of Bern, for support of the "Unfreezing History" project.

Conflicts of Interest: The authors declare no conflict of interest. The funders had no role in the design of the study; in the collection, analyses, or interpretation of data; in the writing of the manuscript; or in the decision to publish the results.

References

1. Lüttge, U.; Kluge, M.; Bauer, G. *Botanik*; Wiley Verlag: Weinheim, Germany, 2005.
2. Kost, B. Die Gewebe der Gefäßpflanzen. In *Strasburger–Lehrbuch der Pflanzenwissenschaften*; Kadereit, J.W., Körner, C., Kost, B., Sonnewald, U., Eds.; Springer: Berlin/Heidelberg, Germany, 2014; pp. 72–96.
3. Chang, Y.-P. *Anatomy of Common North American Pulpwood Barks*; Technical Association of the Pulp and Paper Industry: Tokyo, Japan, 1954.
4. Jensen, W. The Connection between the Anatomical Structure and Chemical Composition and the Properties of Outer Bark of White Birch. *Pappers-och trävarutidskrift för Finl. Suom. Pap.-Ja Puutavaralehti* **1949**, *15*, 113–119.
5. Jensen, W.A. *Botanical Histochemistry: Principles and Practice*; Freeman, W.H.: San Francisco, CA, USA, 1962; Volume 12.
6. Evert, R.F.; Esau, K.; Langenfeld-Heyser, R.; Eichhorn, S.E. *Esaus Pflanzenanatomie: Meristeme, Zellen und Gewebe der Pflanzen- ihre Struktur, Funktion und Entwicklung*; Walter de Gruyter Verlag: Göttingen, Germany, 2009.
7. Klügl, J.; Hafner, A.; Di Pietro, G. On the rolling and plasticization of birch bark (in print). In Proceedings of the 14th ICOM-CC Wet Organic Archaeological Materials (WOAM) Working Group Conference, Portsmouth, UK, 20–24 May 2019.
8. Fortes, M.A.; Rosa, M.E. Growth stresses and strains in cork. *Wood Sci. Technol.* **1992**, *26*, 241–258. [CrossRef]
9. Krahmer, R.L.; Wellons, J.D. Some Anatomical and Chemnical Characteristics of Douglas-Fir Cork. *Wood Sci.* **1973**, *6*, 97–105.
10. Xu, X.; Schneider, E.; Zaremba, C.; Stucky, G.D.; Wudl, F. Modification of the Semitransparent Prunus serrula Bark Film: Making Rubber out of Bark. *Chem. Mater.* **1998**, *10*, 3523–3537. [CrossRef]
11. Groh, B.; Hübner, C.; Lendzian, K. Water and oxygen permeance of phellems isolated from trees: The role of waxes and lenticels. *Planta* **2002**, *215*, 794–801. [CrossRef] [PubMed]
12. Schönherr, J.; Ziegler, H. Water permeability of Betula periderm. *Planta* **1980**, *147*, 345–354. [CrossRef] [PubMed]
13. Shibui, H.; Sano, Y. Structure and formation of phellem of Betula maximowicziana. *Int. Assoc. Wood Anat.* **2018**, *39*, 18–36. [CrossRef]
14. Zajączkowska, U. Cork. *eLS* **2016**, 1–8.
15. Mauseth, J.D. *Botany: An Introduction to Plant Biology*; Jones & Bartlett Learning: Burlington, MA, USA, 2019.
16. Pinto, P.; Sousa, A.; Silvestre, A.; Neto, C.P.; Gandini, A.; Eckerman, C.; Holmbom, B. Quercus suber and Betula pendula outer barks as renewable sources of oleochemicals: A comparative study. *Ind. Crop. Prod.* **2009**, *29*, 126–132. [CrossRef]
17. Ekman, R. The Suberin Monomers and Triterpenoids from the Outer Bark of Betula-verrucosa Ehrh. *Holzforschung* **1983**, *37*, 205–211. [CrossRef]
18. Jensen, W.; Fremer, K.E.; Sierilä, P.; Wartiovaara, V. The Chemistry of Bark. In *The Chemistry of Wood*; Browning, B.L., Ed.; Interscience Publishers: New York, NY, USA; London, UK, 1963.
19. Ferreira, J.P.A.; Quilhó, T.; Pereira, H. Characterization of Betula pendula Outer Bark Regarding Cork and Phloem Components at Chemical and Structural Levels in View of Biorefinery Integration. *J. Wood Chem. Technol.* **2017**, *37*, 10–25. [CrossRef]
20. Koljo, B.; Sitte, P. Das Suberin. In *Die Chemie der Pflanzenzellwand: Ein Beitrag zur Morphologie, Physik, Chemie und Technologie der Cellulose und ihrer Begleiter*; Treiber, E.E., Ed.; Springer: Berlin/Heidelberg, Germany, 1957; pp. 416–432.
21. Kiyoto, S.; Sugiyama, J. Histochemical structure and tensile properties of birch cork cell walls. *Cellulose* **2021**. [CrossRef]
22. Graça, J.; Santos, S. Suberin: A Biopolyester of Plants' Skin. *Macromol. Biosci.* **2007**, *7*, 128–135. [CrossRef] [PubMed]
23. Tse, S.; Dignard, C.; Kata, S.; Henderson, E.J. A study of the light sensitivity of birch bark. *Stud. Conserv.* **2018**, *63*, 423–440. [CrossRef]
24. Monahan, V. *Condition of Artifacts from Ice Patches*; 2008.
25. Klügl, J.; Hafner, A.; Di Pietro, G. Towards a description of the degradation of archaeological birch bark. In Proceedings of the ICOM-CC 18th Triennial Conference Preprints, Copenhagen, Denmark, 4–8 September 2017.
26. Orsini, S.; Ribechini, E.; Modugno, F.; Klügl, J.; Di Pietro, G.; Colombini, M.P. Micromorphological and chemical elucidation of the degradation mechanisms of birch bark archaeological artefacts. *Herit. Sci.* **2015**, *3*, 1–11. [CrossRef]
27. Vasiljeva, N. *Softening, Drying and Consolidation of Birch Bark*; 2018.
28. Klügl, J.; Di Pietro, G. The interaction of water with archaeological and ethnographic birch bark and its effects on swelling, shrinkage and deformations. *Herit. Sci.* **2021**, *9*, 3. [CrossRef]
29. Fletcher, L.; Milner, N.; Taylor, M.; Bamforth, M.; Croft, S.; Little, A.; Pomstra, D.; Robson, H.K.; Knight, B. The Use of Birch Bark. In *Star Carr Volume 2: Studies in Technology, Subsistence and Environment*; Milner, N., Taylor, B., Conneller, C., Eds.; White Rose University Press: York, UK, 2018; pp. 419–534.
30. Hägele-Masnick, J. The birch bark torches from the Kyffhäuser caves. In Proceedings of the 13th ICOM-CC Group on Wet Organic Archaeological Materials Conference, Florence, Italy; 2016; pp. 392–394.
31. Larsen, P.K.; Jensen, L.A.; Ryhl-Svendsen, M.; Padfield, T. The microclimate within a Neolithic passage grave. In Proceedings of the ICOM-CC 18th Triennial Conference, Copenhagen, Denmark, 4–8 September 2017; pp. 109–127.
32. Ward, C.; Giles, D.; Sully, D.; Lee, D.J. The conservation of a group of waterlogged neolithic bark bowls. *Stud. Conserv.* **1996**, *41*, 241–249.

33. Rao, H.; Yang, Y.; Hu, X.; Yu, J.; Jiang, H. Identification of an Ancient Birch Bark Quiver from a Tang Dynasty (A.D. 618–907) Tomb in Xinjiang, Northwest China. *Econ. Bot.* **2017**, *71*, 32–44. [CrossRef]
34. Tintner, J.; Smidt, E.; Aumüller, C.; Martin, P.; Ottner, F.; Wriessnig, K.; Reschreiter, H. Taphonomy of prehistoric bark in a salt environment at the archaeological site in Hallstatt, Upper Austria–An analytical approach based on FTIR spectroscopy. *Vib. Spectrosc.* **2018**, *97*. [CrossRef]
35. Kolattukudy, P.E. Suberin from Plants. In *Biopolymers*; Steinbüchel, A., Yosiharu, D., Eds.; Wiley: Weinheim, Germany, 2002; Volume 3a, pp. 41–73.
36. Agrawal, O.P.; Dhawan, S. Studies on fungal resistance of birch-bark. In Proceedings of the ICOM-CC 7th Triennial Meeting, Copenhagen, Denmark, 10–14 September 1984; pp. 1–3.
37. Ranathunge, K.; Schreiber, L.; Franke, R. Suberin research in the genomics era—New interest for an old polymer. *Plant Sci.* **2011**, *180*, 399–413. [CrossRef]
38. Parameswaran, N.; Wilhelm, G.E. Micromorphology of naturally degraded beech and spruce barks. *Eur. J. For. Pathol.* **1979**, *9*, 103–112. [CrossRef]
39. Rypacek, W. *Biologie Holzzerstörender Pilze*; VEB Gustav Fischer Verlag: Jena, Germany, 1966.
40. Zimmermann, W.; Seemüller, E. Degradation of Raspberry Suberin by Fusarium solani f. sp. Pisi and Armillaria mellea. *J. Phytopathol.* **1984**, *110*, 192–199. [CrossRef]
41. Ofong, A.U.; Pearce, R.B. Suberin degradation by Rosellinia desmazieresii. *Eur. J. For. Pathol.* **1994**, *24*, 316–322. [CrossRef]
42. Fernando, G.; Zimmermann, W.; Kolattukudy, P.E. Suberin-grown *Fusarium solani* f. sp *pisi* generates a cutinase-like esterase which depolymerizes the aliphatic components of suberin. *Physiol. Plant Pathol.* **1984**, *24*, 143–155. [CrossRef]
43. Swift, J.M. Loss of Suberin from Bark tissue rotted by Armillaria mellea. *Nature* **1965**, *207*, 436–437. [CrossRef]
44. Kontkanen, H.; Westerholm-Parvinen, A.; Saloheimo, M.; Bailey, M.; Ratto, M.; Mattila, I.; Mohsina, M.; Kalkkinen, N.; Nakari-Setala, T.; Buchert, J. Novel Coprinopsis cinerea polyesterase that hydrolyzes cutin and suberin. *Appl Env. Microbiol* **2009**, *75*, 2148–2157. [CrossRef] [PubMed]
45. Ferreira, R.; Garcia, H.; Sousa, A.F.; Freire, C.S.R.; Silvestre, A.J.D.; Rebelo, L.P.N.; Pereira, C.S. Isolation of suberin from birch outer bark and cork using ionic liquids: A new source of macromonomers. *Ind. Crop. Prod.* **2013**, *44*, 520–527. [CrossRef]
46. Ferreira, R.; Garcia, H.; Sousa, A.F.; Petkovic, M.; Lamosa, P.; Freire, C.S.R.; Silvestre, A.J.D.; Rebelo, L.P.N.; Pereira, C.S. Suberin isolation from cork using ionic liquids: Characterisation of ensuing products. *New J. Chem.* **2012**, *36*, 2014–2024. [CrossRef]
47. Martins, I.; Hartmann, D.O.; Alves, P.C.; Martins, C.; Garcia, H.; Leclercq, C.C.; Ferreira, R.; He, J.; Renaut, J.; Becker, J.D.; et al. Elucidating how the saprophytic fungus Aspergillus nidulans uses the plant polyester suberin as carbon source. *BMC Genom.* **2014**, *15*, 613. [CrossRef]
48. Ursache, R.; De Jesus Vieira Teixeira, C.; Dénervaud Tendon, V.; Gully, K.; De Bellis, D.; Schmid-Siegert, E.; Grube Andersen, T.; Shekhar, V.; Calderon, S.; Pradervand, S.; et al. GDSL-domain proteins have key roles in suberin polymerization and degradation. *Nat. Plants* **2021**. [CrossRef]
49. Tibbett, M.; Grantham, K.; Sanders, F.E.; Cairney, J.W.G. Induction of cold active acid phosphomonoesterase activity at low temperature in psychrotrophic ectomycorrhizal *Hebeloma* spp. *Mycol. Res.* **1998**, *102*, 1533–1539. [CrossRef]
50. Weinstein, R.N.; Montiel, P.O.; Johnstone, K. Influence of growth temperature on lipid and soluble carbohydrate synthesis by fungi isolated from fellfield soil in the maritime Antarctic. *Mycologia* **2000**, *92*, 222–229. [CrossRef]
51. Duarte, A.W.F.; dos Santos, J.A.; Vianna, M.V.; Vieira, J.M.F.; Mallagutti, V.H.; Inforsato, F.J.; Wentzel, L.C.P.; Lario, L.D.; Rodrigues, A.; Pagnocca, F.C.; et al. Cold-adapted enzymes produced by fungi from terrestrial and marine Antarctic environments. *Crit. Rev. Biotechnol.* **2018**, *38*, 600–619. [CrossRef]
52. Robinson, C.H. Cold adaptation in Arctic and Antarctic fungi. *New Phytol.* **2001**, *151*, 341–353. [CrossRef]
53. Björdal, C. Microbial degradation of waterlogged archaeological wood. *J. Cult. Herit.* **2012**, *13*, 118–122. [CrossRef]
54. Björdal, C.; Nilsson, T. Decomposition of waterlogged archaeological wood. In Proceedings of the 8th ICOM-CC WOAM Conference, Stockholm, Sweden, 11–15 June 2001; pp. 235–247.
55. Pedersen, N.B.; Björdal, C.G.; Jensen, P.; Felby, C. Bacterial Degradation of Archaeological Wood in Anoxic Waterlogged Environments. In *Stability of Complex Carbohydrate Structures: Biofuels, Foods, Vaccines and Shipwrecks*; Harding, S.E., Ed.; RSC Publishing: Cambridge, MA, USA, 2012; pp. 160–187.
56. Blanchette, R.A. A review of microbial deterioration found in archaeological wood from different environments. *Int. Biodeterior. Biodegrad.* **2000**, *46*, 189–204. [CrossRef]
57. Blanchette, R.A.; Nilsson, T.; Daniel, G.; Abad, A. Biological Degradation of Wood. In *Archaeological Wood*; Rowell, R.M., James Barbour, R., Eds.; Advances in Chemistry; American Chemical Society: Washington, DC, USA, 1990; Volume 225, pp. 141–174.
58. Blanchette, R.; Cease, K.; Abad, A.; Burnes, T.; Obst, J. Ultrastructural characterization of wood from Tertiary fossil forest in the Canadian Arctic. *Can. J. Bot.* **1991**, *69*, 560–568. [CrossRef]
59. Blanchette, R.A. Microbial degradation of wood from aquatic and terrestrial environments. In *Cultural Heritage Microbiology: Fundamental Studies in Conservation Science*; Mitchell, R., McNamara, C.J., Eds.; ASM Press: Washington, DC, USA, 2010; pp. 179–218.
60. Blanchette, R.A.; Held, B.W.; Jurgens, J.; Stear, A.; Dupont, C. Fungi attacking historic wood of Fort Conger and the Peary Huts in the High Arctic. *PLoS ONE* **2021**, *16*, e0246049. [CrossRef]

61. Matthiesen, H.; Jensen, J.B.; Gregory, D.; Hollesen, J.; Elberling, B. Degradation of Archaeological Wood Under Freezing and Thawing Conditions—Effects of Permafrost and Climate Change. *Archaeometry* **2014**, *56*, 479–495. [CrossRef]
62. Pedersen, N.B.; Matthiesen, H.; Blanchette, R.; Alfredsen, G.; Held, B.; Westergaard-Nielsen, A.; Hollesen, J. Fungal attack on archaeological wooden artefacts in the Arctic-implications in a changing climate. *Sci. Rep.* **2020**, *10*, 8187. [CrossRef]
63. Hafner, A.; Klügl, J. Neolithisches Bogenfutteral aus Birkenrinde, Holz und Leder. In *Schnidejoch und Lötschenpass. Archäologische Forschungen in den Berner Alpen. Schnidejoch et Lötschenpass. Investigations Archéologiques dans les Alpes Bernoises*; Archäologischer Dienst Bern: Bern, Switzerland, 2015; Volume 2, pp. 15–18.
64. Junkmanns, J.; Klügl, J.; Schoch, W.H.; Di Pietro, G.; Hafner, A. Neolithic and Bronze Age archery equipment from alpine ice-patches: A review on components, construction techniques and functionality. *J. Neolit. Archaeol.* **2019**, *21*, 283–314. [CrossRef]
65. Klügl, J. How to conserve a birch bark bow case from an ice patch? In Proceedings of the 12th ICOM-CC Group on Wet Organic Archaeological Materials Conference, Istanbul, Turkey, 13–17 May 2013; pp. 270–278.
66. Pilø, L.; Finstad, E.; Barrett, J.H. Crossing the ice: An Iron Age to medieval mountain pass at Lendbreen, Norway. *Antiquity* **2020**, *94*, 437–454. [CrossRef]
67. Haeberli, W.; Alean, J. Temperature and accumulation of high altitude firn in the alps. *Ann. Glaciol.* **1985**, *6*, 161–163. [CrossRef]
68. Isaksen, K.; Holmlund, P.; Sollid, J.L.; Harris, C. Three deep Alpine-permafrost boreholes in Svalbard and Scandinavia. *Permafr. Periglac. Process.* **2001**, *12*, 13–25. [CrossRef]
69. Guthruf, K.; Maurer, V.; Ryser, R.; Zeh, M.; Zweifel, N. *Zustand der Kleinseen*; AWA Amt für Wasser und Abfall, Gewässer-und Bodenschutzlabor GBL: Bern, Switzerland, 2015.
70. Ghemawat, M.S. Polychromatic staining with toluidine blue O for studying the host-parasite relationships in wheat leaves of Erysiphe graminis f. sp. tritici. *Physiol. Plant Pathol.* **1977**, *11*, 251–253. [CrossRef]
71. Pedersen, N.B. *Microscopic and Spectroscopic Characterisation of Waterlogged Archaeological Softwood from Anoxic Environments*; University of Copenhagen: Frederiksberg, Denmark, 2015.
72. Sousa, A.F.; Gandini, A.; Caetano, A.; Maria, T.M.; Freire, C.S.; Neto, C.P.; Silvestre, A.J. Unravelling the distinct crystallinity and thermal properties of suberin compounds from Quercus suber and Betula pendula outer barks. *Int. J. Biol. Macromol.* **2016**, *93*, 686–694. [CrossRef] [PubMed]
73. Daniel, G.; Nilsson, T.; Pettersson, B. Poorly and Non-Lignified Regions in the Middle Lamella Cell Corners of Birch (Betula Verrucosa) and Other Wood Species. *IAWA J.* **1991**, *12*, 70–83. [CrossRef]
74. de Lhoneux, B.; Antoine, R.; Côté, W.A. Ultrastructural implications of gamma-irradiation of wood. *Wood Sci. Technol.* **1984**, *18*, 161–176. [CrossRef]
75. Parameswaran, N.; Wilhelm, G.E.; Liese, W. Ultrastructural aspects of beech bark degradation by fungi. *Eur. J. For. Pathol.* **1976**, *6*, 274–286. [CrossRef]
76. Singh, P.A. A review of microbial decay types found in wooden objects of cultural heritage recovered from buried and waterlogged enviromnents. *J. Cult. Herit.* **2012**, *13S*, 16–20. [CrossRef]
77. Beaulieu, C.; Sidibé, A.; Jabloune, R.; Simao-Beaunoir, A.-M.; Lerat, S.; Monga, E.; Bernards, M.A. Physical, Chemical and Proteomic Evidence of Potato Suberin Degradation by the Plant Pathogenic Bacterium Streptomyces scabiei. *Microbes Environ.* **2016**, *31*, 427–434. [CrossRef] [PubMed]
78. Chowdhury, P.; Nag, S.; Ray, A. Degradation of Phenolic Compounds Through UV and Visible-Light-Driven Photocatalysis: Technical and Economic Aspects. *Phenolic Compd. Nat. Sources Importance Appl.* **2017**, *16*, 395–417.
79. Gandini, A.; Pascoal Neto, C.; Silvestre, A.J.D. Suberin: A promising renewable resource for novel macromolecular materials. *Prog. Polym. Sci.* **2006**, *31*, 878–892. [CrossRef]
80. Rosa, M.E.; Pereira, H.; Fortes, M.A. Effects of hot water treatment on the structure and properties of cork. *Wood Fiber Sci.* **1990**, *22*, 149–164.

applied sciences

MDPI

Article

The Efficiency of Biocidal Silica Nanosystems for the Conservation of Stone Monuments: Comparative In Vitro Tests against Epilithic Green Algae

Flavia Bartoli *, Martina Zuena, Armida Sodo and Giulia Caneva

Dipartimento di Scienze, Università degli Studi "Roma Tre", 00146 Roma, Italy; martina.zuena@uniroma3.it (M.Z.); armida.sodo@uniroma3.it (A.S.); giulia.caneva@uniroma3.it (G.C.)

* Correspondence: flavia.bartoli@uniroma3.it; Tel.: +39-06-5733-6374; Fax: +39-06-573-36321

Abstract: In the last decade, worldwide research has focused on innovative natural biocides and the development of organic and inorganic nanomaterials for long-lasting reliability. In this work, the biocide effects of two different biocides encapsulated in two different silica nanosystems for a multifunctional coating have been performed through in vitro tests, by using *Chlorococcum* sp. as a common stone biodeteriogen. Zosteric sodium salt (ZS), a green biocide, was compared with the commercial biocide, 2-mercaptobenzothiazole (MBT), widely used in the treatment of cultural heritage. The analyzed systems are the following: silica nanocapsules (NC) and silica nanoparticles (MNP) not loaded with biocides, two nanosystems loaded with ZS and MBT, and free biocides. The qualitative and quantitative evaluations of biocide efficiency were performed periodically, analyzing pigment autofluorescence to discriminate between active and inactive/dead cells. The analyses showed multiple differences. All the nanocontainers presented an initial reduction in chlorophyll's autofluorescence. For the free biocide, the results highlighted higher efficiency for MBT than ZS. Finally, the nanosystems loaded with the different biocides highlighted a higher activity for nanocontainers loaded with the commercial biocide than the green product, and better efficiency for MNP in comparison with NC.

Keywords: cultural heritage; multifunctional coating; stone biodeterioration; biofilms; biocide; zosteric sodium salt; 2-mercaptobenzothiazole; silica nanosystems

Citation: Bartoli, F.; Zuena, M.; Sodo, A.; Caneva, G. The Efficiency of Biocidal Silica Nanosystems for the Conservation of Stone Monuments: Comparative *In Vitro* Tests against Epilithic Green Algae. *Appl. Sci.* **2021**, *11*, 6804. https://doi.org/10.3390/app11156804

Academic Editors: Maria Filomena Macedo, António Manuel Santos Carriço Portugal, Ana Miller and Ana Catarina Pinheiro

Received: 29 May 2021
Accepted: 21 July 2021
Published: 24 July 2021

1. Introduction

Cultural heritage artworks have an inestimable value that must be protected and preserved to be passed on to future generations. They are affected by different alteration processes and degradation, which increase over time, and in outdoor conditions, they are exposed to a greater extent of atmospheric agents as well as to biological colonization [1–3].

Thanks to technological advancements in recent times, applied techniques and the biocide products used to remove biofilms on monument surfaces have been improved [4–8]. Biocides are chemical substances which are efficient at killing the undesirable microorganisms that can occur on such surfaces, and they have been widely applied in the conservation of cultural heritage [9–11]. However, traditional commercial biocide can show some harmful effects on the environment and human health; furthermore, they can also show some interferences with the substrate, such as staining, changes in color, and chemical interactions [8,10,12]. Application methods and low doses can reduce the mentioned risks. Moreover, the direct application of biocide has relatively short-term effectiveness, and sometimes, the compounds can be removed and dispersed in the environment [2,8,10,13].

In the last decade, worldwide research has concerned the testing of innovative natural biocide and the development of organic and inorganic nanomaterials, associated with multifunctional coatings, for long-lasting reliability, and non-invasive and selective activity [7,8,13–24]. The use of natural active compounds coupled to this nanotechnology seems a good way to pursue an eco-friendly, sustainable, and safe approach for the conservation of cultural heritage, reducing the amount of the bioactive compound and obtaining a satisfactory, long-lasting, antifouling action [15–17,25–27]. The literature reports the synthesis of different nanocontainers and loading techniques [21,25]. However, nanosystems based on mesoporous silica materials, applied in a multifunctional coating, as a controlled release biocide system over time, are reported only in a few cases [19,21,28–30]. In previous works, two different silica-based nanocontainers (Ns), namely a core-shell nanocapsule (NC) and a mesoporous nanoparticle (MNP), have been synthesized and characterized [13,14,20–22]. Previous preliminary in vitro cultures exposed to the nanoparticles showed a high antifouling activity associated with a slight biocidal activity against photosynthesizing microorganisms [13]. However, the efficiency of these Ns in the release control of the engaged biocides and the improvement of biocide efficiency, allowing a reduction in the quantities of biocide required, have not been sufficiently addressed.

In this work, our goal is to test different biocidal agents through in vitro tests against a significant component of the biofilm occurring on stone in conditions of high humidity and low solar radiation, and in parallel, evaluating different coatings' formulations. According to ecological trends and EU regulations [31], we tested natural zosteric sodium salt, a derivative of zosteric acid, as a natural antifoulant compound, in comparison with the commercial biocide 2-mercaptobenzothiazole widely used in the treatment of cultural heritage [9,32,33]. The choice of both biocides and, in particular, of zosteric sodium salt as a green biocide, despite the possibility of a negative effect being linked to sulfate residues [34], is related to the previous results for the nanocontainers' synthesis [13,14,20–22]. Such a first step will need further evaluation before a wider application.

2. Materials and Methods

2.1. Biocidal Description

We selected two different biocides, namely 2-mercaptobenzothiazole and zosteric sodium salt. 2-mercaptobenzothiazole (MBT) is a well-known biocide used in the conservation of cultural heritage as an antifungal and anticryptogamical compound [9]. From the literature data, the biocidal activity of MBT is linked to its heterocyclic structure, and the substituted benzothiazole derivatives bring different biological properties [35,36].

Zosteric sodium salt (ZS) is a natural product antifoulant (NPA), which is less studied, but a promising product in this context. Zosteric acid and its sodium salt are produced and released by *Zostera marina* L., which is a wide-ranging marine flowering plant in the Northern Hemisphere, and its antifouling capability is related to the sulfate group present in the chemical structure [37].

As reported in Ruggiero et al. [20], we synthesized ZS from trans-4-hydroxycinnamic and the sulfur trioxide pyridine complex.

2.2. Encapsulation Step

According to Ruggiero et al. [20,21], we synthesized two different silica nanocontainers, namely core-shell nanocapsules (NC) and mesoporous nanoparticles (MNP) (Figure 1). Briefly, to synthesize the silica nanocapsules (NC), the following compounds were adopted: water, cetyltrimethylammonium bromide (CTAB, Aldrich, Italy) as a cationic surfactant, ammonia solution (NH3aq 30%, Aldrich) as a basic catalyst, tetraethoxysilane (TEOS, Aldrich) as a silica precursor, and diethyl ether (Et2O, Aldrich) as a cosolvent. All the chemicals were analytical grade and were used without further purification.

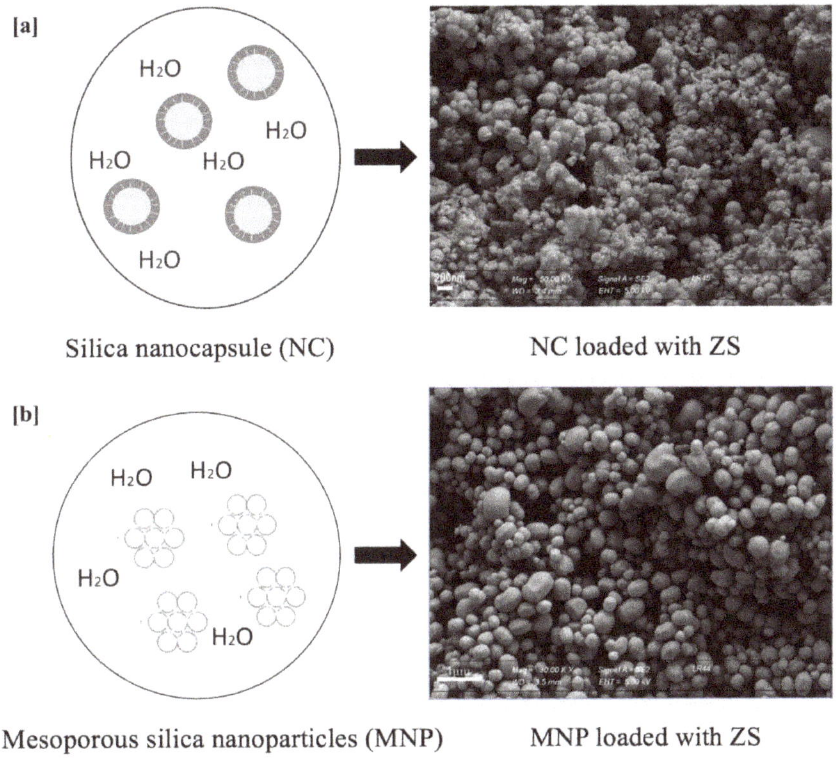

Figure 1. Structure of nanocontainers synthetized: (**a**) Silica Nanocapsule (NC): Scheme, SEM images of NC loaded with Zostera sodium salt (ZS); (**b**) Mesoporous silica nanoparticles (MNP): Scheme, SEM images of NC loaded with Zostera sodium salt (ZS). For the representation of Silica Nanocapsule and Mesoporous silica nanoparticles loaded with MBT, we refer to Ruggiero et al. [21].

A micellar solution was obtained by combining water, surfactant, and the basic catalyst. Alternatively, MBT or ZS were secondly added by using diethyl ether as the cosolvent. Therefore, an oil-in-water mini-emulsion was created due to the presence of diethyl ether, and the TEOS condensates at the mini-emulsion interface to form the inorganic silica network. The subsequent evaporation of the diethyl ether creates the mesoporosity of the shell.

For the synthesis of silica nanoparticles (MNP), the same reagents were used—except for diethyl ether—and the ammonia solution was substituted by NaOH aqueous solution. Differently from the previous synthesis, the biocides were inserted directly into the mixture before the formation of the micellar solution. TEOS plays the role of the template since its molecules polymerize around it.

2.3. In Vitro Tests

To perform in vitro tests, we collected biofilm materials growing on the Aurelian Walls in Rome, as a monument representative of biodeterioration patterns occurring in the Mediterranean area [13,38,39].

Specifically, we performed a survey on the vertical surfaces of the walls in the northern exposure where the biological growth is widely distributed, and not previously treated by restoration activities. Then, we selected and brushed superficially, employing a sterile lancet and a black humid patina. A part of the collected microorganisms was observed by the optical microscope with an immersion objective at 100-magnification (Olympus BX41,

Rome, Italy), following the procedures in UNI [40]. Thanks to the analytic keys of Guiry and Guiry [41], we identified the occurring species, which resulted in a mixed community of green algae, cyanobacteria, and meristematic fungi. Then, we cultured the detected species, according to the suggested nutritional and expositive requirements, in a BG-11 liquid media (Sigma–Aldrich) at 25 °C in conditions of natural sunlight, as previously reported [13]. Indeed, this type of media favored the proliferation of the algal component, and after two months, we obtained an almost pure culture of *Chlorococcum* sp. that we used as target organisms to test the efficiency of our nanosystems.

We tested eight different situations: the two free biocides (MBT and ZS); the two different nanosystems not loaded with biocides (NC, MNP); the two nanosystems loaded with zosteric sodium salt (NC_ZS; MNP_ZS) and with 2-mercaptobenzothiazole (NC_MBT; MNP_MBT). The nanocontainers were loaded with a concentration of biocide related to the loading capability (respectively, NC_Zs = 2.1%; MNP_ZS = 7.8%; NC_MBT = 10%; MNP_MBT 8.2%). We tested, following the protocol fine-tuned in Ruggiero et al. [13], 3 mg of nanocontainers not loaded and loaded with the biocide, added to 1 mL of liquid culture. For the free biocide, we tested a minimum (1 mg) and maximum (3 mg) quantity, emphasizing the concentrations based on those loaded in the nanosystems (~0.01 mg/mL) and the weight put in 1 mL of culture (3 mg).

Biocide efficiency was evaluated periodically, after the 1st, 2nd, 4th, 6th, and 8th week, so we filled 40 test tubes for each product tested in each time control. We observed the control and the treated cultures, in triplicate form for each product, under an optical microscope (Zeiss Axioplan 2) equipped with a photo camera (LEICA DFC 450 C) at 20X magnification in visible and autofluorescent light. Following the methodology of previous works [10,19,42,43], we evaluated the loss of each cell's fluorescence as an indicator of biocidal efficiency. We observed the autofluorescence of cultured microorganisms with multi-channel detection, in blue (450–500 nm), purple (380–450 nm), and green (500–570 nm) wavelengths [44,45]. The photosynthetic microorganisms, in our case green algae, contain chlorophylls, which show autofluorescence at 670–790 nm (excitation, 594 nm) [42–46]. We used multiple channels to observe not only the disappearance of the chlorophyll but also of the other accessory pigments, evaluating the change in chlorophyll fluorescence from red to green until its disappearance. In the monitoring occurring over the weeks, we analyzed the pigment's autofluorescence to discriminate between active cells (the presence of chlorophyll autofluorescence) and inactive/dead cells (between the absence of chlorophyll autofluorescence and complete absence of fluorescence) [42]. In addition to this direct qualitative method to monitor the temporal dynamic of biocidal efficiency, we performed cell counts. We captured the optical image at 20X magnification in visible and fluorescent light. For each product in each week, we observed 3 different optical slides, and for each slide, we captured 3 snapshot images, to have 9 images for each product each week. We analyzed the images through ImageJ software [47,48], and we performed cells counts, discriminating between the total fluorescent cells and the chlorophyll fluorescent cells (red ones). In this way, to normalize the number, we calculated the percentage of cells and the rate of autofluorescence loss.

3. Results

Biocidal Efficiency

The analysis of the three different systems—the free biocides, not-loaded nanocontainers, and loaded nanocontainers—showed multiple differences.

For the free biocide, the optical microscopy images at the 8th week show a higher efficiency of MBT than ZS, which does not highlight evidence of biocidal activity in both concentrations (Figure 2a). Indeed, it is clear in the images of both concentrations of ZS that no fluorescence reduction is evident. Instead, the MBT biocide shows a higher efficiency in 3 mg concentration than 1 mg, as expected, even if the chlorophyll's fluorescence has not completely disappeared by the 8th week.

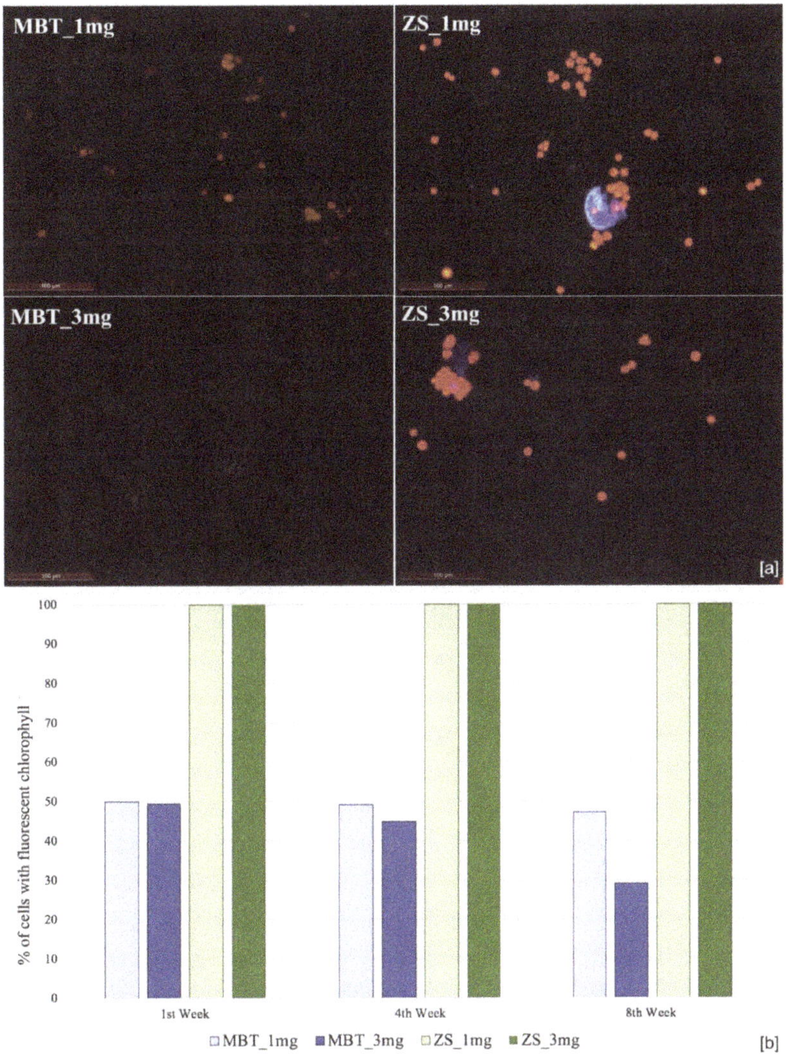

Figure 2. Evaluation of *Zostera* sodium salt (ZS) and 2-mercaptobenzothiazole (MBT) efficiency applied as free biocides in different concentrations (3 mg and 1 mg). (**a**) Optical microscope image in blue channel fluorescent light at the 8th week. The loss of fluorescence of chlorophyll (red) highlights the efficiency of the biocide. (**b**) Graph of red fluorescence percentage of culture treated with different concentrations of both free biocides. The decrease in fluorescent cells' percentage highlights the efficiency of products.

The quantitative analysis, performed with the cell count, confirms the qualitative data obtained by the optical microscope observation. The graph (Figure 2b) highlights that the vitality of the cells treated with MBT is reduced by about 50% after the first week for both concentrations but with different rates of reduction, with a rate of reduction of 0.88% and 4.7% in the 4th week, and 2.7% and 20.37 % in the 8th week for 1 mg and 3 mg, respectively.

All the nanocontainers (loaded and not loaded) present an initial reduction in chlorophyll autofluorescence, which can be noted with a color change of fluorescence from red to green, different to the pure biocide, which, after 1 week, did not show this effect.

The non-loaded nanosystems highlighted a clear reduction in chlorophyll autofluorescence of the photosynthetic microorganisms already after the first week and, almost complete disappearance of chlorophyll fluorescence after the 6th week. Indeed, the optical image at the 8th week shows the complete disappearance of chlorophyll fluorescence and only a few cells with very low green fluorescence. The quantitative analysis of non-loaded nanocontainers highlighted a very high efficiency rate, reducing the vitality of the cells at about 2% in only 4 weeks (Figure 3a).

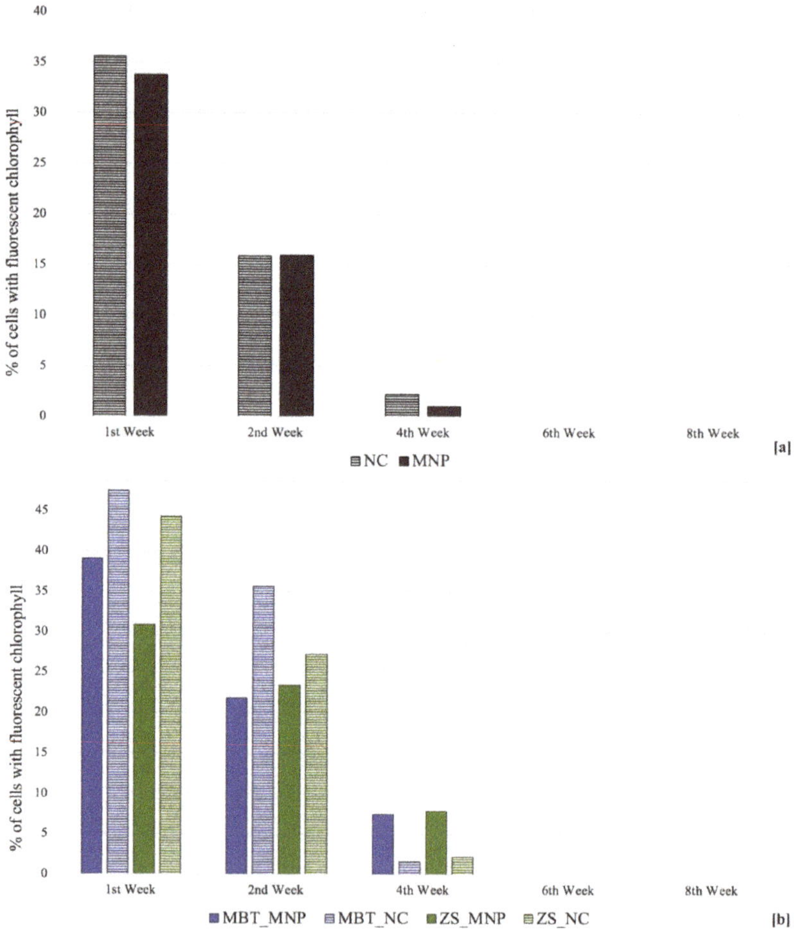

Figure 3. Evaluation of biocide efficiency of nanocontainers not loaded (NC; MNP) and loaded with the biocides (NC_ZS; NC_MBT; MNP_MBT; NC_MBT), where the decrease in fluorescent cells' percentage highlights the efficiency of products. (**a**) Graph of red fluorescence percentage of culture treated with non-loaded nanocontainers. (**b**) Graph of red fluorescence percentage of culture treated with loaded nanocontainers.

Finally, the nanosystems loaded with different biocides highlighted a higher efficiency for the mesoporous systems (MNP) than the nanocapsules (NC). Indeed, in the optical images, it is clear that both MBT-NC and ZS-NC at the 8th week show complete disappearance of red fluorescence but a few cells with low green fluorescence, different to MBT-MNP and ZS_MNP that show complete disappearance of fluorescence.

Moreover, confirming the qualitative data from a quantitative point of view, it is also evident that nanocontainers loaded with the commercial biocide are more efficient than the green product. Indeed, Figure 3b shows that in the 4th week, MBT_MNP has a redution rate of 31.63% in comparison with 23.08% for ZS-MNP.

Comparing the biocide efficiency of pure biocide with the biocide loaded in the nanosystems, it is highlighted that there is a quantity reduction in biocide used of 10 and 12%, respectively, for MBT-NC and MBT-MNP, and 47.6 and 12.6%, respectively, for ZS-NC and ZS-MNP. This reduction is linked to an efficiency increase of 71 % and 50%, respectively, for 3 mg and 1 mg MBT, and of 100% for both nanocontainers with ZS.

Finally, the different efficiencies of the non-loaded nanosystems in comparison with the loaded nanosystems are highlighted by the quantitative analyses of green and red fluorescence (Figure 4), which show the complete disappearance of red cells (chlorophylls' activity) at the 6th week, but differential data resulting for the complete disappearance of fluorescence. Indeed, the quantitative comparison between red and green fluorescence analyses shows the MNP system loaded with both biocides (MBT, ZS) at the 8th week gave rise to a complete disappearance of the fluorescent cells, and the same results occurred at the 6th week for the MNP systems loaded with MBT biocide (Figure 4).

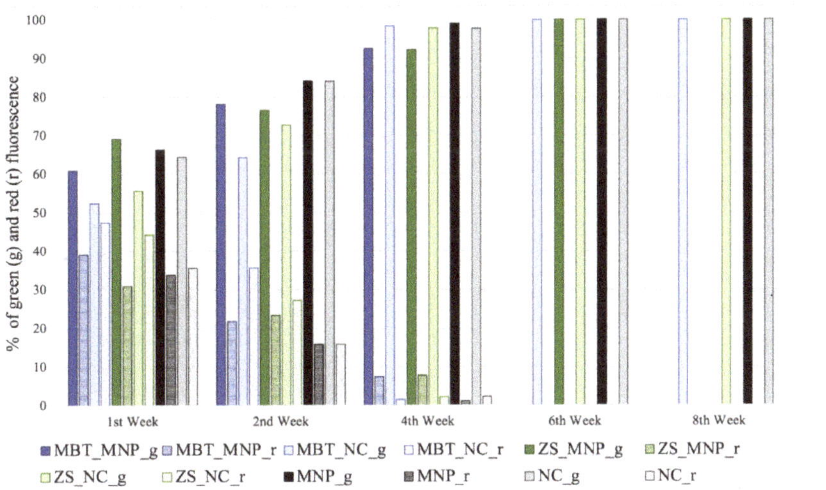

Figure 4. Comparative evaluation between the fluorescence of accessory pigments (green) and the chlorophyll (red) of the nanocontainers not loaded and loaded with biocides. The changing of the cells' fluorescence from red to green and the successive complete disappearance highlights the differential efficiency of products.

4. Discussion

The formulation of nanosystems for a multifunctional coating applicable as a preventive tool against the colonization of microorganisms on stone monuments resulted in a real improvement of chemical treatments with biocides [7,49–52]. As previously stressed in the literature, the multifunctional coating in relation to their formulation and functionality can condition the release of antifouling products loaded in nanosystems [11,13,18,21,53,54]. In this way, it is therefore important as a first step to analyze the biocide efficacy in a controlled environment before incorporating the nanosystems into the coatings.

Our study showed that both the biocide chemical composition and the differential nanostructures can give rise to a differential efficiency in the antifouling activity against green algae. Indeed, the presence of CTAB (cetyl trimethyl ammonium bromide) as a surfactant gives nanocontainers an unequivocal biocidal activity against algal cells. This result could be expected because this product is commonly used in DNA extraction protocol for plants [55]; the negative effect against algal cells seems coherent. The presence of CTAB explains the antifouling activity of the nanosystems in the absence of further biocidal compounds.

However, CTAB is not used as a free biocide due to its toxicity and the possibility of it being released into the environment when directly applied on stone surfaces [56]. According to the EU Regulation [57], CTAB is a cetrimonium bromide ($C_{19}H_{42}BrN$) and it is classified in the safety data sheet as toxic for the environment (very toxic to aquatic life with long-lasting effects, Risk Category 1), and for restorer (if ingested and inhaled, Risk Category 4, or irritating to skin or mucous membranes, Risk Category 2/1), but in the same way, it is completely biodegradable both in water and in soil. Moreover, it does not bioaccumulate in organisms. Furthermore, in this context, CTAB is used in the synthesis procedure of nanocontainers and the amount in excess is removed by centrifugation and the application of multifunctional coatings containing the nanocontainers prevents its release into the environment [19,56,58]. Therefore, in view of an in situ application, even if residue should remain after washing, its quantity and its entrapment inside the coating would not make it toxic for the environment [19,56–58].

Furthermore, our results highlighted that the biocides used by direct applications (Figure 2), although in much higher quantities than in encapsulated biocide (Figure 3b), showed lower antifouling activity. In particular, ZS as a free biocide did not show any relevant biocidal activity against the algal cells at the experimented doses. Noteworthy, these data, which are the first on biocidal efficiency against algal cells, showed different results concerning previously reported data on biocidal efficiency against bacterial [37,59–61] and fungal species [61,62], such as in the case of encapsulated product [63,64]. Despite the promising results of previously detected microorganisms, ZS did not show a broad spectrum of biocidal efficiency. Such data can be related to the mechanisms of biocidal activity of this compound. Indeed, the previous studies demonstrate that this compound does not act by killing the microorganisms but by inhibiting their adhesion, blocking the interaction of the microorganisms with the surface [61,63–65].

Furthermore, the different efficiency between NC for the MNP systems highlighted that the second one presents higher efficiency (Figure 3b). This aspect seems to be related to the structure of the nanocontainers [21,22]. Indeed, the self-assembly method for the two nanosystems is different; the mesoporous system presents hexagonally packed cylindric mesopores whereas the nanocapsule presents a core-shell structure (Figure 1). These different structures seem to explain why in the short period, the MNP resulted more efficiently than the NC, since the biocide compounds result in being less trapped and more quickly released [21,22].

Finally, we also observe that different biodeterioration patterns can highly vary, and they form complex communities, resulting from different environmental and edaphic conditions [9,66–68]. Therefore, a biocidal system that is the best for all conditions does not exist. In any case, the establishment of the efficiency of some different nanosystems against one relevant component of biofilm occurring on stone monuments can be considered a significant result. We believe that such a result could be useful in the case of green algae colonization, occurring on stone surfaces showing high humidity values and low solar radiation. Further in situ tests will give new evaluations which will allow wider testing on other occurring organisms, such as cyanobacteria and meristematic fungi, which are not represented by these in vitro tests.

The potential interference of such coatings on stone surfaces was previously preliminarily assessed. Our results showed that the addition of the silica nanocontainers at 0.05% (*w/v*) in sol did not change the color of the coating and reduced the formation of cracks after drying [22].

Further promising results could also arise, combining systems with nanostructures and enzymes used for cleaning biological coatings, as shown by the recent literature [69].

5. Conclusions

The research, even if limited to in vitro tests, with the selection of one widely recurrent genus of green algae as the cultivable microorganism, confirmed the interest of the silica nanosystems coupled with biocides as a potential protective coating for stone monuments. In particular, nanoparticles, synthesized for the formulation of a multifunctional coating, proved better than nanocontainers as being promising tools in the conservation of stone cultural heritage. Enlarging the testing against other components of biological communities occurring on stone monuments, research must move forward to find the best combination between the type of encapsulated biocide and the most efficient nanosystem structure.

Author Contributions: Conceptualization, F.B.; Methodology and Validation, F.B., M.Z.; Formal Analysis and Investigation, F.B.; Writing, F.B., G.C., M.Z., A.S. Funding Acquisition, G.C., A.S. All authors have read and agreed to the published version of the manuscript.

Funding: This research was funded by Regione Lazio, Italy within the "GRAL- GReen And Long-lasting stone conservation products" project (n. F85F21001710009) financed in the call " Progetto Gruppi di Ricerca 2020".

Institutional Review Board Statement: This study did not involve animals.

Informed Consent Statement: Informed consent was obtained from all subjects involved in the study.

Acknowledgments: The authors acknowledge The Grant of Excellence Departments, MIUR (ARTICOLO 1, COMMI 314—337 LEGGE 232/2016). The authors acknowledge Ludovica Ruggiero for supporting the synthesis procedure of the silica nanocontainers.

Conflicts of Interest: The authors declare no conflict of interest. The funders had no role in the design of the study; in the collection, analyses, or interpretation of data; in the writing of the manuscript, or in the decision to publish the results.

References

1. Videla, H.A.; Herrera, L.K. Biodeterioration and weathering effects on rock decay. *Corros. Rev.* **2004**, *22*, 341–364. [CrossRef]
2. Caneva, G.; Nugari, M.P.; Salvadori, O. *Plant. Biology for Cultural Heritage: Biodeterioration and Conservation*; The Getty Conservation Institute: Los Angeles, CA, USA, 2008.
3. Camuffo, D. *Microclimate for Cultural Heritage: Measurement, Risk Assessment, Conservation, Restoration, and Maintenance of Indoor and Outdoor Monuments*, 3rd ed.; Elsevier: Amsterdam, The Netherlands, 2019.
4. Cappitelli, F.; Villa, F.; Sorlini, C. New environmentally friendly approaches against biodeterioration of outdoor cultural heritage. In *Biocolonization of Stone: Middle Missouri Plains Control and Preventive Village Sites Methods, Proceedings of the MCI Workshop Series*; Elena Charola, A., Ed.; Smithsonian Contributions to Museum Conservation; Smithsonian Institution Scholarly Press: Washington, DC, USA, 2011; pp. 51–58.
5. Baglioni, P.; Chelazzi, D.; Giorgi, R. *Nanotechnologies in the Conservation of Cultural Heritage: A Compendium of Materials and Techniques*; Springer: Dordrecht, The Netherlands; Berlin/Heidelberg, Germany; New York, NY, USA; London, UK, 2014. [CrossRef]
6. Serafini, I.; Ciccola, A. Nanotechnologies and Nanomaterials. Nanotechnologies and Nanomaterials for Diagnostic. *Conserv. Restor. Cult. Herit.* **2019**, 325–380. [CrossRef]
7. Tortora, L.; Di Carlo, G.; Mosquera, M.J.; Ingo, G.M. Nanoscience and Nanomaterials for the Knowledge and Conservation of Cultural Heritage. *Front. Mater. Sci.* **2020**, *7*, 372–374. [CrossRef]
8. Fidanza, M.R.; Caneva, G. Natural biocides for the conservation of stone cultural heritage: A review. *J. Cult. Herit.* **2019**, *38*, 271–286. [CrossRef]
9. Caneva, G.; Nugari, M.P.; Pinna, D.; Salvadori, O. *Il Controllo del Degrado Biologico: I Biocidi nel Restauro dei Materiali Lapidei*; Nardini Editore: Fiesole, Italy, 1996; p. 151.
10. Pinna, D. *Coping with Biological Growth on Stone Heritage Objects: Methods, Products, Applications, and Perspectives*; Apple Academic Press: Waretown, NJ, USA, 2017.
11. Kakakhel, M.A.; Wu, F.; Gu, J.D.; Feng, H.; Shah, K.; Wang, W. Controlling biodeterioration of cultural heritage objects with biocides: A review. *Int. Biodeterior. Biodegrad.* **2019**, *143*, 104721. [CrossRef]
12. Nugari, M.P.; Pallecchi, P.; Pinna, D. Methodological Evaluation of Biocidal Interference with Stone Minerals—Preliminary Laboratory Tests. In Proceedings of the International RILEM/UNESCO Congress on Conservation of Stone and Other Materials, Paris, France, 29 June–1 July 1993; Thiel, M.J., Ed.; E. & F. N. Spon: London, UK, 1993; pp. 295–302.

13. Ruggiero, L.; Bartoli, F.; Fidanza, M.R.; Zurlo, F.; Marconi, E.; Gasperi, T.; Tuti, S.; Crociani, L.; Di Bartolomeo, E.; Caneva, G.; et al. Encapsulation of environmentally-friendly biocides in silica nanosystems for multifunctional coatings. *Appl. Surf. Sci.* **2020**, *514*, 145908. [CrossRef]
14. Ruggiero, L.; Fidanza, M.R.; Iorio, M.; Tortora, L.; Caneva, G.; Ricci, M.A.; Sodo, A. Synthesis and characterization of TEOS coating added with innovative antifouling silica nanocontainers and TiO_2 nanoparticles. *Front. Mater.* **2020**, *7*, 185. [CrossRef]
15. Silva, M.; Rosado, T.; Teixeira, D.; Candeias, A.; Caldeira, A.T. Production of green biocides for cultural heritage. Novel biotechnological solutions. *Int. J. Conserv. Sci.* **2015**, *6*, 519–530.
16. Silva, M.; Salvador, C.; Candeias, M.F.; Teixeira, D.; Candeias, A.; Caldeira, A.T. Toxicological assessment of novel green biocides for cultural heritage. *Int. J. Conserv. Sci.* **2016**, *7*, 265–272.
17. Rotolo, V.; Barresi, G.; Di Carlo, E.; Giordano, A.; Lombardo, G.; Crimi, E.; Costa, E.; Bruno, M.; Palla, F. Plant extracts as green potential strategies to control the biodeterioration of cultural heritage. *Int. J. Conserv. Sci.* **2016**, *7*, 839–846.
18. Dresler, C.; Saladino, M.; Demirbag, C.; Caponetti, E.; Martino, D.F.C.; Alduina, R. Development of controlled release systems of biocides for the conservation of cultural heritage. *Int. Biodeterior. Biodegrad.* **2017**, *125*, 150–156. [CrossRef]
19. Kuznetsova, A.; Domingues, P.M.; Silva, T.; Almeida, A.; Zheludkevich, M.L.; Tedim, J.; Ferreira, M.G.S.; Cunha, A. Antimicrobial activity of 2-mercaptobenzothiazole released from environmentally friendly nanostructured layered double hydroxides. *J. Appl. Microbiol.* **2017**, *122*, 1207–1218. [CrossRef] [PubMed]
20. Ruggiero, L.; Crociani, L.; Zendri, E.; El Habra, N.; Guerriero, P. Incorporation of the zosteric sodium salt in silica nanocapsules: Synthesis and characterization of new fillers for antifouling coatings. *Appl. Surf. Sci.* **2018**, *439*, 705–711. [CrossRef]
21. Ruggiero, L.; Di Bartolomeo, E.; Gasperi, T.; Luisetto, I.; Talone, A.; Zurlo, F.; Peddis, D.; Ricci, M.A.; Sodo, A. Silica nanosystems for active antifouling protection: Nanocapsules and mesoporous nanoparticles in controlled release applications. *J. Alloys Compd.* **2019**, *798*, 144–148. [CrossRef]
22. Ruggiero, L.; Sodo, A.; Cestelli-Guidi, M.; Romani, M.; Sarra, A.; Postorino, P.; Ricci, M.A. Raman and ATR FT-IR investigations of innovative silica nanocontainers loaded with a biocide for stone conservation treatments. *Microchem. J.* **2020**, *155*, 104766. [CrossRef]
23. Arreche, R.; Vázquez, P. Green biocides to control biodeterioration in materials science and the example of preserving World Heritage Monuments. *Curr. Opin. Green Sustain. Chem.* **2020**, 100359. [CrossRef]
24. Palla, F. Biotechnology and Cultural Heritage Conservation. In *Heritage*; IntechOpen: London, UK, 2020. [CrossRef]
25. McCusker, L.; Liebau, F.; Engelhardt, G. Nomenclature of structural and compositional characteristics of ordered microporous and mesoporous materials with inorganic hosts (IUPAC Recommendations 2001). *Pure Appl. Chem.* **2009**, *73*, 381e394. [CrossRef]
26. Mattos, B.D.; Rojas, O.J.; Magalhães, W.L. Biogenic silica nanoparticles loaded with neem bark extract as green, slow-release biocide. *J. Clean. Prod.* **2017**, *142*, 4206–4213. [CrossRef]
27. Campanella, L.; Angeloni, R.; Cibin, F.; Dell'Aglio, E.; Grimaldi, F.; Reale, R.; Vitali, M. Capsulated essential oil in gel spheres for the protection of cellulosic cultural heritage. *Nat. Prod. Res.* **2021**, *35*, 116–123. [CrossRef] [PubMed]
28. Chen, H.; He, J.; Tang, H.; Yan, C. Porous silica nanocapsules and nanospheres: Dynamic self-assembly synthesis and application in controlled release. *Chem. Mater.* **2008**, *20*, 5894–5900. [CrossRef]
29. Popat, A.; Liu, J.; Hu, Q.; Kennedy, M.; Peters, B.; Lu, G.Q.; Qiao, S.Z. Adsorption and release of biocides with mesoporous silica nanoparticles. *Nanoscale* **2012**, *4*, 970e975. [CrossRef]
30. Chan, A.C.; Bravo Cadena, M.; Townley, H.E.; Fricker, M.D.; Thompson, I.P. Effective delivery of volatile biocides employing mesoporous silicates for treating biofilms. *J. R. Soc. Interface* **2017**, *14*, 20160650. [CrossRef]
31. European Parliament. Directive of the European Parliament and of the council of 16 February 1998 concerning the placing of biocidal products on the market. *Off. J. Eur. Commun.* **1998**, *L123*, 1–63.
32. Orlita, A. Microbial biodeterioration of leather and its control: A review. *Int. Biodeterior. Biodegrad.* **2004**, *53*, 157–163. [CrossRef]
33. Nascimbene, J.; Salvadori, O. Lichen recolonization on restored calcareous statues of three Venetian villas. *Int. Biodeterior. Biodegrad.* **2008**, *62*, 313–318. [CrossRef]
34. Jendresen, C.B.; Nielsen, A.T. Production of zosteric acid and other sulfated phenolic biochemicals in microbial cell factories. *Nat. Commun.* **2019**, *10*, 1–10. [CrossRef]
35. Daidone, G.; Maggio, B.; Schillaci, D. Salicylanilide and its heterocyclic analogues. A comparative study of their antimicrobial activity. *Pharmazie* **1990**, *45*, 441–442. [PubMed]
36. Franchini, C.; Muraglia, M.; Corbo, F.; Florio, M.A.; Di Mola, A.; Rosato, A.; Matucci, R.; Nesi, M.; van Bambeke, F.; Vitali, C. Synthesis and Biological Evaluation of 2-Mercapto-1, 3-benzothiazole Derivatives with Potential Antimicrobial Activity. *Arch. Pharm.* **2009**, *342*, 605–613. [CrossRef]
37. Zhang Newby, B.; Cutright, T.; Barrios, C.A.; Xu, Q. Zosteric acid—An effective antifoulant for reducing fresh water bacterial attachment on coatings. *JCT Res.* **2006**, *3*, 69–70.
38. Caneva, G.; De Marco, G.; Dinelli, A.; Vinci, M. The wall vegetation of the roman archaeological areas. *Sci. Technol. Cult. Herit.* **1992**, *1*, 217–226.
39. Cappitelli, F.; Villa, F. Novel Antibiofilm Non-Biocidal Strategies. In *Microorganisms in the Deterioration and Preservation of Cultural Heritage*; Edith, J., Ed.; Springer: Berlin, Germany, 2021; p. 117. [CrossRef]
40. UNI 10923. Beni culturali: Materiali lapidei naturali ed artificiali. In *Allestimento di Preparati Biologici per L'osservazione al Microscopio Ottico*; UNI: Milano, Italy, 2001.

41. Guiry, M.D.; Guiry, G.M. *AlgaeBase Version 4.2. World-Wide Electronic Publication*; National University of Ireland: Maynooth, Ireland, 2007. Available online: http://www.algaebase.org (accessed on 2 February 2020).
42. Ramírez, M.; Hernández-Mariné, M.; Novelo, E.; Roldán, M. Cyanobacteria-containing biofilms from a Mayan monument in Palenque, Mexico. *Biofouling* **2010**, *26*, 399–409. [CrossRef]
43. Hsieh, P.; Pedersen, J.Z.; Bruno, L. Photoinhibition of cyanobacteria and its application in cultural heritage conservation. *Photochem. Photobiol.* **2014**, *90*, 533–543. [CrossRef]
44. García-Plazaola, J.I.; Fernández-Marín, B.; Duke, S.O.; Hernández, A.; López-Arbeloa, F.; Becerril, J.M. Autofluorescence: Biological functions and technical applications. *Plant Sci.* **2015**, *236*, 136–145. [CrossRef]
45. Donaldson, L. Autofluorescence in plants. *Molecules* **2020**, *25*, 2393. [CrossRef] [PubMed]
46. Lamb, J.J.; Røkke, G.; Hohmann-Marriott, M.F. Chlorophyll fluorescence emission spectroscopy of oxygenic organisms at 77 K. *Photosynthetica* **2018**, *56*, 105–124. [CrossRef]
47. Abràmoff, M.D.; Magalhães, P.J.; Ram, S.J. Image processing with ImageJ. *Biophotonics Int.* **2004**, *11*, 36–42.
48. Rasband, W.S. *ImageJ*; US National Institutes of Health: Bethesda, MD, USA, 1997.
49. Colangiuli, D.; Calia, A.; Bianco, N. Novel multifunctional coatings with photocatalytic and hydrophobic properties for the preservation of the stone building heritage. *Constr. Build. Mater.* **2015**, *93*, 189–196. [CrossRef]
50. La Russa, M.F.; Ruffolo, S.A.; Rovella, N.; Belfiore, C.M.; Palermo, A.M.; Guzzi, M.T.; Crisci, G.M. Multifunctional TiO_2 coatings for cultural heritage. *Prog. Org. Coat.* **2012**, *74*, 186–191. [CrossRef]
51. Ruffolo, S.A.; La Russa, M.F. Nanostructured coatings for stone protection: An overview. *Front. Mater. Sci.* **2019**, *6*, 147. [CrossRef]
52. Zuena, M.; Ruggiero, L.; Della Ventura, G.; Bemporad, E.; Ricci, M.A.; Sodo, A. Effectiveness and Compatibility of Nanoparticle Based Multifunctional Coatings on Natural and Man-Made Stones. *Coatings* **2021**, *11*, 480. [CrossRef]
53. Becerra, J.; Mateo, M.; Ortiz, P.; Nicolás, G.; Zaderenko, A.P. Evaluation of the applicability of nano-biocide treatments on limestones used in cultural heritage. *J. Cult. Herit.* **2019**, *38*, 126–135. [CrossRef]
54. Castaldo, R.; de Luna, M.S.; Siviello, C.; Gentile, G.; Lavorgna, M.; Amendola, E.; Cocca, M. On the acid-responsive release of benzotriazole from engineered mesoporous silica nanoparticles for corrosion protection of metal surfaces. *J. Cult. Herit.* **2020**, *44*, 317–324. [CrossRef]
55. Rogers, S.O.; Bendich, A.J. Extraction of DNA from plant tissues. In *Plant Molecular Biology Manual*; Springer: Dordrecht, The Netherlands, 1989; pp. 73–83. [CrossRef]
56. Ray, P.C.; Yu, H.; Fu, P.P. Toxicity and environmental risks of nanomaterials: Challenges and future needs. *J. Environ. Sci. Health C* **2009**, *27*, 1–35. [CrossRef] [PubMed]
57. EU Regulation 1907/2006: Registration, Evaluation, Authorisation and Restriction of Chemicals (REACH). Available online: https://osha.europa.eu/en/legislation/directives/regulation-ec-no-1907-2006-of-the-european-parliament-and-of-the-council (accessed on 23 July 2021).
58. Pastoriza-Santos, I.; Pérez-Juste, J.; Liz-Marzán, L.M. Silica-coating and hydrophobation of CTAB-stabilized gold nanorods. *Chem. Mater.* **2006**, *18*, 2465–2467. [CrossRef]
59. Todd, J.S.; Zimmerman, R.C.; Crews, P.; Alberte, R.S. The AF activity of natural and synthetic phenolic-acid sulfate esters. *Phytochemistry* **1993**, *34*, 401–404. [CrossRef]
60. Villa, F.; Remelli, W.; Forlani, F.; Vitali, A.; Cappitelli, F. Altered expression level of *Escherichia coli* proteins in response to treatment with the antifouling agent zosteric acid sodium salt. *Environ. Microbiol.* **2012**, *14*, 1753–1761. [CrossRef]
61. Vilas-Boas, C.; Sousa, E.; Pinto, M.; Correia-da-Silva, M. An antifouling model from the sea: A review of 25 years of zosteric acid studies. *Biofouling* **2017**, *33*, 927–942. [CrossRef]
62. Villa, F.; Pitts, B.; Stewart, P.S.; Giussani, B.; Roncoroni, S.; Albanese, D.; Giordano, C.; Tunesi, M.; Cappitelli, F. Efficacy of zosteric acid sodium salt on the yeast biofilm model *Candida albicans*. *Microb. Ecol.* **2011**, *62*, 584. [CrossRef]
63. Geiger, T.; Delavy, P.; Hany, R.; Schleuniger, J.; Zinn, M. Encapsulated zosteric acid embedded in poly [3-hydroxyalkanoate] coatings—Protection against biofouling. *Polym. Bull.* **2004**, *52*, 65–72. [CrossRef]
64. Boopalan, M.; Sasikumar, A. Studies on biocide free and biocide loaded zeolite hybrid polymer coatings on zinc phosphated mild steel for the protection of ships hulls from biofouling and corrosion. *Silicon* **2011**, *3*, 207–214. [CrossRef]
65. Cattò, C.; Dell'Orto, S.; Villa, F.; Villa, S.; Gelain, A.; Vitali, A.; Marzano, V.; Baroni, S.; Forlani, F.; CAppitelli, F. Unravelling the Structural and Molecular Basis Responsible for the Anti-Biofilm Activity of Zosteric Acid. *PLoS ONE* **2015**, *10*, e0131519. [CrossRef] [PubMed]
66. Caneva, G.; Salvadori, O.; Ricci, S.; Ceschin, S. Ecological analysis and biodeterioration processes over time at the Hieroglyphic Stairway in the Copán (Honduras) archaeological site. *Plant Biosyst.* **2005**, *139*, 295–310. [CrossRef]
67. Caneva, G.; Bartoli, F.; Savo, V.; Futagami, Y.; Strona, G. Combining statistical tools and ecological assessments in the study of biodeterioration patterns of stone temples in Angkor (Cambodia). *Sci. Rep.* **2016**, *6*, 1–8. [CrossRef] [PubMed]
68. Municchia, A.C.; Bartoli, F.; Taniguchi, Y.; Giordani, P.; Caneva, G. Evaluation of the biodeterioration activity of lichens in the Cave Church of Üzümlü (Cappadocia, Turkey). *Int. Biodeterior. Biodegrad.* **2018**, *127*, 160–169. [CrossRef]
69. Valentini, F.; Diamanti, A.; Carbone, M.; Bauer, E.M.; Palleschi, G. New cleaning strategies based on carbon nanomaterials applied to the deteriorated marble surfaces: A comparative study with enzyme-based treatments. *App. Surf. Sci.* **2012**, *258*, 5965–5980. [CrossRef]

applied sciences

MDPI

Article

Black on White: Microbial Growth Darkens the External Marble of Florence Cathedral

Alba Patrizia Santo [1], Oana Adriana Cuzman [2], Dominique Petrocchi [3], Daniela Pinna [4], Teresa Salvatici [1] and Brunella Perito [3,*]

1 Department of Earth Sciences, University of Florence, Via La Pira 4, 50121 Florence, Italy; alba.santo@unifi.it (A.P.S.); teresa.salvatici@unifi.it (T.S.)
2 Institute of Heritage Science, National Research Council, Via Madonna del Piano 10, Sesto Fiorentino, 50019 Florence, Italy; oanaadriana.cuzman@cnr.it
3 Department of Biology, University of Florence, Via Madonna del Piano, Sesto Fiorentino, 50019 Florence, Italy; dominique.petrocchi@unifi.it
4 Ravenna Campus, University of Bologna, Via Guaccimanni, 42, 48121 Ravenna, Italy; daniela.pinna@unibo.it
* Correspondence: brunella.perito@unifi.it

Abstract: Weathering processes seriously affect the durability of outdoor marble monuments. In urban environments, a very common deterioration phenomenon is the dark discoloration or blackening of marble. This paper describes a multidisciplinary study on the state of conservation of white marbles of the Florence Cathedral and the microbial community involved in their deterioration. The study is focused on the widespread dark discoloration of marble analyzed in two differently exposed sites of the Cathedral. It aims to provide information useful for future interventions to control the microbial growth. By chemical and petrographic analysis, in situ and ex situ microscopy, and cultivation and identification of microorganisms, it was found that (i) the darkening is mainly due to the growth of black fungi and dark cyanobacteria and (ii) the state of conservation of marble and the growth pattern of microorganisms seems to be linked to the microclimatic conditions, in particular to solar radiation exposure. This is the first report on the lithobiontic community inhabiting the Florence Cathedral marbles, with a more detailed investigation of the culturable mycobiota.

Keywords: marble decay; biodeterioration; dark discoloration; stone microbiota; black fungi; cultural heritage conservation

Citation: Santo, A.P.; Cuzman, O.A.; Petrocchi, D.; Pinna, D.; Salvatici, T.; Perito, B. Black on White: Microbial Growth Darkens the External Marble of Florence Cathedral. *Appl. Sci.* **2021**, *11*, 6163. https://doi.org/10.3390/app11136163

Academic Editor: Maria Filomena Macedo

Received: 6 June 2021
Accepted: 30 June 2021
Published: 2 July 2021

1. Introduction

The preservation of stone-built cultural heritage is a major issue in modern societies to be pursued through conservation strategies for its transmission to future generations and the protection of its authenticity [1–3]. Outdoor stone monuments are exposed to several natural abiotic, biotic, and anthropogenic weathering factors. The mineralogy, chemical composition, porosity, surface roughness, water uptake, and state of conservation of the lithotypes as well as the macro- and micro-environmental conditions are key factors affecting the so-called stone bioreceptivity and influencing microbial colonization and growth [3–6]. Although stone represents a harsh habitat to live on/inside, complex and diverse lithobiontic communities develop on rocks coping with large variations in environmental factors such as temperature, water and nutrient availability, electrolyte concentration, and solar radiation [7,8]. Microbial communities usually grow as biofilms adhering to the surface or penetrating inside stone [9,10]. They cause deterioration manifested mainly by aesthetical damages, such as colored patinas or discolorations, but also by severe structural modifications of stones due to biogeophysical and biogeochemical processes that act synergically [2,11].

In urban environments, outdoor monumental and artistic stones are also exposed to air pollution, mainly caused by fossil fuel combustion, which affects chemical weathering, bioreceptivity, and biodeterioration of stones [10,12–14]. A wide range of airborne hydrocarbons and fatty acids settles on stone surfaces [15] and can be utilized by heterotrophic microorganisms for their growth, so that, according to some authors [16], the rock surface in such environments should no longer be regarded as oligotrophic.

Among the natural stones used as constructive and decorative materials since ancient times (e.g., since about 4700 BC in Egypt), marble plays a relevant role, due to its bright white color and translucence. Marble is a metamorphic rock consisting prevalently of calcite or dolomite, and despite its hardness, it is subject to weathering processes that can seriously affect its durability. In outdoor conditions, marble can be damaged by solar irradiation and consequent temperature variations, rain washing, salt crystallization, and atmospheric pollution [17]. In particular, marble is extremely susceptible to acid attack caused by atmospheric compounds such as carbon, sulfur, and nitrogen oxides [18]. Furthermore, outdoor marble monuments can be colonized by several kinds of microorganisms such as bacteria and cyanobacteria, algae, lichens, filamentous, and meristematic fungi, which can cause biodeterioration [5,10,19–21]. Some studies highlight how marble surfaces display the highest degree of microbial growth among monumental stones [22]. Documented biodeterioration phenomena on ancient marble monuments include the production of organic acids by fungi and lichens that chelate metallic cations and dissolve calcite, biomineralization, biopitting, and discolorations [13].

Despite the widespread presence of discolorations on marble, the scientific literature on this subject is limited. Since marble is one of the most used stones by architects and artists of all times [17], its discolorations are worthy of being studied [23], and their specific causes should be clearly identified. The dark discoloration or blackening is often ascribed to the formation of calcium sulfate or gypsum, which contributes to black crusts formation [13,24], but it can be also caused by the colonization of cyanobacteria [25,26] as well as of the so-called black or dematiaceous fungi [27–30].

To plan appropriate conservation measures aimed to control the microbial growth and protect the heritage stone, a correct analysis and diagnosis of each specific biodeterioration aspect is essential [31]. This requires the knowledge of the rock characteristics and the resident microbial community, as well as the interactions between microorganisms and stone [32]. Although biodeterioration phenomena are rarely ascribable to a single microbial component, most studies focus on single taxonomic groups and not on the whole communities that colonize stone surfaces, not considering some of the inter-taxa interactions that may be responsible for deteriorative changes [33].

The city of Florence, known as the cradle of the Renaissance, houses the world's largest concentration of universally renowned artistic and architectural masterpieces. The Historic Centre of Florence is a unique artistic realization, an exceptional testimony of both a medieval city and of a Renaissance one, and a living archive of both European and Italian culture. For these reasons, it was registered in the list of UNESCO World Heritage Sites in 1982. The Cathedral of Santa Maria del Fiore (SMFC; Figure 1) is the most emblematic monument and symbol of Florence. It is an impressive building located in Piazza del Duomo, the beating heart of the city, standing tall over the city, and its conservation is a main issue of worldwide concern. Nonetheless, the exterior of the Cathedral, mainly covered with Apuan marble, shows extended forms of decay, macroscopically visible, consisting of deposits, discolorations, patinas, crusts, erosion, mechanical damage, and granular disaggregation [3,34]. In recent years, the Opera di Santa Maria del Fiore (OSMF)—the institution actively engaged in the protection of the monuments of the complex of Santa Maria del Fiore—started a maintenance program for the cleaning and restoring of all the external façades of the Cathedral. Such interventions, especially those aimed at removing biological patinas, would be better planned with a previous knowledge of the inhabitant lithobiontic community.

The OSMF expressed interest in testing innovative methods to remove patinas through on-site trials on selected external areas, not accessible to visitors, that did not undergo recent restoration interventions. In these areas, marble surfaces showed extended dark discolorations as the main deterioration phenomenon. Therefore, we first investigated the cause of the darkening, then the interaction of microbial communities with marble. We thus had the unique opportunity to carry out a multidisciplinary study on the state of conservation of the SMFC external white marbles and the lithobiontic community involved in their deterioration. The results will be essential to plan adequate interventions to control the microbial colonization.

2. Site Description

The Historic Centre of Florence is an area (latitude 43°46′22″ N; longitude 11°15′25″ E; 50 m at sea level) characterized by a temperate macro-bioclimate with the sub-Mediterranean bioclimatic variant [35]. The average temperature (T), calculated in the period 2010–2020, is about 16.7 °C, with a monthly minimum average of 1.3 °C (February 2012) and a monthly maximum average of 40.5 °C (July 2015). The average winter temperature is about 9.3 °C, while the mean temperature in summer is about 25 °C. The average yearly value of precipitation, mostly occurring in autumn and spring, is 799 mm, and the average relative humidity (RH) is around 61.9%, with maximum peaks of 89–97% in winter. The RH of Florence never dropped below 40% during the reported period. In the climatic zone of SMFC, persistent winds blow all year, mainly from the southeast and northwest at about 1.3–1.6 m/s wind speed [36].

The Cathedral of Santa Maria del Fiore (Figure 1) was built on the site of the previous early-Christian cathedral dedicated to Santa Reparata. Its construction was commissioned by the city council at the end of the 13th century as a symbol of richness and power of the city and lasted from 1296 to 1434. When it was completed, in the 15th century, it was the largest church in the world; today, it is the third-largest church in the world. The first design, in Gothic style, was by the Italian architect Arnolfo di Cambio; successively, other architects, among them Giotto, Andrea Pisano, and Francesco Talenti, tended to its construction, imparting the different styles that we can observe in the current structure. The dome was erected between 1418 and 1434 on the ingenious project of Filippo Brunelleschi. The neo-Gothic façade, on the elaborate design of Emilio de Fabris, was completed only in 1884. The size of the Cathedral, which shows a Latin cross plan, is enormous (8300 m^2 total area); the exterior part is covered with polychrome stone panels consisting of white marble, serpentinite, and red limestone [37–39] coming from Florence's surrounding quarries (Figure 1).

Figure 1. The Cathedral of Santa Maria del Fiore and the study sites. (**a**) The façade; (**b**) the external gallery running around the apses of the SE exposed façade; (**c**) the external gallery on the NW exposed façade; (**d**) a detail of the marble surface of the inner face of the openwork parapet of the SE exposed gallery; (**e**) a detail of the openwork parapet at the NW site showing a widespread dark patina.

The white marbles mainly come from the Apuan Alps, in the Carrara district [37], which represents the world's most important and largest (more than one hundred active quarries) mining area. The Carrara marbles, used since the Ancient Rome time [40,41], represent the best-known stone materials used by architects and artists around the world, from the temple of Apollo Palatinus in Rome to the Marble Arch in London, the Harvard Medical School Building in Boston, and the Buddhist Temple in Singapore [41]. Due to its technical properties, this marble is considered suitable for any type of use, excellent for sculptures—so much that Michelangelo wanted it for his Pietà and David.

In general, marbles, although having the same chemical and mineralogical composition, can show different microstructural characteristics which affect their behavior when subjected to weathering. The microstructural characteristics of Apuan marbles vary depending not only on the different extraction site but also on different areas of the same quarry, according to the complex tectono-metamorphic history that they have undergone [42]. Moreover, in the case of the Cathedral of Florence, Apuan marbles of different provenance were used for the original construction (common white marble quarried from Carrara quarries—the marble studied in the present paper) and the recent 19th century façade (common white marble from Seravezza, Lucca) [37]; the latter displays a more advanced state of decay in comparison to those used in the 15th to 18th centuries [37,43].

3. Materials and Methods

3.1. Sampling

The study was conducted in two differently exposed sites located in the external gallery running around the apses on the upper part of SMFC (Figure 1b,c) at a height of 36 m above ground. They are northwest (NW) and southeast (SE) exposed and characterized by different light/shadow durations (SE area is illuminated longer by direct sun than NW area). The balcony is accessible by an internal staircase. The last intervention on these areas dates back to 1952 and consisted of substitutions of deteriorated marble slabs (B. Agostini, personal communication).

Most in situ observations and sampling were conducted in the fall of 2019. Microscopical observations and microbiological sampling were carried out in different areas of the vertical inner surface of the gallery's openwork parapet (Figure 1d,e).

To perform mineralogical, petrographic, and ex situ microscopic analyses, small fragments already detached from areas with evident forms of degradation (such as fractures and cracks) were collected. The fragments were SMF1, SFMF2, and SMF8 from the NW area, and SMF6, SMF7, and SMF9 from the SE area (Figure S1).

For microbiological analysis, superficial particulate was gently scraped from marble with a sterile spatula (micro-invasive method) and collected into sterile tubes. Three samples of about tens of mg were taken from a surface of about 1000 cm^2 at both the NW and SE sites. Samples were immediately brought to the laboratory and processed.

3.2. In Situ and Ex Situ Microscopy

Biological growth on selected areas was firstly ascertained by naked eye and then was observed under a hand lens (9$\times$ magnification) and a portable light microscope Scalar DG-2A (Tecmet 2000, Corsico, Italy) equipped with an optical zoom 25–200$\times$ and an image capturing system.

The biological growth on and below the surface was examined under transmitted and reflected light (RLM) microscope and scanning electron (SEM) microscope.

The samples were observed under the stereomicroscope Leica DMS300 (Leica Microsystems, Wetzlar, Germany) and the particulate suspensions (Section 3.4) under the ZEISS Axiolab 5 (Zeiss, Oberkochen, Germany) microscope equipped with the video camera Axiocam 208 color (Zeiss, Oberkochen, Germany).

The RLM observations were carried out using the Zeiss Axio Skope.A1 (Zeiss, Oberkochen, Germany) microscope equipped with a video camera (5 megapixel resolution and image analysis software AxioVision) on cross sections (2.5 cm diameter, 1.5 cm height) from both NW and SE areas. The sections were obtained by cutting the samples with a diamond saw. They were stained using the PAS kit (Periodic Acid-Schiff, Sigma Aldrich, St. Louis, MO, USA) to visualize the biological component within the lithic substrate. Since PAS stains carbohydrates, it is used to detect cells as well as extracellular polymeric substances (EPS), and consequently biofilm, on/inside marble.

The petrographic observation was carried out on ultrathin sections (thickness 12–15 μm), using the same Zeiss Axio Scope.A1 (Zeiss, Oberkochen, Germany) microscope with polarized light.

Scanning electron microscopy (SEM) observations were carried out using the Zeiss EVOMA15 microscope (Zeiss, Oberkochen, Germany). The samples were previously coated with both carbon and gold using the Quorum Q150R ES sputter coater (Quorum Technologies, Laughton, UK).

3.3. Chemical Analysis

The surface of the marble samples was scraped and analyzed using the Fourier Transform Infrared Spectroscopy (FT-IR) through ATR mode with a Spectrum 100 FTIR spectrometer (Perkin-Elmer Inc., Norwalk, CT, USA) equipped with a Universal ATR accessory. The acquisition was carried out at room temperature, in the spectral range between 4000 and 350 cm^{-1}, repeating 4 scans with resolution of 4 cm^{-1}. The data were acquired and processed using the Spectrum 100 software.

3.4. Cultivation, Morphological Analysis and Identification of Microorganisms and Lichens

To cultivate green algae and cyanobacteria, aliquots (10 mg) of the collected marble powder were immersed in the liquid nutrient medium BG-11 prepared according to Rippka et al. [44] and adjusted only with 5 mL/L $NaNO_3$ instead of 15 mL in order to allow the growing of nitrogen fixing cyanobacteria, as well. After 2 months, the main phototrophic biodiversity was observed under a Zeiss Axio Scope A1 microscope (Zeiss, Oberkochen, Germany) equipped with a Zeiss Axio Cam ICc3 (Zeiss, Oberkochen, Germany), and the morphological characterization was executed according to Komarek et al. [45] and Bourrelly [46].

In a previous work [3], we performed various lab tests to find suitable conditions for a high recovery of fungal colonies from SMFC marble. Such conditions will be applied in future evaluations of biocide treatments. Among the nutrient media used, Malt Extract Agar (MEA) gave the best results, with a viable titer up to 1.5×10^4 CFU/g at the NW site and an apparently high biodiversity. MEA was also modified by adding 0.1% and 1% marble powder, and 1% aqueous marble extract and no significant differences were observed in viable titer [3]. Consequently, we used MEA at the subsequent sampling. To cultivate fungi, 10 mg of marble particulate, obtained by mixing three samplings from each area, were suspended in 1 mL of Phoshate Buffered Saline (PBS; 8 g/L NaCl, 0.2 g/L KCl, 1.44 g/L Na_2HPO_4, g/L 0.24 KH_2PO_4, pH 7.4) added with 0.001% Tween 80 and vortexed. Then, 0.1 mL of suspension were plated in quadruplicate on MEA (Oxoid) supplemented with 10 μg/mL of chloramphenicol to prevent bacterial growth and then incubated at 30 °C for at least seven days. Viable titer was calculated as mean value of the number of Colony Formant Units (CFUs) per gram of marble particulate. Morphological analysis of colonies was conducted by observing them under the stereomicroscope Olympus SZX9 (Olympus, Tokyo, Japan). Colonies with the same morphological characteristics were grouped in the same morphotype. For each morphotype, at least one strain was re-isolated on MEA and then stored on slant. Identification of fungal morphotypes was based on rDNA analysis. Genomic DNA of isolated strains was extracted using Wizard® Genomic DNA Purification kit (Promega, Madison, WI, USA) following the yeast protocol and modifying the cell lysis step as follows: about 1 cm^2 mycelium from an isolated colony was cut, suspended in 500 μL physiological solution, and centrifuged. The pellet was then suspended in 47.5 mM EDTA + 0.5 mg/mL zymolyase (Sunrise Science Products, Knoxville, TN, USA) and incubated at 37 °C for 45 min. Amplification of the internal transcribed spacer (ITS) region of rRNA genes was performed using primers ITS1D (5′-GTTTCCGTAGGTGAACCTGC-3′) and ITS4 (5′-TCCTCCGCTTATTGATATGC-3′), and PCRBIO Taq Polymerase (PCRBioSystem, London, UK). PCR conditions consisted of an initial denaturation step at 95 °C for 2 min followed by 30 cycles at 95 °C for 30 s, 60 °C for 1 min, 72 °C for 1 min, and a final extension at 72 °C for 5 min. The Sanger sequencing was performed by Bio-Fab Research s.r.l. (Rome, Italy). To identify the isolates, the nucleotide sequences were analyzed by BLAST (Basic Local Alignment Search Tool) using the National Center of Biotechnology Information (NCBI) database [47]. Newly generated ITS rDNA sequences were deposited at the NCBI database under the accession numbers from MW361278 to MW361325.

Lichens were identified directly in situ using the portable stereomicroscope. The identification key of Clauzade et al. [48] was used as a reference. Nomenclature follows Nimis [49].

3.5. Carbonate Solubilization Test

The potential ability of the isolated fungi to solubilize calcite was screened on $CaCO_3$ glucose agar medium (glucose 1%, $CaCO_3$ 0.5%, agar 1.5%; pH adjusted to 8.0 with 1 M HCl). Fungal mycelium was inoculated into the center of the Petri dishes (± 1 cm^2 agar blocks) and incubated at 30 °C for 8 weeks. $CaCO_3$ dissolution ability was evaluated by the presence of a clear zone around the colony.

4. Results

4.1. In Situ Observations

The marble surfaces of both NW and SE areas display a dark discoloration with a higher surface coverage on the NW area (Figure 1d,e).

In situ observations with the stereomicroscope revealed the presence of two patterns of biological colonization: (i) around the grains of the marble but not covering them (Figure 2a), (ii) forming black little spots that cover the marble surface (Figure 2b) and growing up to form patinas more developed at the NW site (Figure 2c). The total effect at the naked eye observation is a strong alteration of the marble color that appears dark gray with widespread black spots (Figures 1 and 2). The dark colonization around grains and the black spots would resemble the growth pattern of cyanobacteria and black fungi (Figure 2a,b). Some lichens preferentially grow on biofilm black spots; the latter seem to behave as pioneer microorganisms, somehow helping the succession to the lichens even if they are able to grow also in polluted atmospheres (Figure 2d).

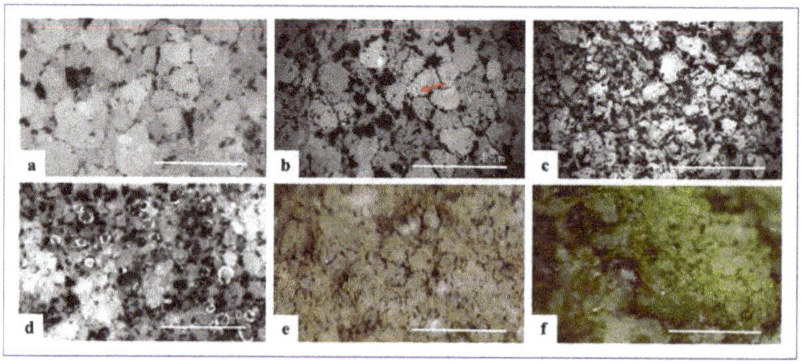

Figure 2. Microscopic observation of the surface of the SMFC external white marbles: (**a**–**d**) in situ observations; (**e**,**f**) ex situ observations of marble fragments. (**a**,**b**) Microbial colonization around the marble grains at the SE (**a**) and NW sites (**b**), little black spots presumably composed of black fungi (arrow) are visible over the marble grains in (**b**). (**c**) Colonization by a more developed biofilm forming a patina over the marble at the NW site. (**d**) Thalli of the lichen *Myriolecis dispersa* growing preferentially on biofilm black spots at the SE site. (**e**) Growth around grains resembling that of cyanobacteria on a marble sample. (**f**) Growth of algae and dark colonies on a marble sample. The scale bar is 1 mm in (**a**–**f**) and 2 mm in (**d**).

4.2. Chemical and Petrographic Analyses

The studied marbles mainly consist of calcite [3]. Accordingly, the obtained FT-IR spectra (Supplementary Figure S2) are characterized by the stretching band around 1420 cm^{-1} and by the absorbance at 871 cm^{-1} and 712 cm^{-1} [50]. In almost all the spectra, also visible is the absorbance at 1033 cm^{-1} (silicates) and the characteristic sharp peaks of gypsum (bending vibrations at 1796 and 1641 cm^{-1} and at 1115 cm^{-1}) whose content is very low. In Supplementary Figure S2, two representative FT-IR spectra (SMF2 and SMF9 from NW and SE area, respectively) are shown.

Marbles observed in ultrathin sections under a petrographic microscope (Figure 3) display the typical polysynthetic twinning of the calcite crystals which do not exhibit a preferred orientation; the grain-size is heterogeneous, mainly in the range 150–200 µm, up to 500 µm.

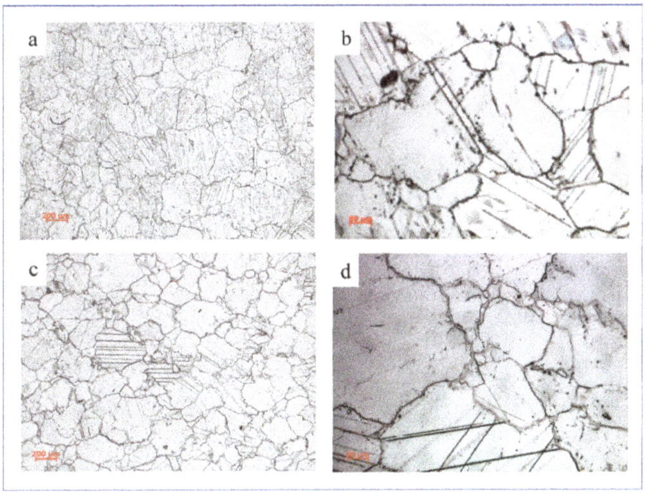

Figure 3. Photomicrographs at different scale of ultra-thin sections of marble in plane-polarized light. (**a**,**b**) NW sample (SMF1); (**c**,**d**) SE sample (SMF6).

The samples collected from the two areas display overall similar characteristics. They have a prevalent heteroblastic or, in some cases, homeoblastic mosaic texture and the grain boundaries show straight to lobate-curved and sutured shapes (Figure 3). All the samples show low macro-porosity; however, where the crystal boundaries are prevalently straight, a slight grain detachment is observed. In particular, the detachment and the presence of fine-grained recrystallized calcite along the grain boundaries are more evident in the samples collected in the SE area (Figure 3d).

4.3. Ex Situ Microscopic Observations

Observation of samples under the stereomicroscope confirmed the biocolonization patterns observed in situ and presumably attributable to cyanobacteria, algae, and fungi (Figure 2e,f). Suspensions of marble particulate used to cultivate fungi were also observed under the light microscope that showed dispersed aggregates of phototrophic microorganisms such as algae, cyanobacteria, and black fungi (Supplementary Figure S3). Observation of cells of black spots picked up with a sterile needle from marble samples showed very similar images of such aggregates (not shown). A few black spots sampled from SMF8 were plated on MEA, in one case leading to the growth of a meristematic colony (strain M1; Section 4.4.3).

RLM observations of cross-sections showed a variable degree of biodeterioration in different samples. The presence of an epilithic biofilm on marble surface was always detected, thicker on the NW samples (Figure 4a,b) whose internal grains, however, showed a better state of conservation (Figure 4a,b,f). An endolithic biofilm developed in the SE samples. It penetrates as up to 3–4 mm in depth, surrounding the internal grains and causing their physical distancing (Figure 4c,d,g,h). In the superficial layer, the biofilm, growing around the grains, causes their detachment (Figure 4e,f). At the magnification used, we could not detect the kind of microbial cells participating to the epilithic or endolithic biofilm detected by PAS, except for the presence of endolithic algae, suggested by green spots successively stained by PAS (Figure 4g,h). Moreover, the presence of black fungi was clearly visible inside some samples (Supplementary Figure S4).

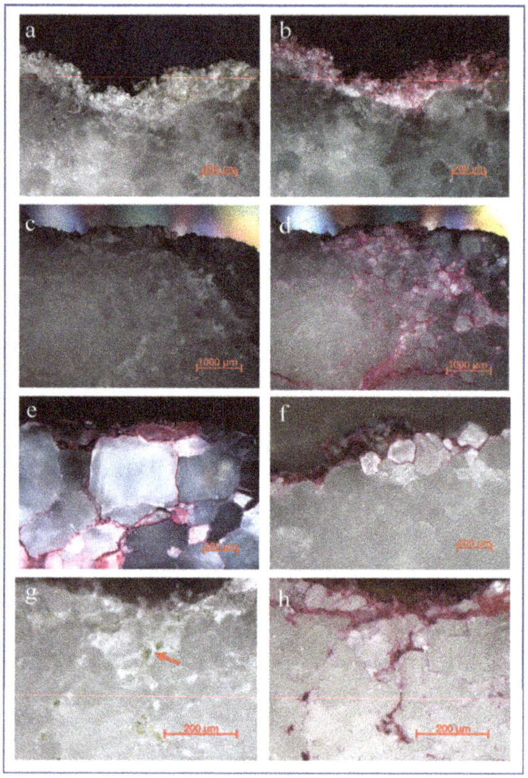

Figure 4. Biological colonization of SMFC marble observed by marble polished cross sections. SMF1 (NW) cross section before (**a**) and after (**b**) PAS staining; SMF6 (SE) cross section before (**c**) and after (**d**) PAS staining; (**e**) a detail of d; (**f**) PAS-stained cross section of SMF2 (NW); SMF9 (SE) cross section before (**g**) and after (**h**) PAS staining, endolithic algae (arrow) are visible in (**g**).

SEM observations confirmed the differences already noted on ultra-thin sections; the SE samples display a more evident grain detachment than NW ones (Figure 5a,b). Moreover, they also revealed the presence of a composite microbial community showing intimate relations with the marble grains. In a few samples, individual colonies are recognizable, as in the case of the cyanobacteria growing among marble grains (Figure 5b,c). In other samples, a well-developed biofilm is visible with differently shaped (filamentous and globular) microorganisms embedded into the slime of EPS; it completely covers the marble (Figure 5d,f) and contributes to grains detachment (Figure 5e).

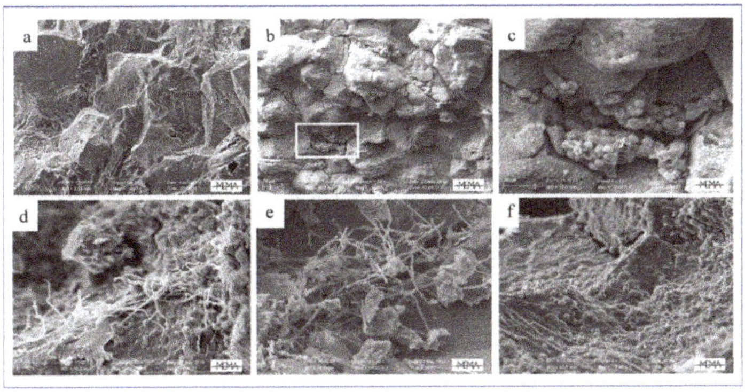

Figure 5. Scanning electron microscope images of SMFC marble. (**a**) NW sample (SMF1); (**b**) SE sample (SMF6), the grain detachment is clearly visible, colonies of unicellular cyanobacteria are scattered around the marble grains. (**c**) A detail of (**b**) (white rectangle), a colony with cells growing in the fissures among grains; (**d**–**f**) NW sample (SMF2), filamentous microorganisms connected through an EPS matrix to a multispecies biofilm adhering to marble (**d**), and detaching marble particulate (**e**). (**f**) Microbial biofilm completely covering marble surface with cells embedded within the EPS matrix.

4.4. Characterization of the Lithobiontic Microbial Community

4.4.1. Lichens

The examined areas were colonized mainly by crustose lichens, just one foliose species was present. Lichen thalli were better developed on the NW facing areas, nevertheless the identified species were the same. Although lichen species richness was extremely low, the abundance of some species was quite high.

Many thalli of *Myriolecis albescens* (Hoffm.) Sliwa, Zhao Xin, and Lumbsch and *Myriolecis dispersa* (Pers.) Sliwa, Zhao Xin, and Lumbsch were present (Figure 2d). They grew preferentially on biofilm black spots. Other species present on the examined areas were *Candelariella aurella* (Hoffm.) Zahlbr., *Flavoplaca citrina* (Hoffm.) Arup, Frödén, and Søching and primordia of thalli of a foliose lichen belonging to *Physcia* genus.

4.4.2. Cyanobacteria and Algae

The most abundant phototrophic microorganisms were coccoid cyanobacteria belonging to the *Chroococcales* group *Gleocapsa* sp. and the unicellular green alga *Chlorococcum* sp. (*Chlorophyta*), being most spread on both the NW and SE facing areas, followed by filamentous cyanobacteria, belonging to the *Oscillatoriales* group. Other few genera were observed in the SE area, such as the cyanobacterium *Aphanocapsa* sp. (*Synechococcales*) and the filamentous green alga *Ulothrix* sp. The microscopic observations of the phototrophic cultures also showed the presence of black fungi associated with the phototrophic cells (Supplementary Figure S3).

4.4.3. Fungi

The fungal viable titer was $1.0 \times 10^4 \pm 4.1 \times 10^3$ CFU/g at the NW and $5.5 \times 10^3 \pm 4.7 \times 10^3$ CFU/g at the SE site, confirming our previous results [3]. Fungi from two different samplings (spring and fall 2019) were investigated for their biodiversity. The morphological analysis of all the colonies identified 28 morphotypes at the NW site and 20 morphotypes at the SE site (Supplementary Figure S5). DNA was extracted and rDNA ITS amplified from all the morphotypes. Results of sequence analysis are reported in Table 1. Although the ITS region is the universal barcode marker currently used for fungi, it has some limitations regarding the identification at species level, depending on the fungal group [51]. Although most of the isolated strains were identified as species, we found ambiguity in species assignment for strains of *Alternaria*, *Cladosporium*, and *Epicoccum*

(Table 1). For this reason, we summarize and discuss the data relative to the genera found. Overall, a total of 21 different genera were identified at the two study sites, with eight genera specific of the NW area and 10 specific of the SE area. Three genera (*Alternaria*, *Cladosporium* and *Epicoccum*) were common to both the areas. *Alternaria* and *Cladosporium* had apparently the highest number of species, all the other genera were represented by only one species, but *Coprinellus* by two.

Table 1. Identification of fungi isolated at the NW and SE study sites of SMFC by rDNA analysis. (**a**) Morphotypes isolated at the NW site. (**b**) Morphotypes isolated at the SE site. In the Morphotype column, bold indicates dematiaceous filamentous fungi; italics indicate dematiaceous meristematic fungi and black yeasts.

Morphotype	Organisms with the Most Similar ITS Sequences	Similarity (%)
(a)		
Nord A	*Parengyodontium album*	99.81
NB	*Dimorphoma saxea*	100
NC	*Alternaria alternata/A. tenuissima*	99.81
ND	*Alternaria tenuissima/A. alternata*	100
NE	*Hyphodermella rosae*	100
NF	*Cladosporium asperulatum/C. xylophilium*	99.80
NG	*Cladosporium halotolerans*	99.61
NH	*Alternaria alternata/A. tenuissima*	99.81
NT-N-1	*Diplodia seriata*	99.82
NT-N-2	*Neoascochyta exitialis*	97.25
NT-N-3	*Alternaria tenuissima*	99.81
NT-N-6	*Pithomyces chartarum* [1]/*Leptosphaerulina chartarum* [1]	99.48
NT-N-7	*Alternaria ethzedia/A. infectoria*	99.64
NT-N-8	*Aspergillus niger/A. welwitschiae*	98.21
NT-N-9	*Alternaria ethzedia*	99.46
NT-N-10	*Cladosporium cladosporioides/C. perangustum*	98.85
NT-N-11	*Epicoccum nigrum/E. layuense*	99.23
NT-N-12	*Alternaria tenuissima/A. alternata*	99.81
NT-N-13	*Leptosphaerulina chartarum*	99.48
NT-N-14	*Aspergillus niger*	99.44
NT-N-15	*Cladosporium asperulatum/C. uredinicola*	99.61
NT-N-17	*Alternaria alternata*	100
NT-N-18	*Alternaria cumini*	99.39
NT-N-19	*Alternaria tenuissima/A. alternata*	100
NT-N-20	*Leptosphaerulina chartarum*	99.49
NT-N-21	*Alternaria alternata/A. malvae*	99.62
NT-N-22	*Pithomyces chartarum* [1]/*Leptosphaerulina chartarum* [1]	99.65
M1	*Lithophila guttulata*	100
(b)		
SA	*Alternaria alternata*	100
SB	*Epicoccum nigrum/E. sorghinum*	99.80
SC	*Phoma conidiogena* [2]/*Didymella glomerata* [2]/*Coniothyrium aleuritis* [2]	100
SD	*Bipolaris coffeana* [3]/*B. austrostipae* [3]/*Cochliobolus cynodontis* [3]	100
SE	*Coprinellus xanthothrix*	99.70
SF	*Arthrinium arundinis*	98.95
SH	*Alternaria ethzedia*	99.63
SI	*Phoma conidiogena* [2]/*Dydymella glomerata* [2]/*Coniothyrium aleuritis* [2]	99.61
SL	*Phoma conidiogena* [2]/*Didymella glomerata* [2]/*Coniothyrium aleuritis* [2]	99.61
SM	*Exophiala capensis*	97.90

Table 1. *Cont.*

Morphotype	Organisms with the Most Similar ITS Sequences	Similarity (%)
SN	*Elaphocordyceps* sp. [4]/*Tolypocladium* sp. [4]	98.85
SO	*Aureobasidium pullulans*	99.64
Sud P	*Aureobasidium pullulans*	99.82
NT-S-1	*Phaeosphaeriopsis pseudoagavacearum*	99.81
NT-S-5	*Dothideomycetes* sp.	99.45
NT-S-6	*Cladosporium ramotenellum/C. puyae*	98.82
NT-S-7	*Alternaria citri*	99.81
NT-S-8	*Cladosporium sinuosum/C. tenellum/C. herbarum*	99.80
NT-S-9	*Paraconiothyrium hawaiiense*	100
NT-S-10	*Coprinellus micaceus*	99.69

[1] *L. chartarum* is the teleomorphic form, *P. chartarum* is the anamorphic form. They were considered as a single species. [2] Some *Phoma* species are the anamorphs and their teleomorphic forms are described in genera *Didymella* and *Coniothyrium*. They were considered as a single species. [3] *Cochliobolus* is the teleomorphic form, *Bipolaris* is the anamorphic form. They were considered as a single species. [4] *Tolypocladium* genus includes some asexual morphs of *Elaphocordyceps* species. They were considered as a single species.

Most of the identified genera (8 out of 11, 72.7% of the genera at NW; 9 out of 13, 69.2% at SE) and species are known as dematiaceous (Table 1). Among these, one strain at NW (M1) and two strains at SE (SM and NT-S-5, the latter based on morphology) were meristematic fungi, while two other strains at SE (SO and SP, both identified as *Aureobasidium pullulans*) showed a yeast-like growth (Supplementary Figure S5).

4.5. $CaCO_3$ Solubilization Test

Twelve fungal strains isolated from SMFC marble and belonging to genera and species frequently detected on marble and limestone were selected to better understand their role in marble deterioration and tested for their ability to solubilize $CaCO_3$. Five strains resulted positive (Table S1, Figure S5): NT-N-8, NT-N-14 (both *Aspergillus niger*), NT-N-10 (*Cladosporium cladosporioides*), SO, and SudP (both *Aureobasidium pullulans*).

5. Discussion

The studied marbles consist of almost pure calcium carbonate; however, trace amounts of silicates and gypsum were revealed, in almost all samples, by the FT-IR analyses. The presence of gypsum is generally imputed to the sulphation process. This consists of the reaction of sulfuric acid, coming from the air pollutant SO_2, with the insoluble calcium carbonate of marble, which is consequently transformed into the sulfate dihydrate or gypsum ($CaSO_4 * 2H_2O$). The process is considered one of the main causes of carbonatic stone deterioration in urban environments and is known to be associated to the blackening of outdoor marble artworks (e.g., [52]). Since gypsum is soluble in water, it is usually washed away in rain-exposed areas, whereas in sheltered areas, its crystals form networks that entrap particles of dirt and airborne pollutants, such as carbonaceous particles, to form black crusts [24]. The Cathedral of Santa Maria del Fiore is exposed to the urban polluted air of Florence and, among pollutants, to SO_2. The annual average values of SO_2 emissions detected in Florence from 2007 to 2012 were around 1–2 $\mu g/m^3$. Since 2014, the annual average values showed a reduction from 3 $\mu g/m^3$ to 1 $\mu g/m^3$ detected in 2019 [53] possibly due to the pedestrianization of this area since October 2009. Due to the long-term SO_2 exposure, the external marbles of the Cathedral display black crusts in several areas [3]; however, they are not present in the studied areas, as demonstrated by microscopic, mineralogical, and chemical investigations. The NW and SE study sites are rain-exposed, thus, soluble salts including gypsum are washed away; this is in agreement with the low amounts of gypsum found by FT-IR. The blackening of marble can thus be imputed to microbial colonization. The dark spots and patches on marble contain in fact dark colonies often present as multispecies aggregates mainly consisting of dematiaceous fungi and dark cyanobacteria, as shown by in situ (Figure 2) and ex situ microscopic observation of samples (Supplementary Figure S3). The presence of black fungi in these aggregates was further confirmed by the growth of a meristematic colony, strain M1, from a black spot. On the other hand, cultivation confirmed the presence of dematiaceous fungi and dark cyanobacteria. Even if dark patinas show different levels of development at the SE and NW sampling sites, they can be attributed to the same biological cause.

Indeed, a complex multi-kingdom microbial community inhabits the SMFC marble. Cyanobacteria, green algae, and lichens have been detected and characterized by morphological criteria. The fungal community was deeper investigated through isolation and identification by molecular methods. Bacteria were also detected [3], and their analysis is the subject of an ongoing work.

Concerning the photoautotrophic community, the biodiversity seems to be very similar in the investigated areas. Lichen species richness is extremely low as it is expected in an area exposed to pollutants as the Cathedral is. Nonetheless, some species (*Myriolecis albescens* and *M. dispersa)* show a high prevalence. This result can be likely due to the pollution resistance of these species as well as to microenvironmental conditions. They grow preferentially on biofilm black spots.

Cyanobacteria and algae are considered primary colonizers of the bare stones. Cyanobacteria, in particular, can easily develop in harsh conditions, and they widely grow on NW and SE surfaces of the SMFC, being important contributors to the darkening alterations. Dark pigments such as mycosporine-like amino acids and carotenoids (common in cyanobacteria and some green algae) and scytonemin (found in sheathed cyanobacteria) contribute to the blackening phenomena even on other monuments [25,26]. The coccoid-sheathed cyanobacteria are dominant on the SMFC marbles, but the green unicellular algae showed a broad presence as well. Both unicellular cyanobacteria and green algae may be part of the symbiotic process in the presence of mycobionts, therefore leading to the formation of lichens. On the other hand, fungi can be prevalent in urban conditions since they utilize the airborne anthropogenic compounds [7].

The isolated fungi showed a high biodiversity, with genera and species richness similar at both study sites. Although some differences in species composition at the NW and SE sites were detected, the strong presence of dematiaceous fungi (more than 70% of the total 21 isolated genera)—in particular, of *Alternaria* and *Cladosporium* (the most abundant genera)—are nevertheless the common denominator of the fungal community at both the study sites. Dematiaceous or black fungi are an artificial heterogeneous group of darkly pigmented fungi with different morphological characteristics (and often pleomorphic behavior) forming dark-brown, green-black, or black colonies due to the production of melanin and melanoid pigments [54]. Among them, meristematic fungi are well known as rock inhabitants able to deal with varying microclimatic conditions. They have cell walls strongly melanized and form small black colonies on and inside the stone, often occurring in close association with lichens [55]. Black fungi are considered as major agents of microbial deterioration of building stones. Their occurrence on marble produces well-documented effects such as aesthetical damage due to darkening (from black spots to black layers completely covering stone) and other color changes, as well as surface erosion and exfoliation [8,20,27,30,32,55–57]. Their activities are favored by the urban "air eutrophication" that increases the occurrence of discolorations and crust formations [30]. Among the isolated dematiaceous fungi, *Alternaria, Aspergillus, Aureobasidium, Cladosporium, Epicoccum*, and *Phoma* are the most frequent genera present as airborne ubiquitous spores growing on stone in urban environments [16], and well known to cause blackening on marble and limestone [30,58]. These genera were also found on darkened areas of the external marble of the Milan Cathedral, where their growth was enhanced by naturally aged acrylic resins used as stone protectives and consolidants [19]. Some of them, particularly *Phoma* and *Alternaria*, are considered as ones of the most damaging organisms that attack and even penetrate the surfaces of stone monuments [29,30,56].

While for the above-mentioned dematiaceous genera the role in stone deterioration is well documented, other isolated fungi would be present as airborne spores occasionally deposited or entrapped in biofilms on marble, with an unclear role in stone colonization.

This would be the case for the plant pathogens *Bipolaris* (strain SD), *Arthrinium* (strain SF), *Hyphodermella* (strain NE), and *Coprinellus* (strains SE and NT-S-10). Concerning *Parengyodontium album* (strain Nord A), it has been frequently isolated from deteriorated materials of cultural heritage, stone included. Although several authors have linked the development of *P. album* on cultural heritage monuments to the presence of insects, its role in biodeterioration is debated [59]. A fungus frequently detected at the NW site is *Pithomyces chartarum* (or its teleomorph *Leptosphaerulina chartarum*; strains NT-N-6, NT-N-13, NT-N-20 and NT-N-22). *Phitomyces* strains are commonly isolated from a wide range of plant material; it has been previously isolated from marble [60], but its possible role in deterioration is unknown. For this reason, *Parengyodontium* and *Pithomyces* (strain NT-N-13) were included among the strains selected for the calcium carbonate dissolution test (Section 4.5). They resulted negative (Table S1), confirming what was found for *P. album* by Trovão et al. [58], and indicating that these fungi do not carry out this kind of chemical attack on the stone. On the other hand, the ability to dissolve calcium carbonate by isolated strains belonging to *Aspergillus*, *Aureobasidium*, and *Cladosporium* shows that such degradative potential is present among the SMFC microbial community members already known as stone deteriogens. This chemical activity is attributed to the production and excretion of organic acids that dissolve the stone carbonates acting as chelators of calcium and other cations. Other members of the SMFC fungal community, such as *Phoma* and *Alternaria*, are known to firmly attach to and penetrate deeper into the marble by a physical attack [30]. The combined action of chemical and mechanical processes allows fungi to actively penetrate marble forming euendolithic communities in the bulk of the substrate [8,32].

Concerning the meristematic fungi, strain SM was identified as *Exophiala capensis*, a species never isolated from stone, and M1 was identified as *Lithophila guttulata*, a fungus previously detected as a new species from marble artworks of the Vatican City [57]. On the contrary, strain NT-S-5 seems to be a not yet identified fungus, since ITS rDNA sequence analysis allowed to classify it up the *Dothideomycetes* class.

Overall, the picture of the fungal community emerging from the current cultivation data shows a prevalence of dematiaceous hyphomycetes with respect to meristematic fungi, according to the climate conditions of Florence. In fact, hyphomycetes including species of *Alternaria*, *Cladosporium*, *Epicoccum*, *Aureobasidium*, and *Phoma* dominate the fungal communities on monuments in moderate and humid climates, while microcolonial black fungi dominate the fungal community in arid and semi-arid environments [55]. However, since we used cultivation conditions favoring the growth of fast-growing fungal strains (Section 3.4), the presence of meristematic fungi on SMFC marble will be better assessed by using more appropriate cultivation conditions for their growth.

Other than biodiversity, the colonization pattern seems to be different at the NW and SE sites. At the NW site, the community grows mainly as a thick epilithic biofilm completely covering the marble surface and strongly adhering to it by EPS (Figures 2, 4 and 5). The biofilm does not invade the inner parts of marble, given the compactness of marble grains (Figure 5). At the SE site, microorganisms grow mainly as epilithic colonies or biofilms along boundaries of marble grains, and as endolithic biofilms that penetrate through and surround the grains' boundaries up to 4 mm in depth (Figures 2, 4 and 5). This penetration pattern relates to the observed detachment of marble grains (Figure 5b). Green unicellular algae, cyanobacteria, and fungi participate in this inward marble colonization due to their small dimension and low nutritive requirements, and also to the white marble transparency that allows the photosynthetic activity. The different growth patterns can be explained by the different microclimatic conditions of the two study areas. Solar radiation and temperature are higher for most of the year on SE-facing surfaces than on NW-facing ones, so the microclimates of the two areas vary during the year. It is widely known that marble thermal weathering is responsible for micro-cracks' formation at the boundaries between grains [17,61,62]. This process is more effective on calcitic marbles than on dolomitic ones due to the anisotropic behavior of calcite crystals. Indeed, when exposed to thermal variations, calcite expands along the crystallographic c-axis and contracts perpendicularly to the same axis causing the detachment of grains, the enlargement of micro-cracks, and the opening of new ones. Experimental studies demonstrated that even day-night thermal excursions can be responsible for the detachment of crystals from their borders and that this process can begin at 40 °C in calcitic marbles [61]. This is the case of the studied marble, consisting of an almost pure calcite, subjected during the summer to temperatures higher than 40 °C. Moreover, the dark microbial patinas on white rock surfaces causes, in addition to the aesthetic damage, a selective absorption of solar radiation that influences the surface thermal behavior and enhances the physical stress leading to crystals' decohesion [2,16]. Such a process would act particularly on the SE-exposed marble, where the irradiation is much stronger and prolonged, causing the distancing of the superficial grains and favoring the penetration of microorganisms. In this way, microorganisms are protected against the irradiation stress and, in turn, cause grains distancing in the inner of marble as a result of biophysical and biochemical processes. On the other hand, NW-exposed marbles have higher water contents and, contrary to SE-facing surfaces, drying occurs at a very slow rate so that NW biofilms may remain wetted for longer periods. This favors a greater epilithic growth, as also indicated by the viable titer of fungi higher at NW than at SE (about double), and the more developed darkening observed at the NW site. According to Diakumaku et al. [30], the major aesthetic damage by fungi may occur when they are spread over the rock surface under favorable conditions (air eutrophication and constant humidity for longer periods) and not forced to grow deep into the crevices.

6. Conclusions

According to the obtained data, the growth of dark cyanobacteria and black fungi is the main cause of the widespread darkening of Santa Maria del Fiore external white marble at both the study sites. The biodiversity of the lithobiontic community seems to be similar in the investigated areas, especially that of the photoautotrophic components. On the contrary, the pattern of microbial colonization on/inside marble seems to be different at the NW and SE sites depending on the distinct climatic conditions of the two study areas, in particular on solar radiation exposure, which influence marble's bioreceptivity.

The results are relevant for the knowledge of the state of conservation of marble building facades in urban environments because they focus on (i) decaying phenomena of ancient marbles exposed in a changing urban environment over centuries without receiving chemical restoration treatments (e.g., synthetic protectives or consolidants); (ii) the darkening in two differently exposed areas with different climatic conditions; and (iii) the broad description of the lithobiontic community (photoautotrophic components and fungi). The acquired knowledge will be used to plan in situ tests with innovative methods to control microbial colonization of NW and SE sides of the Florence Cathedral.

Supplementary Materials: The following are available online at https://www.mdpi.com/article/10.3390/app11136163/s1. Supplementary Figure S1: Marble samples (source), and analyses carried out. Supplementary Figure S2: FT-IR. Supplementary Figure S3: Light microscope observation of marble powder suspensions in PBS (a,b) and cultures in BG-11 medium (c,d). Supplementary Figure S4: Polished cross-sections of SE marble samples. Supplementary Figure S5: Some fungi isolated in this study. Supplementary Table S1: Carbonate dissolution test.

Author Contributions: Conceptualization, A.P.S., O.A.C., D.P. (Daniela Pinna) and B.P.; methodology, A.P.S., D.P. (Daniela Pinna), O.A.C. and B.P.; investigation, A.P.S., B.P., D.P. (Daniela Pinna), D.P. (Dominique Petrocchi), O.A.C. and T.S.; writing—review and editing, A.P.S. and B.P.; supervision, A.P.S. and B.P.; project administration, B.P.; funding acquisition, A.P.S. and B.P. All authors have read and agreed to the published version of the manuscript.

Funding: This research was funded by the Opera di Santa Maria del Fiore (OSMF).

Institutional Review Board Statement: Not applicable.

Informed Consent Statement: Not applicable.

Acknowledgments: The authors are particularly grateful to Archt. Beatrice Agostini, responsible of the OMSF's Restoration, for giving them the opportunity to study the SMFC marbles. It was really exciting to get access to the study area and inspect such precious marble with the view of Florence all around. The authors wish also to thank Archt. Francesca Mannucci for her collaboration and help during the sampling.

Conflicts of Interest: The authors declare no conflict of interest. The funders had no role in the design of the study; in the collection, analyses, or interpretation of data; or in the writing of the manuscript. The funders agree to publish the results.

References

1. Gulotta, D.; Toniolo, L. Conservation of the built heritage: Pilot site approach to design a sustainable process. *Heritage* **2019**, *2*, 797–812. [CrossRef]
2. Favero-Longo, S.E.; Viles, H.A. A review of the nature, role and control of lithobionts on stone cultural heritage: Weighing-up and managing biodeterioration and bioprotection. *World J. Microb. Biot.* **2020**, *36*, 100. [CrossRef] [PubMed]
3. Santo, A.P.; Agostini, B.; Checcucci, A.; Pecchioni, E.; Perito, B. An interdisciplinary study of biodeterioration of the external marbles of Santa Maria del Fiore Cathedral, Florence (IT). *IOP Conf. Ser. Mater. Sci. Engin.* **2020**, *949*, 012085. [CrossRef]
4. Guillitte, O. Bioreceptivity: A new concept for building ecology studies. *Sci. Total Environ.* **1995**, *167*, 215–220. [CrossRef]
5. Miller, A.Z.; Sanmartín, P.; Pereira-Pardo, L.; Dionísio, A.; Saiz-Jimenez, C.; Macedo, M.F.; Prieto, B. Bioreceptivity of building stones: A review. *Sci. Total Environ.* **2012**, *426*, 1–12. [CrossRef]
6. Sanmartín, P.; Miller, A.Z.; Prieto, B.; Viles, H.A. Revisiting and reanalysing the concept of bioreceptivity 25 years on. *Sci. Total Environ.* **2021**, *770*, 145314. [CrossRef]
7. Gorbushina, A.A. Life on the rocks. *Environ. Microbiol.* **2007**, *9*, 1613–1631. [CrossRef] [PubMed]
8. Salvadori, O.; Municchia, A.C. The role of fungi and lichens in the biodeterioration of stone monuments. *Open Conf. Proc. J.* **2016**, *7*, 39–54. [CrossRef]
9. Scheerer, S.; Ortega-Morales, O.; Gaylarde, C. Microbial deterioration of stone monuments—An updated overview. *Adv. Appl. Microbiol.* **2009**, *66*, 97–139.
10. Dakal, T.C.; Cameotra, S.S. Microbially induced deterioration of architectural heritages: Routes and mechanisms involved. *Environ. Sci. Eur.* **2012**, *24*. [CrossRef]
11. Pinna, D.; Salvadori, O. Biodeterioration processes in relation to cultural heritage materials. Stone and related materials. In *Plant Biology for Cultural Heritage*; Caneva, G., Nugari, M.P., Salvadori, O., Eds.; The Getty Conservation Institute: Los Angeles, CA, USA, 2008; pp. 128–143.
12. Nuhoglu, Y.; Oguz, E.; Uslu, H.; Ozbek, A.; Ipekoglu, B.; Ocak, I.; Hasenekoglu, I. The accelerating effects of the microorganisms on biodeterioration of stone monuments under air pollution and continental-cold climatic conditions in Erzurum, Turkey. *Sci. Total Environ.* **2006**, *364*, 272–283. [CrossRef] [PubMed]
13. Negi, A.; Sarethy, I.P. Microbial biodeterioration of cultural heritage: Events, colonization, and analyses. *Microb. Ecol.* **2019**, *32*, 967–982. [CrossRef]
14. Ortega-Morales, O.; Montero-Muñoz, J.L.; Baptista Neto, J.A.; Beech, I.B.; Sunner, J.; Gaylarde, C. Deterioration and microbial colonization of cultural heritage stone buildings in polluted and unpolluted tropical and subtropical climates: A meta-analysis. *Int. Biodet. Biodegr.* **2019**, *143*, 104734. [CrossRef]
15. Zanardini, E.; Abbruscato, P.; Ghedini, N.; Realini, M.; Sorlini, C. Influence of atmospheric pollutants on the biodeterioration of stone. *Int. Biodet. Biodegr.* **2000**, *45*, 35–42. [CrossRef]
16. Sterflinger, K.; Prillinger, H. Molecular taxonomy and biodiversity of rock fungal communities in an urban environment (Vienna, Austria). *Antonie Leeuwenhoek* **2001**, *80*, 275–286. [CrossRef] [PubMed]
17. Siegesmund, S.; Ullemeyer, K.; Weiss, T.; Tschegg, E.K. Physical weathering of marbles caused by anisotropic thermal expansion. *Int. J. Earth Sci.* **2000**, *89*, 170–182. [CrossRef]
18. Pitzurra, L.; Moroni, B.; Nocentini, A.; Sbaraglia, G.; Poli, G.; Bistoni, F. Microbial growth and air pollution in carbonate rock weathering. *Int. Biodet. Biodegr.* **2003**, *52*, 63–68. [CrossRef]
19. Cappitelli, F.; Principi, P.; Pedrazzani, R.; Toniolo, L.; Sorlini, C. Bacterial and fungal deterioration of the Milan Cathedral marble treated with protective synthetic resins. *Geomicrobiol. J.* **2007**, *385*, 172–181. [CrossRef]
20. Marvasi, M.; Donnarumma, F.; Frandi, A.; Mastromei, G.; Sterflinger, K.; Tiano, P.; Perito, B. Black microcolonial fungi as deteriogens of two famous marble statues in Florence, Italy. *Int. Biodet. Biodegr.* **2012**, *68*, 36–44. [CrossRef]
21. Pinna, D.; Galeotti, M.; Perito, B.; Daly, G.; Salvadori, B. In situ long-term monitoring of recolonization by fungi and lichens after innovative and traditional conservative treatments of archaeological stones in Fiesole (Italy). *Int. Biodet. Biodegr.* **2018**, *132*, 49–58. [CrossRef]
22. Gorbushina, A.A.; Lyalikova, N.N.; Vlasov, D.Y.; Khizhnyak, T.V. Microbial communities on the monuments of Moscow and St. Petersburg: Biodiversity and trophic relations. *Microbiology* **2002**, *71*, 350–356. [CrossRef]
23. Pinna, D.; Galeotti, M.; Rizzo, A. Brownish alterations on the marble statues in the church of Orsanmichele in Florence: What is their origin? *Herit. Sci.* **2015**, *3*. [CrossRef]

24. Lazzarini, L.; Laurenzi Tabasso, M. *Il Restauro della Pietra*; UTET Scienze Tecniche: Milano, Italy, 2010.
25. Cappitelli, F.; Salvadori, O.; Albanese, D.; Villa, F.; Sorlini, C. Cyanobacteria cause black staining of the national museum of the American Indian building, Washington, DC, USA. *Biofouling* **2012**, *28*, 257–266. [CrossRef]
26. Gaylarde, C. Influence of environment on microbial colonization of historic stone buildings with emphasis on cyanobacteria. *Heritage* **2020**, *3*, 1469–1482. [CrossRef]
27. Krumbein, W.E.; Urzì, C.E. Biodeterioration processes of monuments as a part of (man-made?) global climate change. In *Conservation of Stone and other Materials, Proceedings of the International RILEM/UNESCO Congress, Paris, 29 June–1 July 1993*; E. & FN Spon Ltd.: Paris, France, 1993; pp. 558–564.
28. Gorbushina, A.A.; Krumbein, W.E.; Hamman, C.H.; Panina, L.; Soukharjevski, S.; Wollenzien, U. Role of black fungi in color change and biodeterioration of antique marbles. *Geomicrobiol. J.* **1993**, *11*, 205–221. [CrossRef]
29. Wollenzien, U.; Hoog, G.S.; Krumbein, W.E.; Urzì, C. On the isolation of microcolonial fungi occurring on and in marble and other calcareous rocks. *Sci. Total Environ.* **1995**, *167*, 287–294. [CrossRef]
30. Diakumaku, E.; Gorbushina, A.A.; Krumbein, W.E.; Panina, L.; Soukharjevski, S. Black fungi in marble and limestones—An aesthetical, chemical and physical problem for the conservation of monuments. *Sci. Total Environ.* **1995**, *167*, 295–304. [CrossRef]
31. Charola, A.E. Stone Deterioration Characterization for its Conservation. *Geonomos* **2016**, *24*, 16–20. [CrossRef]
32. Pinna, D. *Coping with Biological Growth on Stone Heritage Objects. Methods, Products, Applications, and Perspectives*; Apple Academic Press: Oakville, ON, Canada, 2017.
33. Owczarek-Kościelniak, M.; Krzewicka, B.; Piątek, J.; Kołodziejczyk, L.M.; Kapusta, P. Is there a link between the biological colonization of the gravestone and its deterioration? *Int. Biodet. Biodegr.* **2020**, *148*, 104879. [CrossRef]
34. Cuzman, O.A.; Vettori, S.; Fratini, F.; Cantisani, E.; Ciattini, S.; Chelazzi, L.; Ricci, M.; Garzonio, C.A. Alteration of marble stones by red discoloration phenomena. In *Science and Art: A Future for Stone, Proceedings of the 13th International Congress on the Deterioration and Conservation of Stone, Paisley, UK, 6–10 September 2016*; University of the West of Scotland: Glasgow, UK, 2016; Volume 1, pp. 75–82.
35. Pesaresi, S.; Biondi, E.; Casavecchia, S. Bioclimates of Italy. *J. Maps* **2017**, *13*, 955–960. [CrossRef]
36. Available online: http://www.lamma.rete.toscana.it/meteo/osservazioni-e-dati/dati-stazioni; http://www.sir.toscana.it/consistenza-rete (accessed on 14 January 2021).
37. Malesani, P.; Pecchioni, E.; Cantisani, E.; Fratini, F. Geolithology and provenance of materials of some historical buildings and monuments in the centre of Florence (Italy). *Episodes* **2003**, *26*, 250–255. [CrossRef]
38. Santo, A.P.; Pecchioni, E.; Garzonio, C.A. The San Giovanni Baptistery in Florence (Italy): Characterisation of the serpentinite floor. *IOP Conf. Ser. Mater. Sci. Engin.* **2018**, *364*, 012069. [CrossRef]
39. Pecchioni, E.; Magrini, D.; Cantisani, E.; Fratini, F.; Garzonio, C.A.; Nosengo, C.; Santo, A.; Vettori, S. A non-invasive approach for the identification of "red marbles" from Santa Maria Del Fiore Cathedral (Firenze, Italy). *Int. J. Archit. Herit.* **2019**. [CrossRef]
40. Rodolico, F. *Le Pietre delle Città d'Italia*; Le Monnier: Firenze, Italy, 1953.
41. Primavori, P. Carrara marble: A nomination for "Global Heritage Stone Resource" from Italy. *Geol. Soc. London Spec. Publ.* **2017**, *407*, 137–154. [CrossRef]
42. Cantisani, E.; Fratini, F.; Malesani, P.; Molli, G. Mineralogical and petrophysical characterisation of white Apuan marble. *Per. Mineral.* **2005**, *74*, 117–140.
43. Cantisani, E.; Pecchioni, E.; Fratini, F.; Garzonio, C.A.; Malesani, P.; Molli, G. Thermal stress in the Apuan marbles: Relationship between microstructure and petrophysical characteristics. *Int. J. Rock Mech. Min. Sci.* **2009**, *46*, 128–137. [CrossRef]
44. Rippka, R.; Deruelles, J.; Waterbury, J.B.; Herdman, M.; Stanier, R.Y. Generic assignments, strain histories and properties of pure cultures of cyanobacteria. *Microbiology* **1979**, *111*, 1–61. [CrossRef]
45. Komarek, J.; Anagnostidis, K. Cyanoprokaryota 1. Teil Chroococcales. In *Süβwasser-Flora von Mitteleuropa*; Ettl, H., Gärtner, G., Heynig, H., Mollenhauer, D., Eds.; Fisher Press: Stuttgart-Jena, Germany, 1998; Volume 19/1, pp. 1–548.
46. Bourrelly, P. *Les Algues d'Eau Douce. Initiation à la Systématique. Tome I: Les Algues Vertes*; Éditions N. Boubée & Cie: Paris, France, 1966.
47. Available online: https://www.ncbi.nlm.nih.gov/ (accessed on 1 December 2020).
48. Clauzade, G.; Roux, C. *Likenoj de Okcidenta Europo. Illustrita Determinlibro*; Bulletin de la Société Botanique du Centre-Ouest, nouvelle Série. Numéro special 7-1985; Société Botanique du Centre-Ouest: Royan, France, 1985.
49. Nimis, P.L. *The Lichens of Italy. A Second Annotated Catalogue*; EUT Edizioni: Trieste, Italy, 2016.
50. Derrick, M.; Stulik, D.; Landry, J. *Infrared Spectroscopy in Conservation Science*; Getty Publications: Los Angeles, CA, USA, 1999.
51. Badotti, F.; Silva de Oliveira, F.; Garcia, C.F.; Camargos Fonseca, P.L.; Alves Nahum, L.; Oliveira, G.; Góes-Neto, A. Effectiveness of ITS and sub-regions as DNA barcode markers for the identification of Basidiomycota (Fungi). *BMC Microbiol.* **2017**, *17*, 42. [CrossRef]
52. Moropoulou, A.; Bisbikou, K.; Torfs, K.; van Grieken, R.; Zezza, F.; Macri, F. Origin and growth of weathering crusts on ancient marbles in industrial atmosphere. *Atmos. Environ.* **1998**, *32*, 967–982. [CrossRef]
53. Available online: www.arpat.toscana.it (accessed on 11 December 2020).
54. Sterflinger, K. Black yeasts and meristematic fungi: Ecology, diversity and identification. In *Biodiversity and Ecophysiology of Yeasts. The Yeast Handbook*; Péter, G., Rosa, C., Eds.; Springer: Berlin/Heidelberg, Germany, 2006.
55. Sterflinger, K. Fungi: Their role in deterioration of cultural heritage. *Fungal Biol. Rev.* **2010**, *24*, 47–55. [CrossRef]

56. Urzì, C.; Realini, M. Colour changes of Notos calcareous sandstone as related to its colonisation by microorganisms. *Int. Biodet. Biodegr.* **1998**, *42*, 45–54. [CrossRef]
57. Isola, D.; Zucconi, L.; Onofri, S.; Caneva, G.; de Hoog, G.S.; Selbmann, L. Extremotolerant rock inhabiting black fungi from Italian monumental sites. *Fungal Divers.* **2016**, *76*, 75–96. [CrossRef]
58. Trovão, J.; Gil, F.; Catarino, L.; Soares, F.; Tiago, I.; Portugal, A. Analysis of fungal deterioration phenomena in the first Portuguese King tomb using a multi-analytical approach. *Int. Biodet. Biodegr.* **2020**, *149*, 104933. [CrossRef]
59. Leplat, J.; François, A.; Bousta, F. Parengyodontium album, a frequently reported fungal species in the cultural heritage environment. *Fungal Biol. Rev.* **2020**, *34*, 126–135. [CrossRef]
60. Cuzman, O.A.; Olmi, R.; Riminesi, C.; Tiano, P. Preliminary study on controlling black fungi dwelling on stone monuments by using a microwave heating system. *Int. J. Conserv. Sci.* **2013**, *4*, 133–144.
61. Malaga-Starzec, K.; Lindqvist, J.E.; Schouenborg, B. Experimental study on the variation in porosity of marbles as a function of temperature. In *Natural Stone, Weathering Phenomena, Conservation Strategies and Case Studies*; Siegesmund, S., Weiss, T., Vollbrecht, A., Eds.; Geological Society: London, UK, 2002; Special Publications; Volume 205, pp. 81–88.
62. Luque, A.; Ruiz-Agudo, E.; Cultrone, G.; Sebastiàn, E.; Siegesmund, S. Direct observation of microcrack development in marble caused by thermal weathering. *Environ. Earth Sci.* **2011**, *62*, 1375. [CrossRef]

applied sciences

MDPI

Article

Impact of Herbicide Treatments on the Construction Materials in the Roman Wall of Lugo, Spain (UNESCO World Heritage Site)

Beatriz Prieto [1], Patricia Sanmartín [1,*], Javier Cancelo-González [1], Lucía Torres [2,3] and Benita Silva [1]

1 Departamento de Edafoloxía e Química Agrícola, Facultade de Farmacia, Universidade de Santiago de Compostela, 15782 Santiago de Compostela, Spain; beatriz.prieto@usc.es (B.P.); javier.cancelo@usc.es (J.C.-G.); benita.silva@usc.es (B.S.)
2 Departamento de Botánica, Escuela Politécnica Superior, Universidade de Santiago de Compostela, 27002 Lugo, Spain; lu_to_ga@yahoo.es
3 Departamento de Producción Vegetal y Proyectos de Ingeniería, Escuela Politécnica Superior, Universidade de Santiago de Compostela, 27002 Lugo, Spain
* Correspondence: patricia.sanmartin@usc.es; Tel.: +34-881814984

Citation: Prieto, B.; Sanmartín, P.; Cancelo-González, J.; Torres, L.; Silva, B. Impact of Herbicide Treatments on the Construction Materials in the Roman Wall of Lugo, Spain (UNESCO World Heritage Site). *Appl. Sci.* **2021**, *11*, 5276. https://doi.org/10.3390/app11115276

Academic Editors: Maria Filomena Macedo, António Manuel Santos Carriço Portugal, Ana Miller and Ana Catarina Pinheiro

Received: 28 April 2021
Accepted: 4 June 2021
Published: 7 June 2021

Publisher's Note: MDPI stays neutral with regard to jurisdictional claims in published maps and institutional affiliations.

Abstract: Combined laboratory and field research examining the possible alterations caused by herbicide treatments applied to the construction materials (schist and some granite, bound with mortar) in the Roman wall of Lugo (NW Spain), declared a World Heritage site by UNESCO in 2000, was performed in three separate studies in the past 20 years. In the summers of 1998 and 1999, the herbicides glyphosate, sulphosate and glufosinate–ammonium, as well as physical treatments (infrared and burning) were separately applied to different areas of the wall. In the spring of 2016, the oxyfluorfen herbicide Goal Supreme® was applied to test areas. In the winter of 2018, three essential oils, *Origanum vulgare* L., *Thymus zygis* Loefl. ex L., and *Thymus vulgaris* L., were each applied to test areas. Mineralogical modifications in the materials (determined by X-ray diffraction analysis), as well as visible physical changes, such as colour changes, and the appearance of saline residues were evaluated after the treatments. In the 1998/9 trial, glyphosate and both physical treatments triggered changes in the vermiculite clay minerals in the schists, and the physical treatments also caused changes in the kaolinite. None of the treatments caused highly perceptible colour changes. The oxyfluorfen herbicide did not cause any mineralogical alterations in the construction materials, but it did generate an increase in chloride, nitrate and sulphate contents of the granite and a slight darkening of this material. In the most recent study, the only deleterious effect observed was a perceptible increase in lightness and reduction in the yellow component after the application of *Thymus zygis* Loefl. ex L. essential oil to granite.

Keywords: built cultural heritage; cityscape; granite; laboratory and field analysis; maintenance plan; mortar; *Parietaria judaica*; schist; urban area; weed control

1. Introduction

The city of Lugo (Galicia, Northwest Spain) is renowned for its third century Roman wall (Figure 1). The wall represents one of the finest surviving examples of late Roman military fortifications and was declared a World Heritage site by UNESCO in 2000 (for further information, see references [1,2]). In relation to this recognition, efforts have been made in the last 20 years to conserve the construction materials (schist and to a lesser extent granite, bound together with mortar), which are subjected to strong biodeterioration phenomena caused by growth of biodeteriogenic vascular plant species [3]. Biodeterioration affects two-thirds of the world's historical buildings and stone monuments [4], with most of the studies focused on microphyte populations composed of cyanobacteria, algae and lichens (see [5,6], for recent reviews) and few focusing on the role of higher plants [4,7].

Figure 1. Images of the Roman wall of Lugo. (**a**) Photograph of the structure in 1928 (source: ABC, newspaper); (**b**) a recent photograph of the structure.

The propensity of a wall to provide a habitat for biological colonization, i.e., the bioreceptivity of the wall in a broad sense, considering intrinsic and extrinsic factors [8], mainly depends on its size, the construction materials, inclination, exposure and age [9]. The establishment of vascular plants generally requires crevices, fractures and interstices in the wall [10], and the heterogeneity of the building materials, exposure and slope are key factors at this stage [11]. Weed control in monumental walls is of particular importance, because plants can damage monuments with their roots, give the appearance of neglect, obstruct access to sites by visitors and/or conceal the monuments themselves [12]. On walls, the damage caused by root systems of herbaceous species, which tend to spread as pioneers, can have a significant influence on biodeterioration, causing cracking, deformation and detachment [13–15]. In the Roman wall of Lugo, the weeds are removed annually, and herbicide treatments are injected in the roots and stems of vascular plants or by spraying to prevent mechanical damage to the stone. The breakage and loss of cohesion of the structure, rather than aesthetic factors (about which opinions may differ), lead to systematic treatments being applied to weeds on the wall every year.

Mowing and clearing procedures such as cutting are commonly used for weed control. However, these may be insufficient because perennial vegetation rapidly reinfests the same sites because their roots remain alive and regenerate as soon as the climatic and substratum conditions are favourable [16]. This leads to the use of herbicides, among which, nitro-organic compounds (amides, diazines, triazines, piridines and urea derivates) and organophosphorous compounds have been the most commonly used [17], up until the time the present study was carried out. The ideal characteristics of herbicides include low toxicity, low risk of environmental pollution, high efficacy against biodeteriogens, wide range of action, easy to use, no interference with the substratum and low cost [18,19]. In this respect, a low usable dose of herbicide, which can be encouraged by the method of application (for instance, injection in roots instead of spraying), generally entails an actual decrease in the risk to human health and the environment, along with less interference with the substrate [17].

Herbicides can be divided into two broad categories: contact herbicides and systemic herbicides. The former act in the proximity of the areas of entry, while the latter (which are the most commonly used) act after penetration and are transported to other parts of the plant where they act [17]. Historically, glyphosate-based herbicides have been widely used for weed control [17]. In the Roman wall of Lugo, in the late 1990s, chemical herbicides, such as glyphosate, were applied prior to further clearing by manual mowing. Changes in regulations and public health policies regarding the use of glyphosate-based herbicides in urban areas, after the reclassification of glyphosate as a Category 2A compound (probably carcinogenic to humans) by the International Agency for Research on Cancer (IARC) in

2015, led to the search for green conservation strategies. Thus, oxyfluorfen was tested in 2016 and three formulations composed of *Origanum* and *Thymus* spp. essential oils were tested in 2018.

The Roman wall of Lugo fosters a varied biological colonization, including vascular flora (with *Parietaria judaica* or pellitory-of-the-wall currently dominating), mainly on the mortar that binds the ancient stones and the soil and other particles that accumulate on the ledges [3]. *Parietaria judaica*, which is often found in shaded and nitrate-rich microhabitats, such as in rock environments close to bird nests [20], is abundant on the mortar, which is then loosened through the transfer of moisture to the plants [21].

Possible interference with materials has rarely been considered, and few studies, mainly conducted in North America and Italy in the 1990s, reported results of preserving historical stone and masonry by weed control [22–27]. In the present study, which forms part of a wider research project on the best practices for controlling the flora on the Roman wall of Lugo (NW Spain), as well as the assessment of the impact of treatment on *Parietaria judaica* and the analysis of the run-off water from the wall [3], the potential impacts of the treatments on the construction materials of the wall were examined in field and laboratory tests. Samples of schist and mortar (mainly from the wall) were characterized in detail and the impacts on all materials (schist, mortar and granite), considering colour changes, the appearance of saline residues and mineralogical alterations after the herbicidal treatments used in the last 20 years, are reported.

2. Materials and Methods

2.1. Case Study

The Roman wall of Lugo, built between 263 and 276 A.D. to defend the Roman town of Lucus Augusti (nowadays Lugo, Figure 2a) against local tribesmen and Germanic invaders, is particularly well preserved, and the entire perimeter remains intact (Figure 2b).

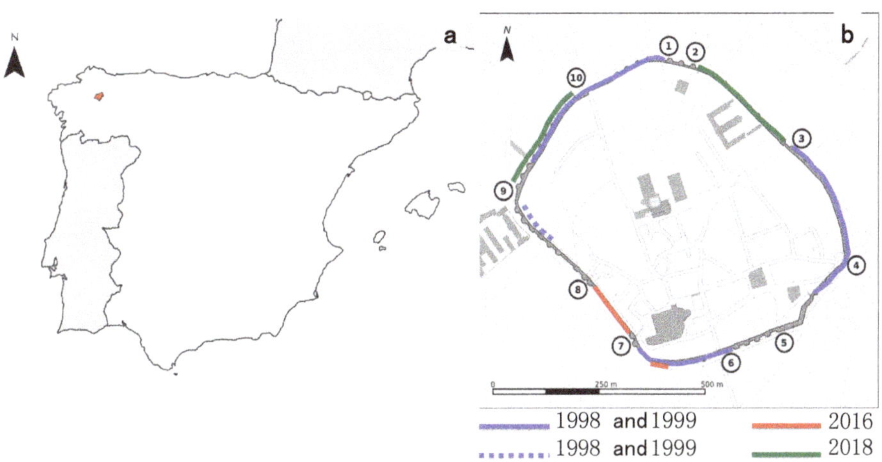

Figure 2. (**a**) ArcGIS Map of the Iberian Peninsula showing the location of the city of Lugo (red area); (**b**) map of the Roman wall of Lugo, showing the test areas of the wall (solid and dashed lines represent, respectively, internal and external faces of the wall) where the treatments were applied over time, and the entrance gates (1: San Fernando Gate, 2: False Gate, 3: Station Gate, 4: San Pedro Gate, 5: Bishop Izquierdo Gate, 6: Bishop Aguirre Gate, 7: Santiago Gate or Pexigo Gate, 8: Gate of El Carmen or Miñá Gate, 9: Bishop Odoario Gate, 10: New Gate).

The wall is around 2120 m (6960 ft) long, encloses an area of 34.4 ha, is about 4.2 m (14 ft) wide and varies in height between 8 m (26 ft) and 12 m (39 ft). The wall consists of internal and external stone faces with a core of earth mixed with gravel, pebbles and worked Roman stone recycled from demolished buildings, cemented with water. The walled enclosure is accessed through 10 gates (Figure 2b), of which five date to Roman times (False Gate [2] and Miñá Gate [8] are the best preserved) and five were added after 1853 in response to urban growth.

At present, the dominant species on the Roman wall is *Parietaria judaica*, followed by *Cymbalaria muralis*, with a more seasonal character and offering fewer problems in its control. In the successive inventories carried out during the last decade, the presence of other less representative species belonging to the genera *Andryala, Epilobium, Oxalis, Soleirolia, Sonchus, Umbilicus* and *Vulpia* have been detected on the wall.

2.2. Characterization of the Construction Materials

The construction materials of the Roman wall of Lugo mainly consist of schist (affected by different degrees of weathering) and to a lesser extent granite, bound together with mortar.

Representative samples of schist were visually classified in 3 categories: weakly weathered, moderately weathered and strongly weathered. Three samples from each of the three categories were characterized petrographically and mineralogically by thin-section analysis (by examination under a light optical microscope) and by X-ray diffraction (XRD) analysis. Powdered samples of particles with sizes less than 50 μm were studied with a PW1710 Philips diffractometer (Amsterdam, the Netherlands) equipped with a PW 1820/00 goniometer and an Enraf Nonius FR590 generator operating at 40 kV and 30 mA. The X-rays were obtained with CuKα-radiation (λ = 1.5406 Å), and the XRD diffractogram patterns were acquired in the angular range of $2 < 2\theta < 65$ with a step size of 0.02° and a measuring time of 2 s per step. The minerals were identified by comparison with the ICSD and COD databases.

Ten mortar samples taken from both inner and outer faces of the wall (five samples of each) were characterized by X-ray diffraction analysis, as previously described for schist samples. The moisture content (expressed as a percentage) of the mortar samples under natural conditions was calculated relative to the dry weight at 110 °C. The aggregate and binder were separated by gradual heating of the samples to a maximum temperature of 550 °C for the time required to achieve total deagglomeration, and the aggregate/binder ratios were determined by gravimetry. The nature and particle size distributions of the aggregates were also determined. Note that these samples do not correspond to original Roman work but to recent interventions. Deeper samples were not obtained, as the objective of the study was to determine the effect of the treatments on the exposed masonry material.

2.3. Experimental Field and Laboratory Trials with Herbicide Treatments

Laboratory- and field-based research trials were carried out in 1998 and 1999, 2016 and 2018 on different areas of the wall, to assess the possible alterations caused by herbicidal treatments applied to the construction materials in the Roman wall of Lugo (Table 1), within a wider project on the implementation of best practices for controlling flora. The main problems caused by the use of herbicides on cultural heritage asset are (1) negative side effects on the substrate and (2) hazards to the operator [28]. Regarding the former, the main problems that may arise are colour modifications, the appearance of saline residues and mineralogical alterations in stone materials, which are those aspects analysed in the present study. Mineralogical (X-ray diffraction analysis) and chemical (appearance of saline residues) modifications in the materials were evaluated after the application of each herbicidal treatment directly on the wall. Otherwise, physical change (colour change) was evaluated after the application of herbicidal treatments on samples in the laboratory.

2.3.1. Herbicidal Application and Sampling

Different herbicide treatments (Table 1) were used in each trial following changes in regulations and public health policies. Treatments acting on developed plants were applied in spring or summer, while those acting on germination were applied in winter.

In order to ensure that the various treatments were not applied to the same material, different parts of the wall were selected for each trial and treatment. This fact, together with the difficulties inherent in sampling monuments, led to differences in the type and number of samples and the time during which monitoring was conducted, as different types of permission had to be obtained for each trial.

Table 1. Summary of the experimental field and laboratory trials with herbicide treatments.

Trial Date	Treatment	Preparation	Application in the Field	Application in the Laboratory
Summer 1998 and summer 1999	Infrared treatment (I) (1) Burn treatment (B) (1)	Not applicable	At 2.5–3.5 bar pressure Application time: 10 min (I) and 1–2 min (B)	Not applicable
	Glyphosate (G) Sulphosate (S) Glufosinate–ammonium (G-a)	For each chemical, an aqueous solution at 2.5% *v/v* plus the wetting agent nonylphenyl polyethylene glycol at 0.15% *v/v*	20 mL/m^2 (G and S) and 30 mL/m^2 (G-a) with a low volume hand-held sprayer	1-h immersion (2)
Spring 2016	Oxyfluorfen	Aqueous solution at 0.5% *v/v*, 1.25% *v/v*, 100% *v/v*	20 mL/m^2 of solution at 1.25% *v/v* with an ultra-low volume sprayer (3)	Solutions at 0.5% *v/v*, 1.25% *v/v* (3) and 100% *v/v* applied by brush
Winter 2018	*Origanum vulgare* L. *Thymus zygis* Loefl. ex L. *Thymus vulgaris* L.	Aqueous solution at 2% *v/v* plus the herbicide adjuvant *Oil Oro* at 2% *v/v*	250 mL/m^2 with an ultra-low volume sprayer (3)	By brush (3)

(1) Not applied in 1999. (2) Applied in 1999. (3) Two applications.

2.3.2. Trials Performed in 1998 and 1999

Two physical treatments, i.e., infrared and burning, and three chemical treatments, i.e., glyphosate (trade name: Roundup Ultra Plus, isopropylamine salt 36% *w/v*, Monsanto, Creve Coeur, MO, USA), sulphosate (trade name: Touchdown, trimethyl-salt with N-(phosphonomethyl)glycine 48% *w/v*, Syngenta Agro, Switzerland) and glufosinate–ammonium (trade name: Finale, glufosinate–ammonium 15% *w/v*, Bayer, Crop Science, Germany), each applied at 2.5% *v/v*, together with a wetting agent (nonylphenyl polyethylene glycol solution 20% *w/v*, Sigma-Aldrich, St. Louis, MO, USA, at 0.15% *v/v*) were tested in the first trial. Each treatment (Table 1) was applied to three test areas (1.5 m × 2.5 m) in the outer face of the wall (15 test areas) and in the inner face of the wall (15 test areas), in the summer of 1998 (Figure 2b). One year later, in the summer of 1999, chemical treatments were again applied in the same test areas. Physical treatments were not repeated, as they gave rise to alterations in the materials (see result section). All test areas included schist with the three different degrees of weathering indicated in Section 2.2, mortar and scarce granite. There was no granite in those test areas where physical treatments were applied.

In order to analyse possible modifications to materials due to herbicidal treatments, one piece of each type of masonry material (granite and weakly, moderately and strongly weathered schist) and one piece of mortar were selected in each test area. A portion of these pieces of approximately 8 cm^3 was obtained at different times: 17 days before and one year after the treatments, i.e., just before the re-application of chemical treatments (summer 1999). Five weeks after re-application, portions from the same pieces were collected.

In addition, two representative, untreated pieces of schist masonry differing in colour and texture and two representative, untreated pieces of granite masonry, each of about 40 cm^3 and differing in colour and texture, were taken from the wall. Each piece was divided into three portions, and glyphosate, sulphosate and glufosinate–ammonium (the three chemical treatments) were applied (Table 1) in order to analyse surface colour modifications three months after the application.

2.3.3. Trials Performed in 2016

In the spring of 2016, oxyfluorfen (trade name: Goal Supreme, 2-chloro-1-(3-ethoxy-4-nitrophenoxy)-4-(trifluoromethyl) benzene 48% *w/v*, Dow AgroSciences, Indianapolis, IN, USA) was applied twice at a concentration of 1.25% *v/v* to three test areas (70–200 m^2) in the outer face of the wall (Figure 2b, Table 1). Before the application of the oxyfluorfen herbicide, one piece of schist masonry, granite masonry and mortar were selected in each test area and a portion of these pieces, of approximately 8 cm^3, was obtained immediately before and six months after the application of the herbicide.

Moreover, three pieces of untreated schist masonry and one piece of untreated granite masonry (present in a low proportion in the wall) were removed for treatment in the laboratory and analysis of the change in colour induced three months after the application. Each piece of approximately 40 cm^3 was divided in three portions and the oxyfluorfen herbicide was applied, as follows: (1) twice at 1.25% *v/v* to simulate the treatment carried out on the wall to three portion samples of granite and schist, (2) at 0.5% *v/v* to represent a low level of application to three portion samples of schist, and (3) at 100% *v/v* to represent the maximum effect that the herbicide could cause, if, e.g., the product was spilled on the materials by accident, to three portion samples of schist.

2.3.4. Trials Performed in 2018

In winter 2018, *Origanum vulgare* L., *Thymus zygis* Loefl. ex L., and *Thymus vulgaris* L. essential oils (for chemical composition, see Table 2) were each applied at 2% *v/v*, together with herbicide adjuvant marketed under the trade name *Oil Oro* (paraffinic oil surfactant blend/emulsifier at 83%, Químicas Oro, Spain) at 2% *v/v*, to the outer part of the wall (Figure 2b, Table 1).

Table 2. Chemical composition of the three essential oils determined by gas chromatography, with detection by mass spectrometry (GC/MS), flame ionization detection (GC/FID), or proton nuclear magnetic resonance (NMR) of the oils in deuterated chloroform. Data provided by the producer (Esencias Martínez Lozano, Murcia, Spain).

Compound (1)	*Origanum vulgare* L.	*Thymus zygis* Loefl. ex L.	*Thymus vulgaris* L.
α-pinene	0.91 (2)	1.26 (2)	-
α-thuyene	1.05	0.73	-
β-myrcene	1.50	1.62	3.50 (3)
α-terpinene	1.12	1,45	8.14
p-cymene	6.34	19.23	2.68
limonene	-	-	3.01
1–8-cineole	-	-	1.30
γ-terpinene	4.66	8.06	-
linalool	1.30	4,71	81.37
camphor	-	0.79	-
borneol	-	1.30	-
4-terpineol	0.78	1.04	-
thymol	3.80	49.39	-
carvacrol	70.24	2.76	-
β-caryophyllene	2.23	1.48	-

(1) Compounds detected by mass spectrometry analysis and retention times (in percentage ≥ 0.5%), or proton nuclear magnetic resonance (NMR). (2) Percentages calculated from GC/FID without correction factor. (3) Percentages relative to the total of the compounds detected by NMR. The spectra obtained were compared with those of the pure compounds obtained from the Sigma-Aldrich website (https://www.sigmaaldrich.com (accessed on 5 June 2021)), the "Spectral Database for Organic Compounds, SDBS" and the base SciFinder.

Each treatment was applied (Table 1) to two test areas of 9 m^2 (1.2 m × 7.5 m). In each test area, two pieces each of schist masonry and granite masonry were obtained (four masonry pieces of each material); a sample of these pieces of about 8 cm^3 was taken immediately before and 3 months after treatment in order to analyse mineralogical and chemically induced modifications.

In addition, three untreated granite masonry pieces and three untreated schist masonry pieces, each of about 40 cm^3, were taken from the wall. Each piece was divided in three portions and *Origanum vulgare* L., *Thymus zygis* Loefl. ex L., and *Thymus vulgaris* L. essential oils were applied (Table 1) to test any colour change, determined in the laboratory three months after the application.

2.3.5. Analysis of Interactions between the Herbicidal Treatments and the Substrate

The mineralogy of the samples before and after the interventions on the wall was determined by X-ray diffraction (XRD) analysis, with the same device and conditions described in Section 2.2.

In order to analyse the presence of soluble salts due to herbicidal treatments, the ion content (Cl^-, NO_3^-, SO_4^{-2}, HCO_3^- and PO_4^{-3}) and cation content (NH_4^+, only in the first trial) were measured in samples of the wall taken before and after the application of the herbicide. The soluble salts were extracted during 24 h from 1 g of sample ground to a particle size of less than 4 mm with 10 mL of ultrapure water. The extracts were filtered (0.45μm) and the ions present were analysed. The chloride (mg Cl^-/g), nitrate (mg NO_3^-/g), sulphate (mg SO_4^{-2}/g) and phosphate (mg PO_4^{-3}/g) contents were determined in a 930 Compact IC Flex (Metrohm, Switzerland) ion chromatography system. The ammonium ion content (mg NH_4^+/g) was determined with a selective ORION 720 electrode. The bicarbonate content (mg HCO_3^-/g) was analysed by potentiometric titration (Compact titrometer, Crison, Barcelona, Spain).

The colour change was recorded immediately before and three months after herbicide application, with a Konica Minolta tristimulus colourimeter equipped with a CR-300 measuring head (1998 and 1999) or with a portable spectrophotometer (Konica Minolta CM-700d) equipped with CMS100w (SpectraMagicTM NX) software (2016 and 2018). The measurement conditions in both devices were illuminant D65, observer 2° and target area, 8-mm diameter. A number of readings (in proportion to the size of sample) were made at different randomly selected zones on each sample [29]. The measurements were made in the CIELAB colour space [30] and expressed as L*, a*, and b*. Partial colour differences (ΔL*, Δa*, and Δb*), i.e., before and after treatment, represent the change in lightness (ΔL*), from black (negative value) to white (positive value), the change in the green–red component (Δa*), from greenness (negative value) to redness (positive value), and the change in the blue–yellow component (Δb*), from blueness (negative value) to yellowness (positive value). The total or global colour difference (ΔE^*_{ab}) was determined according to UNE-EN 15886 [31] by using the following equation:

$$\Delta E^*_{ab} = (\Delta L^{*2} + \Delta a^{*2} + \Delta b^{*2})^{1/2}, \quad (1)$$

To determine any perceptible colour changes in these three colour attributes and in the global colour, a threshold of 3 CIELAB units was assumed to indicate a notable change in colour [32–34].

3. Results

3.1. Characterization of Construction Materials

The three schist samples (weakly, moderately and strongly weathered) can be described as micaceous quartz schist with marked foliation, composed of micaceous minerals (biotite, muscovite and chlorite), quartz and accessory minerals (Figure 3, Table 3).

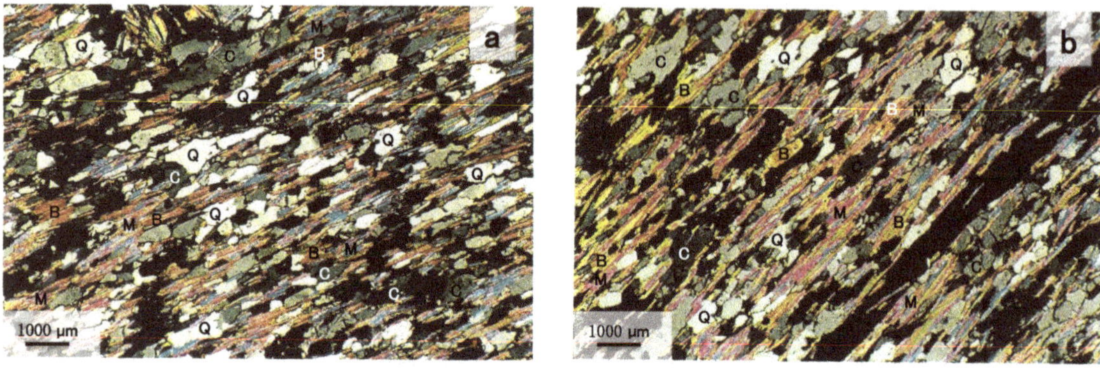

Figure 3. Thin sections of the weakly (**a**) and strongly (**b**) weathered schist samples. 'Q', 'C', 'B', and 'M' denote the quartz, chlorite, biotite and muscovite grains, respectively, and the pores and foliation appear in black.

Table 3. Mineral composition of the schist by modal analysis. The values are expressed as mean ± error (in percentage) of three independent thin sections from three subsamples from each category.

Mineral	Degree of Weathering of Schist		
	Weak	Moderate	Strong
Quartz	31.6 ± 4.1	37.0 ± 2.8	30.3 ± 5.0
Biotite	19.7 ± 2.8	26.1 ± 2.4	37.9 ± 3.6
Muscovite	28.9 ± 2.2	17.4 ± 1.8	18.2 ± 1.5
Chlorite	14.5 ± 1.1	13.0 ±1.0	10.6 ± 0.9
Accessory minerals	5.3 ± 0.3	6.5 ± 0.6	3.0 ± 0.7

Thin-section analysis of the weakly weathered schist showed that the biotite was formed by crystals of a tabular–laminar habit (0.1–0.8 mm) with brown–yellow pleochroism and frequent pleochroic halos; no alteration processes were observed (Figure 3a). Muscovite, with crystals of 0.06 to 0.15 mm, was interwoven with biotite. Isolated chlorite crystals (0.10–0.45 mm) were scarce; however, chlorite blasts (1.2 × 0.7 mm)–(1.6 × 1.0 mm) were frequent, constituted in turn by aggregates of chlorite crystals, quartz crystals, biotites and some iron ore. Quartz occurred in allotriomorphic crystals that tended to be elongated and occupied the intercrystalline positions of micas. These crystals varied in size from 0.05 to 0.25 mm, although larger crystals were also observed (0.5 × 0.10 mm). Accessory minerals were frequent, particularly apatite in rounded and elongated crystals (0.05–0.15 mm), zircon in small, rounded crystals (0.03 mm) as inclusion in biotite, opaque iron minerals in small sections of sizes from 0.05–0.01 mm and feldspars forming part of the chlorite blasts.

Thin-section analysis of the moderately weathered schist showed that the biotite was formed by crystals of a tabular–laminar habit (0.10–0.25 mm) with distinct pleochroism (from brown–yellow to white) bordering the quartz crystals. No obvious alteration processes were observed in biotite crystals. Muscovite occurred in colourless tabular crystals associated with biotite. The crystals varied in size from 0.12 to 0.20 mm. Chlorite appeared in tabular crystals ranging from 0.12 mm to 0.25 mm, forming aggregates of chlorites or chlorite blasts (of sizes from 1.0 × 0.9 mm–2.4 × 1.6 mm) made up of aggregates of chlorite and quartz crystals. Rock foliation tended to deform around the grains. The following accessory minerals, which were relatively frequent and represent a source of metamorphic granitic–pegmatitic materials, were detected: apatite, in rounded and elongated crystals (0.08–0.15 mm); tourmaline, in sub-rounded crystals (0.05–0.15 mm); zircon, in small, rounded crystals (0.03–0.05 mm); iron minerals, in small sections (0.02–0.08 mm); and, feldspars forming part of the chlorite blasts.

Thin-section analysis of the strongly weathered schist (Figure 3b) showed that the biotite was formed by crystals of a tabular–laminar habit (0.12–0.5 mm) with brown–yellow pleochroism; no alteration processes were observed, only the slight segregation of iron oxyhydroxides. Muscovite crystals of 0.10 to 0.20 mm were interwoven with biotite. Chlorite appeared as well-developed tabular crystals (0.15–0.40 mm) generally oriented transversely to the foliation. Although present in isolated cysts, chlorite usually occurred as aggregates or blasts (1.25 × 0.30 mm), formed in turn by aggregates of crystals and some degree of laminar development. Quartz occurred in allotriomorphic crystals occupying the intercrystalline positions of micas. The quartz crystals varied in size from 0.07 mm to 0.50 mm. Although the crystals were allotriomorphic, elongated sections oriented according to the foliation predominated. As accessory minerals, small opaque iron oxide crystals (0.05–0.10 mm) predominated in individual sections, and some iron hydroxides scattered in the micaceous matrix gave the sample an overall reddish-brown colour. Some feldspar crystals were also observed forming part of the chlorite blasts.

The schist samples analysed are therefore almost identical in mineralogical composition and structure. They probably correspond to the Precambrian schist formation of Villalba [35], a small town near the city of Lugo. The samples are formed by weakly weathered minerals, such as quartz and muscovite (Table 3), only containing biotite as an easily weathered component. Biotite determines the degree of weathering in the different samples, as indicated by the iron oxyhydroxides secreted, which give the most altered rock samples a browner colour, and by the greater porosity, with holes of a planar type, i.e., fissures that follow the direction of foliation of the rock samples (Figure 3b).

In addition to these results, X-ray diffraction analysis showed that some of the micaceous minerals are interstratified, i.e., mica–chlorite, mica–vermiculite, and vermiculite–chlorite. This is an interesting finding because it is precisely the interstratified minerals which may be most susceptible to being modified by the treatments.

Regarding the mortar samples, calcite ($CaCO_3$) and crushed quarry material containing quartz, feldspar and mica minerals (from granite) and chlorite (from schist) were identified in all samples, confirming that the mortars are lime-based. Some samples also contained large amounts of kaolinite, indicating a fine inert charge such as mud or clay, common in old mortars. Furthermore, the samples were quite coherent, which indicates the complete carbonation of the lime (calcium oxide, CaO). Under natural conditions, the moisture content, which varies between 0.7% and 1.90%, is not related to the position of the mortar in the outer or in the inner part of the wall (Table 4). In general, except for sample 6I, the mortars contain very little binder, especially those from the outer face, where the percentage of binder varied between 6.36% and 11.47%. Mortars from the inner face contained larger amounts of binder, between 10.13% and 21.24%. The difference between mortar samples from both faces of the wall was also observed in the particle size distribution of the aggregates (Table 4).

Table 4. Moisture content, particle size distribution of the aggregates and aggregate/binder ratio of the mortar samples.

Sample [(1)]	M [(2)]	Aggregate [(3)]						Binder [(3)]	Aggregate/ Binder Ratio
		>4 mm	4–2 mm	2–1 mm	1–0.5 mm	0.5–0.25 mm	0.25–0.10 mm		
1O	1.40	0.00	15.50	31.60	26.90	15.58	10.42	11.47	7.72
2O	0.70	8.39	23.35	37.66	18.94	6.39	5.27	8.75	10.42
3O	1.68	5.06	16.06	32.68	31.74	9.35	5.12	8.34	10.99
4O	1.63	3.47	20.85	34.26	30.13	7.36	3.93	6.36	14.72
5O	1.66	7.25	19.72	31.94	28.21	7.88	4.99	8.18	11.22
6I	1.90	0.88	14.93	26.95	25.91	18.52	12.81	21.24	3.70
7I	0.96	7.24	36.14	23.47	16.11	9.49	7.55	10.13	8.87
8I	1.05	8.10	25.44	24.14	19.96	12.71	9.66	11.73	7.52
9I	1.65	1.30	21.82	26.87	22.78	17.11	10.11	14.65	5.82
10I	1.37	0.00	17.17	28.47	23.51	17.87	12.98	16.45	5.08

(1) O: from the outer face of the wall. I: from the inner face of the wall. (2) M: moisture content (in percentage). (3) Percentage.

3.2. Impact of Herbicide Treatment on the Construction Materials

In the first trial (1998 and 1999), of the three chemical treatments tested, only glyphosate caused detectable mineralogical alteration of the schist samples, triggering the disappearance of vermiculite (Table 5), as the signal disappeared at 1.4 mm in the X-ray diffractograms at the same time as the signal increased by around 1.0 mm. Glyphosate can cause the complexation of iron and aluminum, both of which are present in vermiculite; if this occurs, it would be detected by a change in the interlaminar spacing of the vermiculite, which would shift from 1.4 mm to around 1.0 mm, overlapping with the signal of biotite and muscovite. No mineralogical changes were detected in granite after the application of the chemicals.

Table 5. Changes in the ion content and the mineral composition. +: change detected; −: no change detected; n.c.: not considered; d.k.: disappearance of kaolinite; d.v.: disappearance of vermiculite.

		I	B	G	S	G-a	O	*O. v.*	*T. z.*	*T. v.*
Cl^-	Schist	n.c.	n.c.	−	−	−	+	−	−	−
	Granite	n.c.	n.c.	−	−	−	+	−	−	−
	Mortar	n.c.	n.c.	−	−	−	+	−	−	−
NO_3^-	Schist	n.c.	n.c.	−	−	−	+	−	−	−
	Granite	n.c.	n.c.	−	−	−	+	−	−	−
	Mortar	n.c.	n.c.	−	−	−	+	−	−	−
SO_4^{-2}	Schist	n.c.	n.c.	−	−	−	+	−	−	−
	Granite	n.c.	n.c.	−	−	−	+	−	−	−
	Mortar	n.c.	n.c.	−	−	−	+	−	−	−
HCO_3^-	Schist	n.c.	n.c.	−	−	−	n.c	−	−	−
	Granite	n.c.	n.c.	−	−	−	n.c	−	−	−
	Mortar	n.c.	n.c.	−	−	−	+	−	−	−
PO_4^{-3}	Schist	n.c.	n.c.	−	−	−	-	n.c.	n.c.	n.c.
	Granite	n.c.	n.c.	−	−	−	-	n.c.	n.c.	n.c.
	Mortar	n.c.	n.c.	−	−	−	-	n.c.	n.c.	n.c.
NH_4^+	Schist	n.c.	n.c.	−	−	−	n.c.	n.c.	n.c.	n.c.
	Granite	n.c.	n.c.	−	−	−	n.c.	n.c.	n.c.	n.c.
	Mortar	n.c.	n.c.	−	−	−	n.c.	n.c.	n.c.	n.c.
Minerals	Schist	+(d.k.;d.v.)	+(d.k.;d.v.)	+(d.v.)	−	−	−	−	−	−
	Granite	n.c.	n.c.	-	−	−	−	−	−	−
	Mortar	−	−	-	−	−	−	n.c.	n.c.	n.c.

I: infrared, B: burn, G: glyphosate, S: sulphosate, G-a: glufosinate–ammonium, O: oxyfluorfen, *O. v.*: *Origanum vulgare* L., *T. z.*: *Thymus zygis* Loefl. ex L., *T. v.*: *Thymus vulgaris* L.

Both physical infrared and burning treatments caused notable mineralogical changes in schist, such as the disappearance of kaolinite and vermiculite (Table 5), as indicated in the X-ray diffractograms. The high temperatures (above 800 °C) reached in both treatments led to the dehydroxylation of kaolinite and the loss of water from the interlaminar space of the vermiculite.

Regarding the mortars, none of the treatments caused mineral alterations (Table 5). This is probably due to the fact that the mortars are very hardened, with (as indicated in Section 3.1) all the lime (CaO) converted into calcite ($CaCO_3$), a thermally highly stable mineral.

Regarding the salt content, we expected an increase in the ammonium salt after treatment with glyphosate and glufosinate–ammonium and an increase in the formation of sulphates after the application of sulphosate; however, none of these changes occurred. Before and after the application of the three chemical treatments, the ion contents were low, and no changes were detected (Table 5).

After one year, none of the three chemical treatments caused further mineralogical changes. It was also confirmed that the herbicides glyphosate, sulphosate and glufosinate–ammonium did not induce saline residues on the wall materials.

In the laboratory tests, the colour of the samples varied with the treatments (Table 6). The colour changes were more notable in schist than in granite. In the schist samples, sulphosate caused a perceptible change (with an average value of around 4 CIELAB units) in all three colour parameters. All treatments reddened the schist (Δa*) and led to a noticeable reduction in the yellow component (Δb*). In granite, perceptible changes were detected in the b* parameter after the application of glyphosate, indicating a perceptible reduction (with an average value around 3 CIELAB units) in the yellow component of the stone colour due to the treatment. Furthermore, granite samples darkened perceptively after the application of sulphosate and glufosinate–ammonium (Table 6).

Table 6. Partial changes in lightness (ΔL*), the green–red colour component (Δa*) and the blue–yellow colour component (Δb*), and total colour change (ΔE^*_{ab}). According to the values obtained and taking a threshold of 3 CIELAB units, perceptible changes (≥3 CIELAB units) are indicated in bold. n.c.: not considered.

		G	S	G-a	O (0.5%)	O (1.25%)	O (100%)	*O. v.*	*T. z.*	*T. v.*
ΔL*	Schist	−1.13 ± 0.58	**−4.03 ± 0.81**	−2.11 ± 0.35	−0.60 ± 0.87	+0.17 ± 1.00	**−17.20 ± 5.28**	−0.96 ± 0.88	−0.67 ± 0.34	−0.70 ± 0.06
	Granite	−1.52 ± 0.34	**−3.55 ± 0.93**	**−3.68 ± 0.74**	n.c.	**−2.86 ± 0.19**	n.c.	+1.16 ± 0.58	**+4.75 ± 1.13**	+1.50 ± 0.49
Δa*	Schist	**+4.26 ± 0.63**	**+4.53 ± 0.82**	**+4.02 ± 0.82**	−0.02 ± 0.13	+0.10 ± 0.17	**+7.68 ± 2.42**	+0.02 ± 0.08	−0.02 ± 0.07	+0.00 ± 0.12
	Granite	−0.94 ± 0.29	−0.91 ± 0.45	−0.71 ± 0.25	n.c.	−0.32 ± 0.87	n.c.	−0.63 ± 0.84	−1.21 ± 0.43	−0.43 ± 0.24
Δb*	Schist	**−5.54 ± 0.91**	**−4.90 ± 0.93**	**−4.34 ± 0.86**	−0.10 ± 0.61	+0.52 ± 0.88	**+12.09 ± 5.57**	+0.12 ± 0.04	−0.26 ± 0.18	−0.48 ± 0.65
	Granite	**−3.22 ± 0.67**	−1.40 ± 0.33	−1.80 ± 0.51	n.c.	−0.67 ± 1.70	n.c.	+0.97 ± 0.23	**−3.56 ± 0.91**	−1.24 ± 0.76
ΔE^*_{ab}	Schist	**7.01 ± 0.45**	**7.76 ± 0.79**	**6.24 ± 0.81**	0.61 ± 0.48	0.56 ± 0.65	**22.53 ± 7.48**	0.97 ± 0.87	0.72 ± 0.38	0.85 ± 0.28
	Granite	**3.64 ± 0.51**	**3.87 ± 0.60**	**4.08 ± 0.35**	n.c.	**3.14 ± 0.32**	n.c.	1.70 ± 0.84	**6.06 ± 1.51**	1.92 ± 0.75

G: glyphosate, S: sulphosate, G-a: glufosinate–ammonium, O: oxyfluorfen, *O. v.*: *Origanum vulgare* L., *T. z.*: *Thymus zygis* Loefl. ex L., *T. v.*: *Thymus vulgaris* L.

In the second trial (2016), no mineralogical changes were detected in the construction materials in the wall after the application of oxyfluorfen. Moreover, in general, the amounts of chlorides, nitrates and sulphates increased in the construction materials after the treatment. The first two are produced during the degradation of the herbicide. In addition, the amount of soluble bicarbonates increased in the mortar, probably due to an increase in the pH provoked by the herbicide.

Regarding the colour change determined in laboratory tests (Table 6), the application of the herbicide at maximum strength (100%) produced highly perceptible changes, darkening, reddening and yellowing of the schist samples. Colour changes in a* were three times higher and changes in b* were six times higher than the threshold of 3 CIELAB units, and changes in L* exceeded 20 CIELAB units. Thus, total colour changes (ΔE^*_{ab}) of 30 CIELAB units were reached. The application of herbicide at 1.25% (twice) produced a barely perceptible darkening in the granite and no perceptible colour changes in the schist. No colour change was observed when oxyfluorfen was applied at the lowest concentration (0.5%) to schist samples.

Finally, in the third trial (2018), there were no mineralogical changes or changes in the salt content. The only deleterious effect occurred after the application of *Thymus zygis* Loefl. ex L. essential oil to the granites, which caused a perceptible increase in lightness and a perceptible reduction in the yellow component, of slightly more than 3 CIELAB units, which resulted in a total colour change (ΔE^*_{ab}) of around 6 CIELAB units (Table 6).

4. Discussion

In the last two decades, there has been a major shift in thinking with regard to the herbicidal treatments applied to cultural heritage. The treatments have changed from invasive procedures such as controlled burning and flooding [22] and the use of chemical herbicides containing harmful compounds to the present use of natural, eco-innovative compounds with a low risk to human health, e.g., essential oils and quorum sensing inhibitors [36,37].

The efficacy of herbicides or other protective measures has been tested numerous times against target plants, but as already noted, the question regarding how the treatments affect masonry materials has rarely been examined. In this respect, although herbicides are commonly used to prevent the growth of weeds, very few products have been tested or developed specifically for use on historic masonry materials. Indeed, most products are borrowed from agricultural and landscape applications.

The first published investigation into whether chemical treatments affect masonry is from the late 1970s [25] and the next major study of the effects of a herbicide on historic masonry was carried out 1989 [26]. In the former, brick and mortar samples (clay brick, clay tile, terracotta, cinder block, concrete block, adobe) were soaked in two different herbicides (Roundup® and Garlon®4) for one week. In the latter, Roundup®, Weed-B-Gon®, and borax were applied to brick, limestone, concrete and granite. Both studies concluded that the cumulative use of herbicides could cause mechanical damage and staining on stone and mortars. In a later comprehensive study, in 1999, in which a glyphosate-basedherbicide (Round-up®) was tested [22], it was concluded that glyphosate causes three types of damage to masonry. It first attacks calcareous stone by acid dissolution. Secondly, the compound and the solvent (in this case water) introduce or redeposit soluble salts. Thirdly, in the presence of calcium, the compound forms insoluble salts. In acid rocks, such as schist, glyphosate also caused mineralogical changes, such as the disappearance of vermiculite, as confirmed in the present study. However, the most important problem with glyphosate is the doubt regarding the danger it poses to human health. According to several studies, the potentially carcinogenic effects of glyphosate are complex in nature [38]. Although considered a probable carcinogen by the IARC since 2015, many regulatory agencies, including the European Food Safety Authority and European Chemicals Agency, continue to hold that glyphosate is unlikely to pose a carcinogenic risk [38].

In Galicia (NW Spain), during the 1990s, chemicals such as Neo Desogen, Paragon Invisible, Hyvar-X, Sanit-S and the cleaning agent AB-57 (a calcium dissolving solvent) were used to remove lichens and algae from granite monuments. The deleterious effects of these compounds are similar to those caused by Goal Supreme® in the present study. Hence, although there was no mineralogical alteration, there was an appreciable colour change, in addition to saline efflorescence after treatment [39]. Important colour changes of more than 30 CIELAB units were reported by Tretiach et al. [24] after the application of biocide (Koretrel©) to grey granite. However, the change in colour was reversible because after washing the sample in running tap water for 48 h, the colour change measured was less than one CIELAB unit. In the present study, the colour change induced can be considered more stable over time, as it was observed up to three months after application.

The use of chemicals will be increasingly questioned in terms of both human safety and environmentally friendliness. However, this attention should also be extended in the field of the conservation of stone monuments and buildings, to determine the short-term and long-term potential damage to historic masonry materials caused by herbicides or other protective measures. Indeed, according to a recent review on natural biocides for the conservation of stone cultural heritage, only 15 of 94 evaluated the interference of the substances with the materials [36]. This additional information can be considered a very important factor in the decision-making process. According to Pinna [40], chlorine-containing compounds and hydrogen peroxide-based biocides are now avoided in the cultural heritage field because they can interact negatively with stone materials; in addition, oxidizing metal ions (e.g., iron) are strongly oxidizing chemicals that can cause rust or black stains, even though they are often stabilized by adding acids. To the best of our knowledge, this type of damage has not yet been observed after the use of essential oils, such as those used in the third trial carried out on the Roman wall of Lugo. Considering the lower toxicity of the essential oils tested in this work, together with the absence of deleterious effects on the building materials in the Roman wall of Lugo, these products may represent a promising alternative to chemical compounds for biodeterioration control, although it cannot be guaranteed that the following application of the same products will not cause a noticeable colour change. The manufacturing cost could also be a major obstacle to the application of these natural compounds (mainly on very large structures such as the Roman wall of Lugo), which may be much more expensive to produce than the conventional treatments currently in main use.

5. Conclusions

In this study, trends in the use of herbicides in the last 20 years were examined from the point of view of how the compounds have affected masonry material. Physical (infrared and burning) treatments based on the application of high temperatures (above 800 °C) were completely ruled out after a first application in 1998, because they caused irreversible mineralogical alterations in the schist, the main construction material in the Roman wall of Lugo. Among the other treatments, only oxyfluorfen left salt residues, and all treatments caused perceptible colour changes. As the colour of the wall is already heterogeneous owing to differences in the colour of the various materials and pieces of the same material, the perceptible colour changes caused by treatments should not represent a problem regarding conservation of the monument.

Furthermore, considering the restrictions on the use of chemicals in parks and public gardens (where parts of the Roman wall of Lugo occur), herbicidal treatments should focus on environmentally friendly technologies with low risk to human health and should not damage the construction materials. In this respect, *Origanum vulgare* L. and *Thymus vulgaris* L. essential oils seem to be most appropriate, as they did not provoke any mineralogical alterations or changes in the salt content and original colour.

Author Contributions: Conceptualization, B.P. and B.S.; methodology, B.P., P.S., J.C.-G., L.T. and B.S.; formal analysis, B.P., P.S., J.C.-G., L.T. and B.S.; investigation, B.P., P.S. and B.S.; writing—original draft preparation, P.S.; writing—review, B.P. and P.S.; supervision, B.P. and B.S. All authors have read and agreed to the published version of the manuscript.

Funding: The results of the different studies presented here were financially supported by the following contracts: "Evaluación de diferentes métodos de control de las malas hierbas de la muralla de Lugo. Xunta de Galicia. Consellería de Cultura, Comunicación Social e Turismo. 1998 and 1999"; "Seguimiento de los tratamientos de control de la flora vascular en la muralla romana de Lugo. Restauraciones y Construcciones Luis J. Sánchez SA. 2015–2017"; "Realizar el seguimiento del segundo tratamiento fitosanitario 2017 y realizar un ensayo de herbicidas naturales en las paredes da muralla. Restauraciones y Construcciones Luis J. Sánchez SA. 2017–2018"; "Realizar el seguimiento de la limpieza manual de las paredes de la muralla romana de Lugo en el verano-otoño de 2019 y la aplicación en las paredes de la muralla romana de Lugo de herbicidas naturales potenciales para controlar la germinación de Parietaria Judaica L. Restauraciones y Construcciones Luis J. Sánchez SA. 2019–2020". The authors are also grateful for financial support from the Xunta de Galicia (grant ED431C 2018/32).

Institutional Review Board Statement: Not applicable.

Informed Consent Statement: Not applicable.

Data Availability Statement: Not applicable.

Acknowledgments: The authors are grateful to Antonio Rigueiro Rodríguez and Elvira Antonia Díaz Vizcaíno, who collaborated in the programming, planning and monitoring of the activities contemplated in the projects listed in the section above.

Conflicts of Interest: The authors declare no conflict of interest.

References

1. Faraci, G. Ensuring the Conservative Process: The Roman Walls of Lugo Maintenance Plan. *Conserv. Manag. Archaeol. Sites* **2016**, *18*, 411–421. [CrossRef]
2. Prieto, B.; Sanmartín, P.; Pereira-Pardo, L.; Silva, B. Recovery of the traditional colours of painted woodwork in the Historical Centre of Lugo (NW Spain). *J. Cult. Herit.* **2011**, *12*, 279–286. [CrossRef]
3. Torres García, L. Flora vascular de la muralla de Lugo: Control del crecimiento y germinación de la dominante *Parietaria judaica* L. Ph.D. Thesis, Universidade de Santiago de Compostela, Lugo, Spain, 2018.
4. Motti, R.; Bonanomi, G.; Stinca, A. Deteriogenic Flora of the Phlegraean Fields Archaeological Park: Ecological Analysis and Management Guidelines. *Nord. J. Bot.* **2020**, *38*, 1–11. [CrossRef]
5. Favero-Longo, S.E.; Viles, H.A. A review of the nature, role and control of lithobionts on stone cultural heritage: Weighing-up and managing biodeterioration and bioprotection. *World J. Microbiol. Biotechnol.* **2020**, *36*, 100. [CrossRef]
6. Gaylarde, C. Influence of Environment on Microbial Colonization of Historic Stone Buildings with Emphasis on Cyanobacteria. *Heritage* **2020**, *3*, 1469–1482. [CrossRef]
7. Motti, R.; Bonanomi, G. Vascular plant colonisation of four castles in southern Italy: Effects of substrate bioreceptivity, local environment factors and current management. *Int. Biodeterior. Biodegrad.* **2018**, *133*, 26–33. [CrossRef]
8. Sanmartín, P.; Miller, A.Z.; Prieto, B.; Viles, H.A. Revisiting and reanalysing the concept of bioreceptivity 25 years on. *Sci. Total Environ.* **2021**, *770*, 145314. [CrossRef]
9. Francis, R.A. Wall ecology: A frontier for urban biodiversity and ecological engineering. *Progr. Phys. Geogr.* **2011**, *35*, 43–63. [CrossRef]
10. Fisher, G.G. Weed damage to materials and structures. *Int. Biodeterior. Bullettin* **1972**, *8*, 101–103.
11. Motti, R.; Bonanomi, G.; Stinca, A. Biodeteriogens at a southern Italian heritage site: Analysis and management of vascular flora on the walls of Villa Rufolo. *Int. Biodeterior. Biodegrad.* **2021**, *162*, 105252. [CrossRef]
12. Caneva, G.; Ceschin, S.; De Marco, G. Mapping the risk of damage from tree roots for the conservation of archaeological sites: The case of the Domus Aurea, Rome. *Conserv. Manag. Archaeol. Sites* **2006**, *7*, 163–170. [CrossRef]
13. Caneva, G.; Galotta, G.; Cancellieri, L.; Savo, V. Tree roots and damages in the Jewish catacombs of Villa Torlonia (Roma). *J. Cult. Herit.* **2009**, *10*, 53–62. [CrossRef]
14. Kanellou, E.; Economou, G.; Papafotiou, M.; Ntoulas, N.; Lyra, D.; Kartsonas, E.; Knezevic, S. Flame weeding at archaeological sites of the Mediterranean region. *Weed Technol.* **2017**, *31*, 396–403. [CrossRef]
15. Trotta, G.; Savo, V.; Cicinelli, E.; Carboni, M.; Caneva, G. Colonization and damages of *Ailanthus altissima* (Mill.) Swingle on archaeological structures: Evidence from the Aurelian Walls in Rome (Italy). *Int. Biodeterior. Biodegrad.* **2020**, *153*, 105054. [CrossRef]
16. Mouga, T.; Almeida, M. Neutralisation of herbicides: Effects on wall vegetation. *Int. Biodeterior. Biodegrad.* **1997**, *40*, 141–149. [CrossRef]
17. Caneva, G.; Nugari, M.P.; Salvadori, O. *Plant Biology for Cultural Heritage: Biodeterioration and Conservation*; The Getty Conservation Institute: Los Angeles, CA, USA, 2008.
18. Introduction to Weeds and Herbicides. Available online: https://extension.psu.edu/introduction-to-weeds-and-herbicides (accessed on 29 May 2021).
19. Caneva, G.; De Marco, G. *Il Controllo Della Vegetazione Nelle Zone Archeologiche e Monumentali, Atti del Convegno 'Manutenzione e Conservazione del Costruito, fra Tradizione e Innovazione'*; Libreria Progetto: Bressanone, Italy, 1986; pp. 553–570.
20. Ceschin, S.; Bartoli, F.; Salerno, G.; Zuccarello, V.; Caneva, G. Natural habitats of typical plants growing on ruins of Roman archaeological sites (Rome, Italy). *Plant Biosyst.* **2016**, *150*, 866–875. [CrossRef]
21. Honeyborne, D.B. Weathering and decay of masonry. In *Conservation of Building and Decorative Stone*; Ashurst, J., Dimes, F.G., Eds.; Butteworth: Guildford, UK, 1990; Volume 1, pp. 153–184.
22. Dewey, C.C. An Investigation into the Effects of an Herbicide on Historic Masonry Materials. Master's Thesis, University of Pennsylvania, Philadelphia, PA, USA, 1999.
23. Altieri, A.; Coladonato, M.; Lonati, G.; Malagodi, M.; Nugari, M.P.; Salvadori, O. Effects of biocidal treatments on some Italian lithotypes samples. In Proceedings of the 4th International Symposium on the Conservation of Monument in the Mediterraneum, Rhodes, Greece, 6–11 May 1997; Moropoulou, A., Zezza, F., Kollias, E., Papachristodoulou, Eds.; Technical Chamber of Greece: Athens, Greece, 1997; Volume 3, pp. 31–40.
24. Tretiach, M.; Crisafulli, P.; Imai, N.; Kashiwadani, H.; Moon, K.; Wada, H.; Salvadori, O. Efficacy of a biocide tested on selected lichens and its effects on their substrata. *Int. Biodeterior. Biodegrad.* **2007**, *59*, 44–54. [CrossRef]
25. James, E.F. *The Effects of Herbicides on Masonry*; National Technical Information Service: Springfield, VA, USA, 1978.
26. Cook, L. The Effects of Herbicides on Masonry: Products, Choices and Testing. Master's Thesis, Columbia University, New York, NY, USA, 1989.
27. Tiano, P.; Caneva, G. Procedures for the elimination of vegetal biodeteriogens from stone monuments. In Proceedings of the ICOM 8th Triennial Meeting, Sydney, Australia, 6–11 September 1987; pp. 1201–1205.
28. Mishra, A.K.; Jain, K.K.; Garg, K.L. Role of higher plants in the deterioration of historic buildings. *Sci. Total Environ.* **1995**, *167*, 375–392. [CrossRef]

29. Prieto, B.; Sanmartín, P.; Silva, B.; Martinez-Verdú, F. Measuring the color of granite rocks. A proposed procedure. *Color Res. Appl.* **2010**, *35*, 368–375. [CrossRef]
30. CIE S014-4/E; *Colorimetry Part 4: CIE 1976 L*a*b* Colour Space*; Commission Internationale de L'eclairage, CIE Central: Bureau, Vienna, 2007.
31. UNE-EN 15886; *Conservation of Cultural Property—Test Methods—Colour Measurement of Surfaces*; Asociación Española de Normalización y Certificación: Madrid, Spain, 2011.
32. Berns, R.S. *Billmeyer and Saltzman's Principles of Color Technology*, 3rd ed.; Wiley: New York, NY, USA, 2000.
33. Völz, H.G. *Industrial Color Testing*; Wiley–VCH: Weinheim, Germany, 2001.
34. Sanmartín, P.; Chorro, E.; Vázquez-Nion, D.; Martínez-Verdú, F.M.; Prieto, B. Conversion of a digital camera into a non-contact colorimeter for use in stone cultural heritage: The application case to Spanish granites. *Meas. J. Int. Meas. Confed.* **2014**, *56*, 194–202. [CrossRef]
35. Capdevila, R. Le Métamorphisme Régional Progressif et les Granites Dans le Segment Hercynien de Galice Nord Orientale (NW de l'Espagne). Ph.D. Thesis, University of Montpellier, Montpellier, France, 1969.
36. Fidanza, M.R.; Caneva, G. Natural biocides for the conservation of stone cultural heritage: A review. *J. Cult. Herit.* **2019**, *38*, 271–286. [CrossRef]
37. Raveau, R.; Fontaine, J.; Lounès-Hadj Sahraoui, A. Essential Oils as Potential Alternative Biocontrol Products against Plant Pathogens and Weeds: A Review. *Foods* **2020**, *9*, 365. [CrossRef] [PubMed]
38. Davoren, M.J.; Schiestl, R.H. Glyphosate-based herbicides and cancer risk: A post-IARC decision review of potential mechanisms, policy and avenues of research. *Carcinogenesis* **2018**, *39*, 1207–1215. [CrossRef] [PubMed]
39. Prieto, B.; Rivas, T.; Silva, B. The effect of selected biocides on granites colonized by lichens. In *Biodeterioration and Biodegradation*, 1st ed.; Bousher, A., Chandra, M., Edyvean, I.R., Eds.; Inst. Chemistry: Rugby, UK, 1995; Volume 9, pp. 204–209.
40. Pinna, D. *Coping with Biological Growth on Stone Heritage Objects: Methods, Products, Applications, and Perspectives*; Apple Academic Press: Waretown, NJ, USA; CRC Press, Taylor and Francis Group: Boca Raton, FL, USA, 2017.

applied sciences

MDPI

Article

Inhibitory Effect of Cinnamaldehyde on Main Destructive Microorganisms of Nanhai No. 1 Shipwreck

Xinduo Huang [1], Yeqing Han [1], Jing Du [2], Peifeng Guo [1], Yu Wang [1], Kaixuan Ma [1], Naisheng Li [2], Zhiguo Zhang [2], Yue Li [3,*] and Jiao Pan [1,*]

1 Key Laboratory of Molecular Microbiology and Technology of the Ministry of Education, Department of Microbiology, College of Life Sciences, Nankai University, Tianjin 300071, China; 2120191084@mail.nankai.edu.cn (X.H.); 2120191000@mail.nankai.edu.cn (Y.H.); Gary19981114@163.com (P.G.); 2120201036@mail.nankai.edu.cn (Y.W.); makaixuan16@163.com (K.M.)

2 National Center of Archaeology, Beijing 100013, China; ldusts@163.com (J.D.); lineas@126.com (N.L.); zzgwys@126.com (Z.Z.)

3 State Key Laboratory of Component-Based Chinese Medicine, Ministry of Education Key Laboratory of Pharmacology of Traditional Chinese Medicine Formulae, Institute of Traditional Chinese Medicine, Tianjin University of Traditional Chinese Medicine, Tianjin 301617, China

* Correspondence: liyue2018@tjutcm.edu.cn (Y.L.); panjiaonk@nankai.edu.cn (J.P.)

Abstract: Nanhai No. 1, a shipwreck in the Southern Song Dynasty, China, has a history of more than 800 years. It was salvaged in 2007 and is now on display in the Guangdong Maritime Silk Road Museum. Due to the fact that the hull is a wooden cultural relic and exposed to the air, the biological corrosion and biodegradation caused by microorganisms are key problems of hull protection. At present, the antimicrobial agent Euxyl® K100 (isothiazolinone) has a significant antimicrobial effect in the field, but it has a certain negative impact on the environment and archeologists. In order to reduce the use of chemical antimicrobial agents, we evaluated the inhibitory effects of cinnamaldehyde on the main destructive microorganisms of Nanhai No. 1. Cinnamaldehyde is the main active component of cinnamon, and has broad-spectrum antimicrobial properties. The paper diffusion method, gas diffusion method and minimum inhibitory concentration experiment were used to detect the inhibitory effects of cinnamaldehyde on the main microorganisms of Nanhai No. 1. We found that cinnamaldehyde had significant inhibitory effects on *Bacillus tequilensis* NK-NH5, *Bacillus megaterium* NK-NH10, *Bacillus velezensis* NK-NH11, *Bacillus* sp. NK-NH15, *Bacillus* sp. NK-NH16, *Bacillus* sp. NK-NH17, *Fusarium solani* NK-NH1 and *Scedosporium apiospermum* NK.W1-3. At the same time, cinnamaldehyde had more inhibitory effects on fungi than bacteria. Finally, we verified that cinnamaldehyde can effectively inhibit the growth of microorganisms in water, for storing the scattered wood blocks of the Nanhai No. 1 hull through laboratory simulation experiments. Cinnamaldehyde, as an environment-friendly antimicrobial agent, is of great significance to protecting water-saturated wooden relics from microbial corrosion and degradation in the future.

Keywords: Nanhai No. 1 shipwreck; cinnamaldehyde; antimicrobial activity; biodegradation

Citation: Huang, X.; Han, Y.; Du, J.; Guo, P.; Wang, Y.; Ma, K.; Li, N.; Zhang, Z.; Li, Y.; Pan, J. Inhibitory Effect of Cinnamaldehyde on Main Destructive Microorganisms of Nanhai No. 1 Shipwreck. *Appl. Sci.* **2021**, *11*, 5262. https://doi.org/10.3390/app11115262

Academic Editor: Maria Filomena Macedo

Received: 9 April 2021
Accepted: 21 May 2021
Published: 5 June 2021

1. Introduction

Nanhai No. 1, a Chinese Song Dynasty shipwreck, was discovered in 1987 near the Chuanshan archipelago in the South China Sea, Guangdong Province. In 2007, Nanhai No. 1 was salvaged and displayed in "the Crystal Palace" of the Guangdong Maritime Silk Road Museum. In 2012, archaeological excavation was officially started [1], and a large number of precious cultural relics such as gold, silver, bronze, iron, lacquer and porcelain were found in the wreck. The discovery of Nanhai No. 1 wreck is of great significance to Chinese underwater archaeology, and also provides important clues for the study of Maritime Silk Road.

The hull and the cultural relics of Nanhai No.1 are the key protected objects. With the excavation of Nanhai No.1, the exposed area of the hull increases gradually. The cultural relics need to go through the process of desalination, desulfurization and iron removal, so they are soaked in a specific buffer for a long time [2]. Due to the high humidity and high temperature in the storage environment of Nanhai No. 1, and the fact that wooden cultural relics are natural organic matter, this provides suitable conditions for the growth and reproduction of microorganisms [3]. Microorganisms are very harmful to wooden relics. Some bacteria and fungi can destroy wood by degrading cellulose, hemicellulose and lignin. For example, the ability of microorganisms such as *Trichoderma viride, Trichoderma reesei, Bacteroides cellulosolvens, Bacteroides succinogenes* to degrade cellulose has been reported [4]. This will cause structural and mechanical changes of wooden cultural relics, and cause irreversible impact on precious cultural artifacts [5–7].

The main microorganisms of Nanhai No. 1 isolated by our previous research group include: *Fusarium solani* NK-NH1 [8], *Bacillus tequilensis* NK-NH5, *Bacillus megaterium* NK-NH10, *Bacillus velezensis* NK-NH11 [9] and *Scedosporium apiospermum* NK.W1-3 [10]. In addition, we have recently identified bacteria on the hull and surrounding sea mud including *Bacillus* sp. NK-NH15, *Bacillus* sp. NK-NH16, *Bacillus* sp. NK-NH17 (Table S1).

At present, Euxyl® K100 is the main antimicrobial agent used in Nanhai No.1 microbial control. Although it can effectively inhibit the growth of microorganisms, it is a little harmful to the environment and human body. Methylchloroisothiazolinone (MCI) and methylisothiazolinone (MI), the main components of Euxyl® K100, are widely used in cosmetic products [11]. MI and MCI can cause allergic contact dermatitis [12]. At the same time, MI has a certain volatility; it will slowly volatilize into the air, affecting the surrounding environment, and then cause harm to the contacts [13]. Therefore, it is an important task to find environment-friendly and health-friendly antimicrobial agents. Plant-derived antimicrobial agents are relatively safe and friendly to the human body and environment [14,15]. At present, there are many reports that plant-derived antimicrobial agents are used in the actual control of microbial diseases [16]. Our research goal is to apply plant-derived antimicrobial agents to the microbial control of the Nanhai No. 1 shipwreck.

Cinnamaldehyde is the main component of cinnamon essential oil, which is a yellow oily liquid [17]. The antiinflammatory, antioxidation, antiulcer, antibacterial, hypoglycemic and hypolipidemic properties of cinnamaldehyde have been reported [18]. Cinnamaldehyde has been widely used in food, medicine, cosmetics and other fields [19]. Previous studies have shown that cinnamaldehyde has good inhibitory effect on *Aspergillus, Fusarium, Penicillium, Rhizopus* and other fungi [20–22], *Escherichia coli, Bacillus subtilis, Staphylococcus* spp. and other bacteria [23,24]. In this study, we first tested the cellulose degradation ability of the main destructive microorganisms of Nanhai No. 1, then tested the inhibition effect of cinnamaldehyde on these main destructive microorganisms. In order to play a practical role in the microbial control of Nanhai No. 1, the simulated experiment was performed in the laboratory, and the inhibitory effect of cinnamaldehyde on microorganisms of the wood blocks of Naihai No.1 was evaluated.

2. Materials and Methods

2.1. Main Destructive Microorganisms of Nanhai No. 1 Shipwreck

In the previous work, we isolated and purified a variety of main destructive microorganisms from the samples of the Nanhai No.1 shipwreck. The details are as follows: *Bacillus tequilensis* NK-NH5, *Bacillus megaterium* NK-NH10, *Bacillus velezensis* NK-NH11 were isolated from the water samples of lacquerware plates [9]. *Bacillus* sp. NK-NH15, *Bacillus* sp. NK-NH16, and *Bacillus* sp. NK-NH17 were isolated from the hull wood and sea mud. *Fusarium solani* NK-NH1 is a fungus isolated from the wood of the shipwreck. *Scedosporium apiospermum* NK.W1-3 is a fungus isolated from water samples of wood storage [8,10]. All microorganisms were frozen at −80 °C.

2.2. Revitalizing the Microorganisms

The strain was taken out from the refrigerator (−80 °C) and inoculated in Luria Broth (LB) agar medium. After one day's culture at 37 °C, a single colony was selected and inoculated in the new LB agar medium again. The strain was cultured at 37 °C for one day for subsequent experiments; the fungi were inoculated in potato dextrose agar (PDA) medium and cultured at 28 °C for 3 days, then a single colony was selected and inoculated in the new PDA medium again and cultured at 28 °C for 3 days for subsequent experiments.

2.3. Determination of Cellulase Activity in Destructive Microorganisms

Two types of carboxylmethylcellulose (CMC) agar media were prepared to evaluate the ability of the microorganisms to degrade cellulose: (1) CMC agar media for bacteria: CMC Na 15.0 g, NaCl 5.0 g, KH_2PO_4 1.0 g, $MgSO_4$ 0.2 g, peptone 10.0 g, yeast extract 5.0 g, agar 18.0 g, and distilled water 1 L; (2) CMC agar media for fungi: $NaNO_3$ 2 g, K_2HPO_4 1 g, $MgSO_4$ 0.5 g, KCl 0.5 g, CMC Na 2 g, peptone 2 g, agar 17 g and 1 L distilled water. The bacteria were put into the center of the plate, cultured at 28 °C for 4 days, and then the plate was dyed with 1 g/L Congo red solution. After 15 min, the dye was discarded, then 1 mol/L NaCl solution was added for washing, and 15 min later, the sodium chloride solution was discarded and the colony diameter and transparent circle diameter were determined [25]; a fungal disk with a diameter of 7.5 mm was cut from the edge of the active colony. Then, the disk was transferred to a CMC agar plate, cultured at 28 °C for 4 days, then 5 mL iodine-potassium iodide solution (2.0 g potassium iodide, 1.0 g iodine, 300 mL double distilled water) was added, and incubated in the dark at room temperature for 5 min to determine the colony diameter and transparent circle diameter [26]. According to the ratio of Hyaline circles (H) to colony diameter (D), the cellulase activity of the tested bacteria can be determined preliminarily. The higher the H/D value, the stronger the ability of cellulose degradation. All media were autoclaved at 121 °C for 20 min.

2.4. Preparation of Bacterial Culture Suspension and Spore Suspension of Fungi

The bacteria in LB agar medium were inoculated into LB liquid medium and cultured overnight at 180 rpm/min and then the bacterial standard curve was prepared. The final concentration of the bacterial solution was diluted to 10^6 CFU/mL. Then, 5 mL 1/1000 Tween-80 solution was added into the purified single colony plate (fungi), the fungal spores were scraped with a glass rod, the spore solution was filtered out through sterile gauze, the number of spores was detected with a blood cell counting plate, and the final concentration of spore suspension was diluted to 1×10^4 CFU/mL. This suspension was used in the subsequent antimicrobial experiment.

2.5. Disk Diffusion Experiment

The inhibition of cinnamaldehyde (Rhawn, Shanghai, China) was determined by disk diffusion method. Cinnamaldehyde was dissolved in the mixture of 10/1000 dimethyl sulfoxide (DMSO) and 1/1000 Tween-80 and mixed evenly to make the final concentration of cinnamaldehyde 100 mg/mL and 50 mg/mL respectively. 0.5% Euxyl® K100 (Schülke, Norderstedt, Germany) was used as the positive control group, and the mixed solution of 10/1000 DMSO and 1/1000 Tween-80 was used as the negative control group. For bacteria, 100 μL bacterial culture solution was added to the LB agar plate and then spread evenly with a glass spreading rod. Four pieces of sterile filter paper with diameters of 7 mm were placed on each plate, then 15 μL of 50 mg/mL cinnamaldehyde, 100 mg/mL cinnamaldehyde, 0.5% Euxyl® K100, mixed solution of 10/1000 DMSO and 1/1000 Tween-80 were added on the corresponding filter paper sheets. After incubation at 37 °C for 24 h, the size of inhibition zone was observed. The larger the diameter of the inhibition zone was, the stronger the inhibitory effect of the drug on bacteria was. For fungi, 100 μL spore suspension was added to a PDA plate and evenly coated with a sterile cotton swab. A piece of filter paper was placed on each plate and 15 μL drug or mixed solution was dropped.

After being cultured at 28 °C for 3 days, the size of the inhibition zone was observed. All materials (media, cotton swabs, etc.) were autoclaved at 121 °C for 20 min.

2.6. Antimicrobial Experiment of Cinnamaldehyde Volatile Gas

The antimicrobial effect of cinnamaldehyde volatile gas was determined by dichotomy plate. LB agar medium (bacteria) or PDA medium (fungi) was poured into one side of a split plate. After the medium solidified, 100 μL bacterial culture suspension or spore suspension of fungi were taken and coated evenly with a sterile cotton swab on one side of the plate. The air concentration of cinnamaldehyde was 250 μL/L when 20 μL cinnamaldehyde was added to the other side and the volume of the 90 mm culture dish was about 80 mL. After 24 h at 37 °C, the inhibition area was observed and the inhibition rate was calculated. Mixed solvents were used as negative control, and Euxyl® K100 as positive control.

$$\text{Inhibition rate} = \text{inhibition area}/(\text{colony area} + \text{inhibition area}) \times 100\%$$

2.7. Determination of Minimum Inhibitory Concentration of Cinnamaldehyde

Minimum inhibitory concentration assays were performed in 96 well plates. Cinnamaldehyde and Euxyl® K100 were diluted by the double dilution method using a mixed solution of 10/1000 (DMSO) and 1/1000 Tween 80. The concentrations of cinnamaldehyde ranged from 0.003125 mg/mL to 2 mg/mL and the concentrations of Euxyl® K100 ranged from 0.003906 mg/mL to 5 mg/mL. Different concentrations of drugs (50 μL) and bacterial culture suspension or spore suspension of fungi (150 μL) were added to each well. Then, 150 μL bacterial culture suspension/spore suspension and 50 μL mixed with solution of 20/1000 DMSO and 1/1000 Tween-80 were added to the wells as negative control, Euxyl® K100 as positive control, 150 μL medium and 50 μL mixed solution of 10/1000 DMSO and 1/1000 Tween-80 were added to the wells as blank control. The 96 well plates with fungi were kept at a constant temperature of 28 °C for 36 h, and the 96 well plates with bacteria were kept at a constant temperature of 37 °C for 24 h. At the end of incubation, the growth of microorganisms was determined by measuring the absorbance of each well at 595 nm (fungi)/600 nm (bacteria) with a multifunctional microplate reader, calculating the minimum inhibitory concentration, and each test was done in three replicates.

2.8. Laboratory Simulation Experiments

Five pieces of Nanhai No. 1 hull wood samples (length: 3 cm; width: 2 cm; height: 1 cm) were taken and placed in five boxes with volumes of 5 L. Two liters of distilled water was added into each box, and 500 μL of each bacterial culture suspension (*Bacillus* sp. (NK-NH15), *Bacillus* sp. (NK-NH16), *Bacillus* sp. (NK-NH17), *B. megaterium* (NK-NH10), *B. velezensis* (NK-NH11), *B. tequilensis* (NK-NH5) was inoculated in the boxes. Five groups (N, C1, C2, C3, C4) were set up. Cinnamaldehyde (0 g, 0.1 g, 0.4 g, 1 g, 2.5 g) was added sequentially to ensure the final concentration of cinnamaldehyde in the solution was 0, 50, 200, 500, 1250 μg/mL, respectively; after stirring evenly, boxes were sealed with a lid and allowed to stand for 14 days at an ambient temperature of 18–25 °C. The annual temperatures in the Marine Silk Road Museum are between 16.4–30.7 °C.

2.9. Determination of Colony Forming Units(CFU)

At the end of the laboratory simulation experiment, the water samples were stirred well with a glass rod, and 1 mL of water sample was taken out of each box for the determination of the total amount of bacteria. Water samples from each group were subjected to ten-fold gradient dilution to give a final concentration of 10^0–10^{-3}. 100 μL of water samples at each concentration were taken and dropped onto LB agar plates, respectively, and spread well with a sterile coating rod before being incubated for 24 h. Colony forming units were counted. Each concentration was repeated three times.

2.10. Quantitative Real-Time PCR

After the laboratory simulation experiment, the water samples in each box were passed through a suction filter device for bacteria collection, the microorganisms were cut off on a 0.22 μM filter membrane, and then DNeasy powersoil Kit (Qiagen, Germany) was employed to extract the total DNA in the water samples. The total biomass of bacteria was determined by qPCR using primers Eub338(5′-ACTCCTACGGGAGGCAGCAG-3′)/Eub518(5′-ATTACCGCGGCTGCTGG-3′) targeted on the 16S rRNA gene [27]. The standard curve was constructed by drawing the logarithm of the concentration of seven gradient dilutions of genomic DNA and the threshold period (Ct value) produced by qPCR analysis. *E. coli* genomic DNA was used as the standard template for quantitative detection of bacteria. QPCR was performed in a Step One Plus ™ real-time PCR system using faststart universal SYBR Green master (Rox) (Roche, Switzerland). Each PCR reaction included 1 μL DNA template, 2 μL 10 mM bacterial primers Eub338 / Eub518, 10 μL SYBR green mix, and 7 μL H_2O for a total of 20 μL. The cycling program consisted of denaturation at 95 °C for 10 min, followed by 40 cycles of 95 °C for 15 s, 54 °C for 30 s, 72 °C for 20 s. Melting curve analysis was established by increasing the temperature from 60 to 95 °C.

3. Results

3.1. Cellulolytic Enzyme Activities of Hull Main Microorganisms

To examine the extent of microbial influence on the hull, we determined the cellulose degrading capacity of these destructive microbes. As shown in Figure 1, all of these strains were able to degrade cellulose to generate degradation circles (H/D: 1.1–2.8), except *B. megaterium* (NK-NH10) which could not. Comparing the cellulose degradation ability of several strains of bacteria, *Bacillus* sp. (NK-NH17), *B. tequilensis* (NK-NH5) showed the highest degradation ability (H/D > 2.5), *Bacillus* sp. (NK-NH 15), *Bacillus* sp. (NK-NH16), and *B. velezensis* (NK-NH11) also showed a high capacity for cellulose degradation (H/D = 1.5–2.5). *F. solani* (NK-NH1) and *S. apiospermum* (NK. W1-3) had weak degradation ability (H/D < 1.5). These destructive microorganisms, by degrading cellulose of the hull or other organic matter cultural relics, can cause irreversible damage to these cultural relics. The control of microorganisms is important for the conservation of Nanhai No. 1.

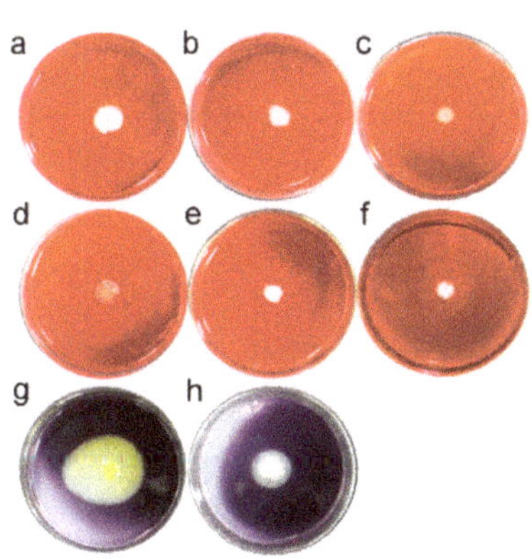

(A)

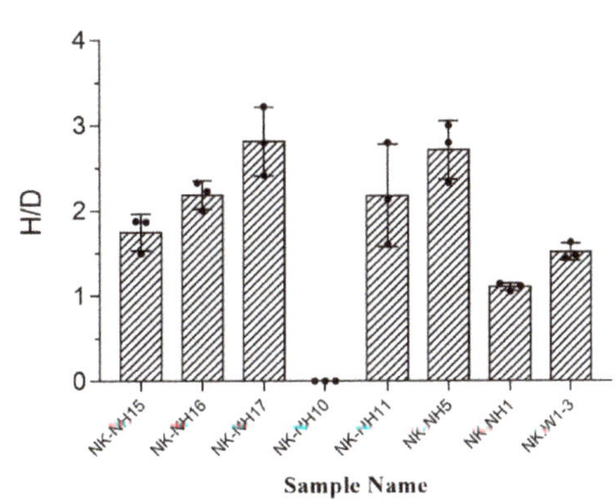

(B)

Figure 1. Cellulase activity of the main destructive microorganisms in Nanhai No. 1. (**A**) Hyaline circles and colony growth on CMC plates of the strains to be tested. (**a**) *Bacillus* sp. (NK-NH15), (**b**) *Bacillus* sp. (NK-NH16), (**c**) *Bacillus* sp. (NK-NH17), (**d**) *B. megaterium*(NK-NH10), (**e**) *B. velezensis* (NK-NH11), (**f**) *B. tequilensis* (NK-NH5), (**g**) *F. solani* (NK-NH1), (**h**) *S. apiospermum* (NK.W1-3). (**B**) The diameter of the colonies and the surrounding hyaline circles on the CMC plates were determined, denoted by D and H, respectively. The ratio of H and D (H/D) can give a preliminary indication of the cellulase activity of each colony.

3.2. Determination of Antimicrobial Activity of Cinnamaldehyde by Disc Diffusion Method

To determine the inhibitory activity of cinnamaldehyde against the tested microorganisms, we first employed the disk diffusion method. As shown in Figure 2, different sizes of inhibition zones were all produced around the filter paper sheets containing cinnamaldehyde, which illustrated that, at a certain concentration, cinnamaldehyde had inhibitory activities on the strains to be tested. As the concentration of cinnamaldehyde increased, the more obvious of an inhibitory effect was observed. Cinnamaldehyde at the concentration of 100 mg/mL showed strong inhibitory ability against *Bacillus* sp. (NK-NH15), *Bacillus* sp. (NK-NH16), *Bacillus* sp. (NK-NH17), *B. megaterium*(NK-NH10), *B. velezensis* (NK-NH11), *B. tequilensis* (NK-NH5), *F. solani* (NK-NH1), and *S. apiospermum* (NK.W1-3), and the diameter of each inhibition zone was 2.01, 2.51, 2.08, 1.89, 2.34, 2.55, 4.77, and 9 cm, respectively. The antimicrobial effect of 100 mg/mL cinnamaldehyde was similar to that of 0.5% Euxyl® K100. The inhibitory ability of cinnamaldehyde on fungi was stronger than that of bacteria. We found that *S. apiospermum* (NK.W1-3) was very sensitive to cinnamaldehyde. Cinnamaldehyde of 50 mg/mL could completely inhibit the growth of *S. apiospermum* (NK.W1-3), but 0.5% Euxyl® K100 could not.

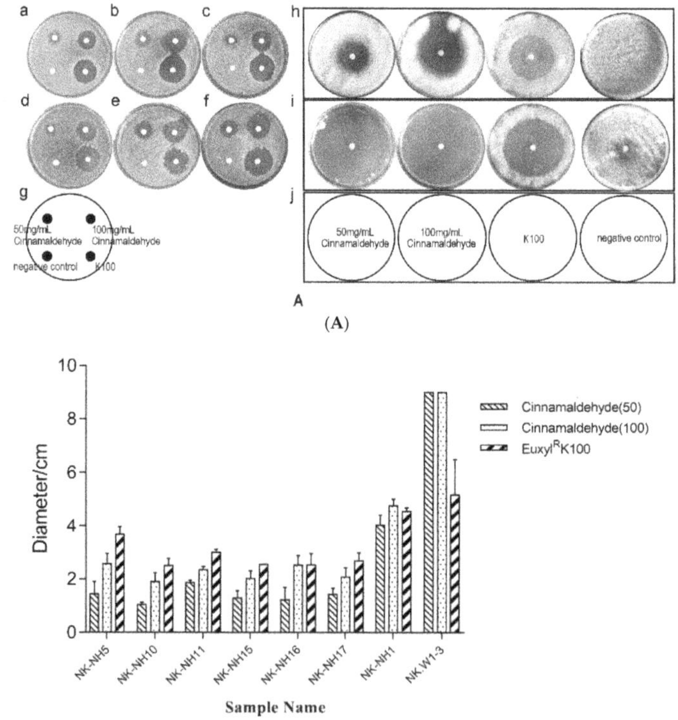

(B)

Figure 2. Inhibitory activities of cinnamaldehyde against the major destructive microorganisms of Nanhai No. 1. (**A**) Drugs created a zone of inhibition for different strains. (**a**) *Bacillus* sp. (NK-NH15), (**b**) *Bacillus* sp. (NK-NH16), (**c**) *Bacillus* sp. (NK-NH17), (**d**) *B. megaterium*(NK-NH10), (**e**) *B. velezensis* (NK-NH11), (**f**) *B. tequilensis* (NK-NH5), (**g**,**j**) drug distribution, (**h**) *F. solani* (NK-NH1), and (**i**) *S. apiospermum* (NK.W1-3); (**B**) cinnamaldehyde (50), cinnamaldehyde at concentration of 50 mg/mL; cinnamaldehyde (100), cinnamaldehyde at concentration of 100 mg/mL; Euxyl® K100, the concentration of Euxyl® K100 was 0.5%.

3.3. Determination of Antimicrobial Activity of Cinnamaldehyde Volatile Gas in Air

To determine the inhibitory activity of cinnamaldehyde volatile gas in air against destructive microorganisms, we completed the experiment using dichotomy plates. As shown in Figure 3, when the volatile gas concentration of cinnamaldehyde is 250 μL/L, cinnamaldehyde exerted antimicrobial activity on all the tested strains, while Euxyl® K100 did not (Figure S1). Specifically, cinnamaldehyde exhibited the strongest inhibitory effect against *S. apiospermum* (NK.W1-3) with an inhibition rate of 100%, and against *F. solani* (NK-NH1) with an inhibition rate of 45%.

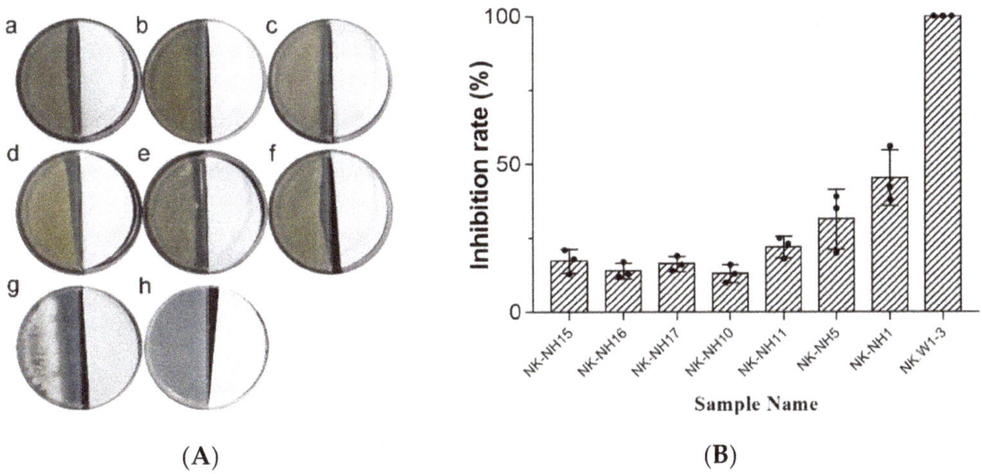

Figure 3. Antimicrobial activity of cinnamaldehyde volatile gas in air (**A**) microbial growth on the dichotomy plate; left (bacteria), right (cinnamaldehyde 250 ul/L). (**a**) *Bacillus* sp. (NK-NH15), (**b**) *Bacillus* sp. (NK-NH16), (**c**) *Bacillus* sp. (NK-NH17), (**d**) *B. megaterium*(NK-NH10), (**e**) *B. velezensis* (NK-NH11), (**f**) *B. tequilensis* (NK-NH5), (**g**) *F. solani* (NK-NH1), and (**h**) *S. apiospermum* (NK.W1-3). (**B**) Inhibition rates of cinnamaldehyde against eight microorganisms to be tested.

Similar to the results of the disk diffusion method, the inhibitory ability of cinnamaldehyde volatile gas to bacteria is relatively weak, and for *Bacillus* sp.(NK-NH15), *Bacillus* sp.(NK-NH16), *Bacillus* sp.(NK-NH17), *B. megaterium*(NK-NH10), *B. velezensis* (NK-NH11), and *B. tequilensis* (NK-NH5), the inhibition rate of bacteria is 17%, 14%, 17%, 13%, 22%, and 31%, respectively.

3.4. MIC Determination of Cinnamaldehyde

To compare the inhibitory ability of cinnamaldehyde, the minimum inhibitory concentration of eight destructive microorganisms was measured. As shown in Table 1, cinnamaldehyde for *Bacillus* sp. (NK-NH15), *Bacillus* sp. (NK-NH16), *Bacillus* sp. (NK-NH17), *B. megaterium* (NK-NH10), *B. velezensis* (NK-NH11), *B. tequilensis* (NK-NH5) had inhibitory activity with respective MIC values of 1 mg/mL, 2 mg/mL, 0.125 mg/mL, 0.5 mg/mL and 1 mg/mL. Inhibition of Euxyl® K100 was somewhat more effective. *F. solani* (NK-NH1) and *S. apiospermum* (NK.W1-3) were more sensitive to cinnamaldehyde with MICs of 0.250, 0.0625 mg/mL, respectively. *S. apiospermum* (NK.W1-3) was more sensitive to cinnamaldehyde than Euxyl® K100 (0.07813 mg/mL).

Table 1. MIC values of cinnamaldehyde against destructive microorganisms of Nanhai No.1 (mg/mL).

Strains	Cinnamaldehyde	Euxyl® K100
Bacillus sp. (NK-NH15)	1	0.07813
Bacillus sp. (NK-NH16)	2	0.03906
Bacillus sp. (NK-NH17)	1	0.03906
B. megaterium (NK-NH10)	0.125	0.07813
B. velezensis (NK-NH11)	0.5	0.03906
B. tequilensis (NK-NH5)	1	0.03906
F.solani (NK-NH1)	0.25	0.1563
S. apiospermum (NK.W1-3)	0.0625	0.07813

3.5. Cinnamaldehyde Inhibitory of Major Destructive Microbes of Hull during Laboratory Simulation Experiment

To verify the inhibitory effect exerted by cinnamaldehyde in practical applications, we simulated the protection status of hull wood from Nanhai No. 1. Shipwreck hull scattered wood was put in deionized water for moisture stabilization treatment and preliminary desalination treatment in the tank. Bacterial suspension of *Bacillus* sp. (NK-NH15), *Bacillus* sp. (NK-NH16), *Bacillus* sp. (NK-NH17), *B. megaterium* (NK-NH10), *B. velezensis* (NK-NH11) and *B. tequilensis* (NK-NH5) was added in the water, then covered and the contact of wood with air was reduced (Figure S2). After standing for 14 days, the total number of bacteria in the water samples was determined. As shown in Table 2, the total number of bacteria in water samples without cinnamaldehyde was $4.27 \times 10^5 \pm 2.04 \times 10^4$ CFU/mL. Cinnamaldehyde could inhibit microbial growth in water, and as the concentration of cinnamaldehyde rose, the number of microorganisms in the water continuously decreased. When the cinnamaldehyde concentration reached the highest value of 1.25 mg/mL, the total number of colonies in the water was $1.2 \times 10^1 \pm 5.10$ CFU/mL.

Table 2. Inhibitory effects of cinnamaldehyde on bacteria in water samples containing shipwreck wood.

	N	C1	C2	C3	C4
CFU/mL	$4.27 \times 10^5 \pm 2.04 \times 10^4$	$1.67 \times 10^5 \pm 9.02 \times 10^3$	$1.2 \times 10^4 \pm 5.89 \times 10^2$	$5.07 \times 10^2 \pm 3.86$	$1.2 \times 10^1 \pm 5.10$

At the same time, the method of qPCR was applied to determine the biomass of bacteria in water samples. The correlation coefficient of the standard curve for *E. coli* was >0.99. (Figure 4A). As shown in Figure 4B, the biomass of bacteria showed a decreasing trend with increasing concentrations of cinnamaldehyde. When the concentration of cinnamaldehyde was 1.25 mg/mL, the concentration of bacterial DNA was 1.34 ± 0.09 ng/μL, which was significantly lower than that of the negative control (584.65 ± 52.26 ng/μL).

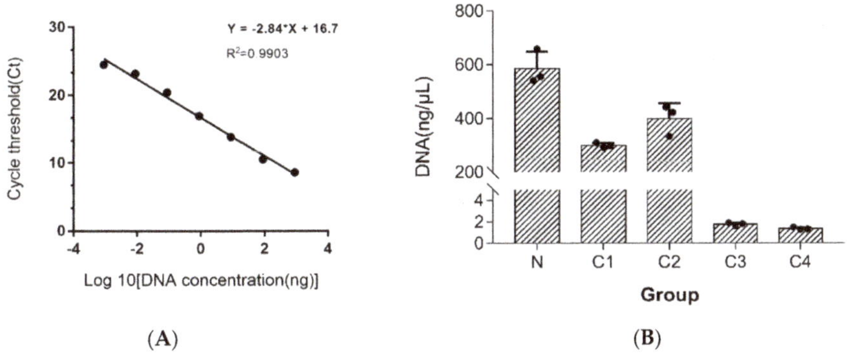

Figure 4. Quantification of bacterial biomass of samples by qPCR. (**A**) Standard curve for *E. coli* quantification using bacterial primers Eub338/Eub518. (**B**) Comparison of bacterial amounts. N: Water samples without cinnamaldehyde addition; C1: cinnamaldehyde concentration at 50 μg/mL; C2: cinnamaldehyde concentration at 200 μg/mL; C3: cinnamaldehyde concentration at 500 μg/mL; and C4: cinnamaldehyde concentration at 1250 μg/mL.

4. Discussion

The present archaeological digging work of Nanhai No. 1 is close to its end, with the hull chronically exposed to air and with high hull humidity (74.3%), which provides a suitable growth environment for microbial growth. Microorganisms, on the one hand, can affect the aesthetics of cultural relics, and on the other hand can cause harm to the cultural relics themselves. So, in the process of cultural relic protection, the prevention and control of microorganisms is of great significance. Previous studies have found that Nanhai No. 1 is affected by multiple microorganisms. *F. solani* (NK-NH1) is the major destructive fungus on ship wood, accounting for 90% of the total fungal load on the hull, which presents white plaques [8]. *S. apiospermum* (NK. W1-3) is a major destructive fungus in water samples stored in shipwreck wood. The genus *Scedosporium* accounted for 72.28% of the total fungi in water samples. *B. tequilensis* (NK-NH5), *B. velezensis* (NK-NH11), *B. megaterium* (NK-NH10) were the main bacteria in water samples storing wooden lacquer plates, *Bacillus* sp. (NK-NH15), *Bacillus* sp. (NK-NH16), *Bacillus* sp. (NK-NH17) were the main harmful bacteria isolated on ship wood. These destructive microorganisms can degrade cellulose and lignin, causing irreversible effects on hulls. Currently, the main measures for controlling microbial degradation of Nanhai No. 1 are to moisturize and inhibit the hulls by spraying them with distilled water containing 0.5% Euxyl® K100. Although the process can be effective in controlling microbial degradation of the hull, once the use of bacteriostatic agents is discontinued, the microorganism undergoes substantial reproductive growth. At the same time, Euxyl® K100 was found to cause some negative effects on the archaeological and relic protection workers' bodies. As such, finding safer and more environmentally friendly ways to inhibit these destructive microorganisms is very important. The aim of this study was to investigate the inhibitory activity of cinnamaldehyde against the destructive microorganisms of Nanhai No. 1. We first show that these destructive microorganisms are potentially harmful to hulls. Cellulose degradation experiments found that except for *B. velezensis* (NK-NH11), other microorganisms to be tested had the ability to degrade cellulose. The prevention and treatment of destructive microorganisms of Nanhai No. 1 should take both fungi and bacteria into account. Meanwhile we determined the antimicrobial activity of cinnamaldehyde. Previous studies have reported that cinnamaldehyde has a broad spectrum of antimicrobial activity [17,19,28,29]. We first preliminarily measured the inhibition of destructive microorganisms of Nanhai No. 1 by the disk diffusion method and found that cinnamaldehyde, at a certain concentration, could effectively inhibit microbial growth, which was comparable to the effects of Euxyl® K100 at 0.5%. At the same time, we found that fungi were more sensitive to cinnamaldehyde. Taking into account the volatile properties of cinnamaldehyde, we measured the inhibition effect of cinnamaldehyde volatile gas in air. Similarly to the paper disk diffusion method results, airborne cinnamaldehyde can effectively inhibit microbial growth, although the inhibition effect is more obvious on fungi. Cinnamaldehyde, through its volatile properties, exerts antimicrobial ability, which can be used on the inhibition of destructive microorganisms of cultural relics in practical work. This provided us with ideas for the future practical application of cinnamaldehyde in the protection of cultural relics, and cinnamaldehyde can be slowly released in a closed vessel to achieve the effect of sustained antimicrobial activity. Edible films containing cinnamaldehyde have been reported to be effective in inhibiting *Salmonella* on vegetables sealed plastic bags [30]. In the results of MIC, cinnamaldehyde showed obvious antimicrobial activity, but lower than Euxyl® K100. For the antimicrobial mechanism of cinnamaldehyde has been intensively studied, and it has been reported that cinnamaldehyde can cause changes in the fatty acids of cell membranes and alter its structure, thus prompting the binding of cinnamaldehyde or other compounds to cells [31]. It also provides ideas for our later experiments: through the combined use of multiple antimicrobial agents, so as to achieve the effect of reducing the hazard and improving the drug efficacy.

The antimicrobial activity of cinnamaldehyde has been largely evaluated in laboratory experiments, while it is also widely used in the food industry [32]. The ability of antimicrobial agents against microorganisms is closely related to the host, the environment, and so on. To evaluate the feasibility of cinnamaldehyde application for microbial degradation control in Nanhai No. 1, we simulated the protection status of the scattered wood of Nanhai No. 1 and added Cinnamaldehyde to observe its bacteriostatic effect. The results showed that cinnamaldehyde could inhibit the growth of bacteria in water, and the concentration of bacteria in water decreased as the concentration of cinnamaldehyde increased. When the concentration of cinnamaldehyde is 0.5 mg/mL, it can inhibit the reproduction of most bacteria. The application of cinnamaldehyde to the protection of wooden cultural relics in water will provide a new way to develop environmentally friendly, safe and inexpensive bacteriostatic agents of plant origin in the future. In the following work, we will pay more attention to the practical use and effect of cinnamaldehyde in field protection, and the combined use of cinnamaldehyde with other antimicrobial agents is also of interest. In this study, cinnamaldehyde was found to inhibit the main destructive fungi of Nanhai No. 1, which is comparable to the Euxyl® K100 effect used for conservation on-site and also has some inhibitory activity against bacteria. Cinnamaldehyde not only showed antimicrobial ability in laboratory experiments, but also effectively inhibited the reproductive growth of bacteria in simulated experiments, indicating that cinnamaldehyde has a promising application in the prevention and control of cultural relic biodegradation.

Supplementary Materials: The following are available online at https://www.mdpi.com/article/10.3390/app11115262/s1, Figure S1: Microbial growth on the dichotomy plate, Figure S2: The state of shipwreck hull scattered wood in water tank. Table S1: Molecular identification of strains isolated from the hull, sea mud.

Author Contributions: Conceptualization, J.P. and Y.L.; methodology, X.H. and Y.H.; software, X.H. and Y.H.; validation, P.G., Y.W. and K.M.; resources, J.D., N.L. and Z.Z.; data curation, X.H.; writing—original draft preparation, X.H.; writing—review and editing, J.P. and Y.L.; visualization, X.H.; supervision, Y.H.; project administration, J.P.; funding acquisition, J.P. and Y.L. All authors have read and agreed to the published version of the manuscript.

Funding: This work was supported by Matching Funds of Natural Science Foundation of Tianjin (Grant No. 19JCZDJC33700 to J.P.); National Key R&D Program of China (Grant No. 2020YFC1521800 to Z.Z. and J.D.); The National Natural Science Foundation of China (Grant No. 82073832 to Y.L.); Tianjin Natural Science Fund for Distinguished Young Scholars (Grant No. 20JCJQJC00070 to Y.L.); Tianjin Municipal Education Commission Scientific Research Project (Natural Science, Grant No. 2019ZD11 to Y.L.).

Institutional Review Board Statement: Not applicable.

Informed Consent Statement: Not applicable.

Data Availability Statement: Not applicable.

Acknowledgments: We gratefully acknowledge the assistance of Dawa Shen from Chinese Academy of Cultural Heritage and Guanglan Xi from National Center of Archaeology.

Conflicts of Interest: The authors declare no conflict of interest.

References

1. Yong, C. Brief Summery on Excavation of Nanhai NO.1 Shipwreck. *Stud. Nat. Cult. Herit.* **2019**, *4*, 14–20.
2. Chen, J.; Huang, X.; Chen, X.; Chen, Z. Corrosion Type and Conservation of Archaeological Waterlogged Wood. *Mater. Rev.* **2015**.
3. BjöRdal, C.G. Microbial degradation of waterlogged archaeological wood. *J. Cult. Herit.* **2012**, *13*, S118–S122. [CrossRef]
4. Bhat, M.; Bhat, S. Cellulose Degrading Enzymes and Their Potential Industrial Applications. *Biotechnol. Adv.* **1997**, *15*, 583–620. [CrossRef]
5. Fazio, A.T.; Papinutti, L.; Gómez, B.A.; Parera, S.D.; Romero, A.R.; Siracusano, G.; Maier, M.S. Fungal deterioration of a Jesuit South American polychrome wood sculpture. *Int. Biodeterior. Biodegrad.* **2010**, *64*, 694–701. [CrossRef]
6. Geoffrey, F.; Daniel, T.; Nilsson, A.P.S. Degradation of lignocellulosics by unique tunnel-forming bacteria. *Can. J. Microbiol.* **1987**, *33*, 943–948.
7. Blanchette, R.A. A review of microbial deterioration found in archaeological wood from different environments. *Int. Biodeterior. Biodegrad.* **2000**, *46*, 189–204. [CrossRef]
8. Liu, Z.; Fu, T.; Hu, C.; Shen, D.; Macchioni, N.; Sozzi, L.; Chen, Y.; Liu, J.; Tian, X.; Ge, Q.; et al. Microbial community analysis and biodeterioration of waterlogged archaeological wood from the Nanhai No. 1 shipwreck during storage. *Sci Rep.* **2018**, *8*, 7170. [CrossRef] [PubMed]
9. Yin, L.; Jia, Y.; Wang, M.; Yu, H.; Jing, Y.; Hu, C.; Zhang, F.; Sun, M.; Liu, Z.; Chen, Y.; et al. Bacterial and Biodeterioration Analysis of the Waterlogged Wooden Lacquer Plates from the Nanhai No. 1 Shipwreck. *Appl. Sci.* **2019**, *9*, 653. [CrossRef]
10. Han, Y.; Huang, X.; Wang, Y.; Du, J.; Ma, K.; Chen, Y.; Li, N.; Zhang, Z.; Pan, J. Fungal Community and Biodeterioration Analysis of Hull Wood and Its Storage Environment of the Nanhai No. 1 Shipwreck. *Front. Microbiol.* **2020**, *11*, 609475. [CrossRef]
11. Aerts, O.; Meert, H.; Goossens, A.; Janssens, S.; Lambert, J.; Apers, S. Methylisothiazolinone in selected consumer products in Belgium: Adding fuel to the fire? *Contact Dermat.* **2015**, *73*, 142–149. [CrossRef]
12. Marrero-Alemán, G.; Santana, P.S.; Liuti, F.; Hernández, N.; López-Jiménez, E.; Borrego, L. The Role of Cleaning Products in Epidemic Allergic Contact Dermatitis to Methylchloroisothiazolinone/Methylisothiazolinone. *Dermatitis* **2018**, *29*, 77–80. [CrossRef] [PubMed]
13. Lundov, M.D.; Kolarik, B.; Bossi, R.; Gunnarsen, L.; Johansen, J.D. Emission of Isothiazolinones from Water-Based Paints. *Environ. Sci. Technol.* **2014**, *48*, 6989–6994. [CrossRef] [PubMed]
14. Liu, S.Q.; Zhang, Y.; Liao, X.L.; Bai, L.Y. Research Status and Application Prospects of Botanical Pesticides in China. *Hunan Agric. Sci.* **2016**, *2*, 115–119.
15. Pavela, R.; Benelli, G. Essential Oils as Ecofriendly Biopesticides? Challenges and Constraints. *Trends Plant. Sci* **2016**, *21*, 1000–1007. [CrossRef] [PubMed]
16. Yoon, M.Y.; Cha, B.; Kim, J.C. Recent trends in studies on botanical fungicides in agriculture. *Plant. Pathol. J.* **2013**, *29*, 1–9. [CrossRef] [PubMed]
17. Doyle, A.A.; Stephens, J.C. A review of cinnamaldehyde and its derivatives as antibacterial agents. *Fitoterapia* **2019**, *139*, 104405. [CrossRef]
18. Hariri, M.; Ghiasvand, R. Cinnamon and Chronic Diseases. *Adv. Exp. Med. Biol.* **2016**, *929*, 1–24. [PubMed]
19. Shreaz, S.; Wani, W.A.; Behbehani, J.M.; Raja, V.; Irshad, M.; Karched, M.; Ali, I.; Siddiqi, W.A.; Hun, L.T. Cinnamaldehyde and its derivatives, a novel class of antifungal agents. *Fitoterapia* **2016**, *112*, 116–131. [CrossRef]
20. Wei, J.; Bi, Y.; Xue, H.; Wang, Y.; Zong, Y.; Prusky, D. Antifungal activity of cinnamaldehyde against Fusarium sambucinum involves inhibition of ergosterol biosynthesis. *J. Appl. Microbiol.* **2020**, *129*, 256–265. [CrossRef]
21. Suwanamornlert, P.; Sangchote, S.; Chinsirikul, W.; Sane, A.; Chonhenchob, V. Antifungal activity of plant-derived compounds and their synergism against major postharvest pathogens of longan fruit in vitro. *Int. J. Food Microbiol.* **2018**, *271*, 8–14. [CrossRef] [PubMed]
22. Sun, Q.; Li, J.; Sun, Y.; Chen, Q.; Zhang, L.; Le, T. The antifungal effects of cinnamaldehyde against Aspergillus niger and its application in bread preservation. *Food Chem.* **2020**, *317*, 126405. [CrossRef] [PubMed]
23. Nazzaro, F.; Fratianni, F.; De Martino, L.; Coppola, R.; De Feo, V. Effect of essential oils on pathogenic bacteria. *Pharmaceuticals* **2013**, *6*, 1451–1474. [CrossRef] [PubMed]
24. Vasconcelos, N.G.; Croda, J.; Simionatto, S. Antibacterial mechanisms of cinnamon and its constituents: A review. *Microb. Pathog.* **2018**, *120*, 198–203. [CrossRef]

25. Ahmad, B.; Nigar, S.; Shah, S.S.; Bashir, S.; Ali, J.; Yousaf, S.; Bangash, J.A. Isolation and Identification of Cellulose Degrading Bacteria from Municipal Waste and Their Screening for Potential Antimicrobial Activity. *World Appl. Sci. J.* **2013**, *27*, 1420–1426. [CrossRef]
26. Kasana, R.; Salwan, R.; Dhar, H.; Dutt, S.; Gulati, A. A rapid and easy method for the detection of microbial cellulases on agar plates using gram's iodine. *Curr. Microbiol.* **2008**, *57*, 503–507. [CrossRef]
27. Fierer, N.; Jackson, J.A.; Vilgalys, R.; Jackson, R.B. Assessment of soil microbial community structure by use of taxon-specific quantitative PCR assays. *Appl. Environ. Microbiol.* **2005**, *71*, 4117–4120. [CrossRef]
28. Cheng, S.S.; Liu, J.Y.; Chang, E.H.; Chang, S.T. Antifungal activity of cinnamaldehyde and eugenol congeners against wood-rot fungi. *Bioresour. Technol.* **2008**, *99*, 5145–5149. [CrossRef]
29. Ma, Y.N.; Chen, C.J.; Li, Q.; Wang, W.; Xu, F.R.; Cheng, Y.X.; Dong, X. Fungicidal Activity of Essential Oils from Cinnamomum cassia against the Pathogenic Fungi of Panax notoginseng Diseases. *Chem. Biodivers.* **2019**, *16*, e1900416. [CrossRef]
30. Zhu, L.; Olsen, C.; McHugh, T.; Friedman, M.; Jaroni, D.; Ravishankar, S. Apple, carrot, and hibiscus edible films containing the plant antimicrobials carvacrol and cinnamaldehyde inactivate Salmonella Newport on organic leafy greens in sealed plastic bags. *J. Food Sci.* **2014**, *79*, M61–M66. [CrossRef]
31. Di Pasqua, R.; Betts, G.; Hoskins, N.; Edwards, M.; Ercolini, D.; Mauriello, G. Membrane toxicity of antimicrobial compounds from essential oils. *J. Agric. Food Chem.* **2007**, *55*, 4863–4870. [CrossRef] [PubMed]
32. Friedman, M. Chemistry, Antimicrobial Mechanisms, and Antibiotic Activities of Cinnamaldehyde against Pathogenic Bacteria in Animal Feeds and Human Foods. *J. Agric. Food Chem.* **2017**, *65*, 10406–10423. [CrossRef]

applied sciences

MDPI

Article

Role of Exposure on the Microbial Consortiums on Historical Rural Granite Buildings

Elsa Fuentes [1], Rafael Carballeira [2] and Beatriz Prieto [1,*]

[1] Departamento de Edafoloxía e Química agrícola, Facultad de Farmacia, Universidade Santiago de Compostela, 15782 Santiago de Compostela, Spain; elsa.fuentes.alonso@usc.es

[2] Centro de Investigacións Científicas Avanzadas (CICA), Facultade de Ciencias, Universidade da Coruña, 15008 A Coruña, Spain; r.carballeira@udc.es

* Correspondence: beatriz.prieto@usc.es

Abstract: Local granite has been used throughout history in Galicia (NW Spain), forming the basis of much of the region's architecture. Like any other rock, granite provides an ecological niche for a multitude of organisms that form biofilms that can affect the physical integrity of the stone. In this study, for the first time, characterization of the microbial consortium forming biofilms that developed on historical rural granite buildings is carried out using a combination of culture-dependent and next generation sequencing (NGS) techniques. Results pointed to differences in biofilm composition on the studied rural granite buildings and that of previously analyzed urban granite buildings, especially in terms of abundance of cyanobacteria and lichenized fungi. Exposure was corroborated as an important factor, controlling both the diversity and abundance of microorganisms on walls, with environmental factors associated with a northern orientation favoring a higher diversity of fungi and green algae, and environmental factors associated with the west orientation determining the abundance of lichenized fungi. The orientation also affected the distribution of green algae, with one of the two most abundant species, *Trentepohlia cf. umbrina*, colonizing north-facing walls, while the other, *Desmococcus olivaceus*, predominated on west-facing walls.

Keywords: biofilm; microbial community; algae; cyanobacteria; fungi; bacteria; next generation sequencing; granite; cultural heritage; Trentepohlia

Citation: Fuentes, E.; Carballeira, R.; Prieto, B. Role of Exposure on the Microbial Consortiums on Historical Rural Granite Buildings. *Appl. Sci.* **2021**, *11*, 3786. https://doi.org/10.3390/app11093786

Academic Editor: Giovanna Traina

Received: 25 March 2021
Accepted: 20 April 2021
Published: 22 April 2021

1. Introduction

Granite is one of the most commonly used types of stone in construction worldwide and is known to have been in use since Egyptian times [1]. This igneous rock is widely used in construction because of its sustainability and durability, being a hard rock that is resistant to both heat and ultraviolet radiation. Nonetheless, all materials exposed to the environment gradually deteriorate as a result of physical, chemical, and biological processes [2,3]. Interactions between the atmosphere and stone invariably lead to the formation of altered surface layers, which eventually results in damage to the original stone. This damage is of great concern in the case of stones in unique buildings and monuments [4]. Information about the biodeterioration of historical granite buildings is very limited relative to what is known about buildings constructed with limestone or sandstone [5], but it is known that the main factors leading to the granite biodeterioration process are the presence of soluble salts [6] and biodeterioration caused by colonizing organisms [7–9].

Microbial colonization alters the stone substrate causing physical–chemical and aesthetic damage. These negative effects have been widely studied [10–15]. However, the idea of the protective effect of organisms on stone has recently gained relevance as their function as consolidants, binding particles, and protecting stone from erosion or aggressive substances, as well as from other abiotic factors, has been demonstrated [16–18]. Both deteriorative and protective actions can be caused by a single species, but can also be due to the combined, inter-related effects of various different microorganisms, which also depends on the environmental conditions [19–21]. It is therefore important to understand the complexity, composition, and structure of the community inhabiting the stone of cultural heritage buildings, in order to foresee the risk deterioration degree and, thus, be able to carry out the most appropriate conservation tasks.

Due to the interactions between organisms and substrate, with proved positive and negative effects, the development of techniques allowing the study and identification of microbial communities has gained importance as a key issue in the heritage conservation field. In this sense, different techniques have been used over time to identify organisms forming biofilms. Traditionally, their isolation, identification, and study have been carried out by means of classical culture techniques. These culture-dependent techniques have the advantage of being able to perform biochemical and physiological studies in parallel to identification; however, they greatly underestimate existing diversity [22] and favor the rapid growth of opportunistic species, making it difficult to identify other microorganisms that may play an important role [23]. With the emergence of metagenomics and, in particular, next generation sequencing (NGS) techniques, detection capacity has increased exponentially, thus improving the potential and level of detail of microbial ecology studies [24]. NGS techniques enable large amounts of data to be obtained more quickly than with any of the previous methods. Thus, the combination of these modern techniques together with traditional culture-dependent methods would enable the production of more data and a better understanding of specific findings [25]. Numerous studies on the microbial colonization of historical monuments have been carried out using these modern techniques, especially on sandstone, marble, and limestone substrates. In the case of historical granite buildings, studies characterizing microbial communities are scarce [12,26–28] and most of them have been carried out using traditional methods [29–31], with only one using NGS techniques [26]. In this study, biofilms (algae, cyanobacteria, and fungi) on historical granite buildings in Santiago de Compostela were characterized; however, bacterial communities, which are important at early stages of biocolonization, were not studied.

In this study, we investigated the microbiological colonization of three granite-built churches located in rural areas of Galicia, in the northwestern part of the Iberian Peninsula. Colonization was characterized using a combination of traditional and NGS methods (culture dependent methods for phototrophic organisms and NGS methods for fungi and bacterial organisms). The biodiversity was determined using the Shannon diversity index. Moreover, taking into account that the orientation of buildings is one of the factors that most strongly influences the development and composition of subaerial algae biofilms, as it determines the amount of light and the availability of water on the substrate [32,33], the role of exposure was also considered.

2. Materials and Methods

The three churches under study, Sta. María de Vilar (SMV) (43°8′1.2″ N, 7°54′30.1″ W; 466 m), San Cibrao de Barreiro (BAR) (42°48′25.4″ N, 7°55′20.6″ W; 509 m), and San Cosme de Rocha (ROC) (43°02′30.7″ N, 7°50′59.7″ W; 532 m) were selected as typical examples of granite constructions in rural Galicia (in the NW of the Iberian Peninsula). The churches (hereafter referred to as SMV, BAR and ROC churches) are located in rural areas, far from major cities, with low levels of pollution (Figure 1). The churches are located along a 35 km transect, from north to south, and each church is surrounded by a small graveyard and local vegetation. The region has an Atlantic climate, with a mean relative humidity of 80% and a mean yearly rainfall of 1180 mm. The mean number of days per month with rainfall is 13 and the mean temperature is 12 °C. The wind rose (the dominant frequencies of wind direction) showed clear northeastern and southwestern wind dominances (https://www.meteogalicia.gal/; Accessed: 21 January 2021).

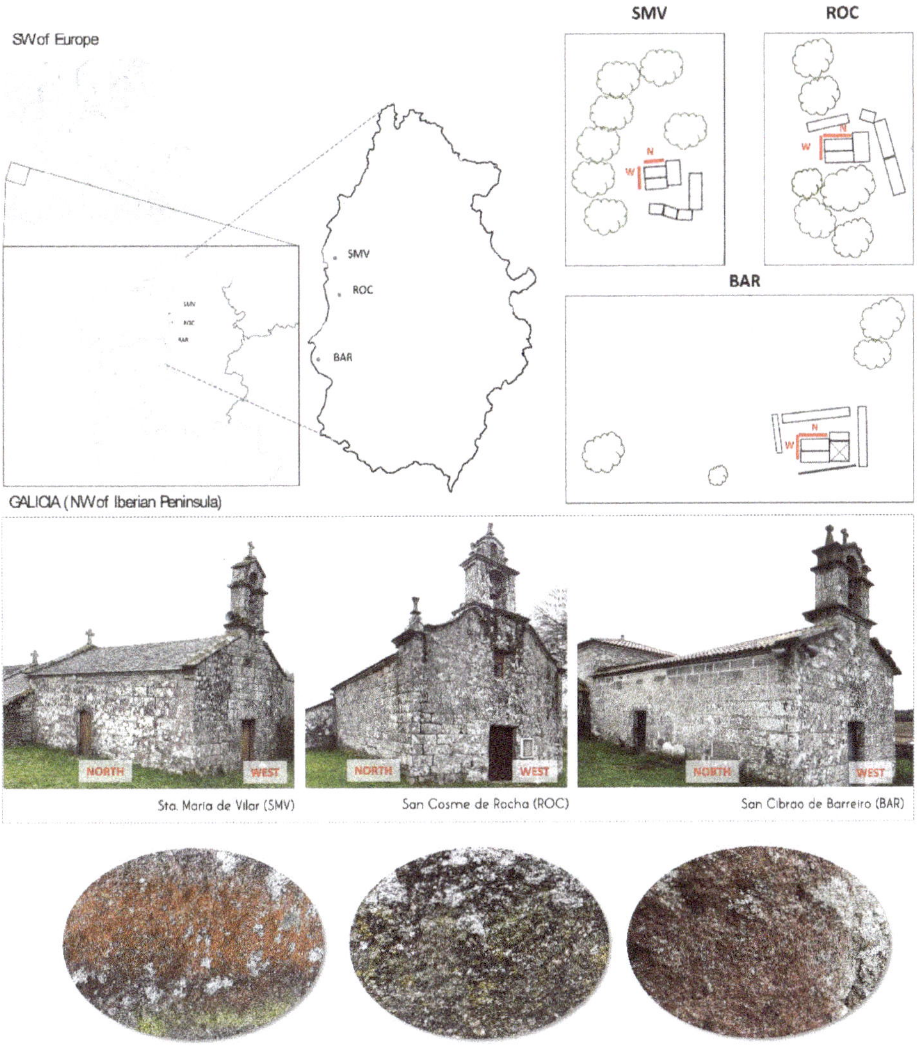

Figure 1. Location and photographs of the north and west façades of the SMV, ROC and BAR churches (Galicia, in the NW of the Iberian Peninsula) and the different biopatinas.

The church walls are almost completely covered by lichens and biofilms are only present on the north- and west-facing walls. Representative samples of the subaerial biofilms were obtained by scraping approximately 5 mg from three areas of the north (N)- and west (W)-facing walls of each church.

Phototrophs (i.e., algae and cyanobacteria) were studied by means of conventional morphological examination under light microscopy, since it allows for a reliable identification and for the estimation of the abundance of the different taxa. Bacteria and fungi were studied using next generation sequencing, since growing different taxonomic groups to obtain representative data requires great effort, and not all taxa grow in culture.

2.1. *Cyanobacteria and Algae Morphological Identification*

A subsample of 0.5 mg of each direct scraping was disaggregated and resuspended in 1 mL of BG11 culture medium for microscopic examination of the algal and cyanobacterial communities present on the N- and W-facing walls of each church. The subaerial biofilm species were examined under an optical microscope (Nikon Eclipse E600 equipped with an E-Plan 40x objective, N.A. 0.65), with differential interference optical contrast (Nomarski). Photomicrographs were taken with a digital camera (AxioCam ICc5 Zeiss). The algal communities were characterized using a semi-quantitative approach in which 5 aliquots of 150 μL of each sample, resuspended in medium, were examined by counting a total of 100 fields (40× magnification) and a minimum of 1000 cells. The percentages of each species present were calculated from the total number of cells observed; the species detected by culture were assigned 1% as a record of their presence in the total count. Agglomerative hierarchical analysis of the percentages of each species/taxon was carried out, and Ward's criterion was applied. The different samples of subaerial biofilm were compared in relation to the affinities between the communities by means of a Euclidean distance matrix of the dissimilarity coefficient, following the community classification criteria of Gauch and Whittaker [34].

BG11 media [35], with and without nitrogen ($BG11_0$), were prepared to promote growth of cyanobacteria (Cyanoprokaryota). Specifically, the $BG11_0$ medium was used to promote nitrogen-fixing cyanobacteria and their dormant stages. BBM medium [36,37] was used to promote the growth of green algae and the different phases of their life cycle in order to confirm their taxonomic identification. Cultures were maintained in sterile controlled laboratory conditions at a constant temperature of 23 °C, a 12:12 h light/dark photoperiod and a photon irradiance of 20 μmol m^{-2} s^{-1}. Cultures were maintained for 6 months and examined every 2 months to confirm taxonomic identifications based on morphometric characteristics and life history stages not presented in the initial sample.

Species identification and nomenclature were mainly based on Rifón et al. [31,38]. The following articles were also consulted: Škaloud [39], for green algae (Chlorophyta); Komárek [40], for cyanobacteria (Cyanophyta/Cyanoprokariota); and Lange-Bertalot and Hoffman [41] for diatoms (Bacillariophyta).

2.2. *Bacteria and Fungi Next Generation Sequencing Analysis*

2.2.1. DNA Extraction and PCR Amplification.

Total genomic DNA was isolated from 0.25 g of each individual biofilm sample with the QIAamp® Power® Fecal DNA Kit (Qiagen Sciences Inc., Germantown, MD, USA) following the manufacturer's protocol. DNA was quantified and quality-checked using a Nanodrop spectrophotometer. A 5-ng aliquot of total genomic DNA was used as a template for the specific PCR amplification of hypervariable V3-V4 regions of the 16S rRNA bacterial gene and of the ITS II region between the 5.8S rDNA and LSU of fungal ribosomal DNA in order to detect respectively bacterial and fungal species in the biofilm samples. PCR amplification of V3–V4 hypervariable regions in the 16S rDNA gene consisted of 25 cycles with 55 °C as annealing temperature, as described by [42]. PCR amplification conditions for the ITS region consisted of 40 cycles with 55 °C as the annealing temperature. The PCR analysis was performed at Lifesequencing-ADM (Valencia, Spain).

2.2.2. Illumina Sequencing of 16S V3-V4 and 5.8S rDNA-LSU Amplicons

PCR amplification products of variable regions V3 –V4 in the 16S rRNA gene were obtained with fusion primers S-D-Bact-0341b-S-17 (Illumina adaptors + 5′ CCTAC-GGGNGG CWGCAG 3′) and S-D-Bact-0785-a-A-21 (Illumina adaptors + 5′ GAC-TACHVGGGTATCT AATCC 3′). PCR amplification products of the ITS3-ITS4 region were obtained with fusion primers ITS3F (Illumina adaptors + 5′ GCATCGATGAA-GAACGCAGC 3′) and ITS4R (Illumina adaptors + 5′ GCATATCAATAAGCGGAG-GA3′).

Amplicon multiplexing and sequencing was carried out using a dual indexing tag-tailed design with 8-nt indices, with the Nextera XT Index Kit v2. Paired-end sequencing of 16S and ITS PCR amplicon libraries was performed with the MiSeq Reagent Kit v3, and 600 cycles were performed to produce 300 paired end sequences and the Illumina MiSeq. Library preparation and sequencing was performed at Lifesequencing-ADM (Valencia, Spain).

2.2.3. Bioinformatic Analysis

The 16S and ITS rRNA raw sequence data were first demultiplexed and then processed using PEAR V.0.9.1 (http://www.exelixis-lab.org/web/software/pear; Accessed: 6 October 2020) with default parameters, except for the overlap between the sequences of each end, which was fixed at 70 nts. This finally produced a single file with all overlapping sequences. Trimming steps were applied to remove adaptors and low-quality and short sequences. Primers for the target overlapping sequences were trimmed using CUTADAPT v.1.8.1 software. Sequences with quality values lower than Phred Q20 were excluded using BBMap version 3.38 (https://jgi.doe.gov/data-and-tools/bbtools/bb-tools-user-guide/bbmap-guide/; Accessed: 6 October 2020). Finally, only sequences longer than 200 nts were retained, as shorter-length sequences are more likely to be incorrectly assigned to certain taxonomic groups.

For 16S rRNA only, chimeric reads were identified and excluded using Chimera UCHIME53 [43]. High-quality reads were taxonomically identified using operational taxonomic units (OTUs), and samples were assigned to the same group when reads showed 97% similarity (clustering at 97% homology was carried out using CD-HIT; Fu et al., 2012).

Based on the OTU approach, the alpha-diversity of the biofilm samples was estimated in terms of community richness (Chao-1) and diversity (Shannon). The longest read of each cluster was used as a reference for taxonomic classification, which was conducted using a local BLAST search with the blast-2.2.26+ algorithm (http://blast.ncbi.nlm.nih.gov/Blast.cgi; Accessed: 6 October 2020), default settings and the NCBI database for 16s rRNA. In the final step, the taxonomic path of the reference sequence from each cluster was mapped to the additional reads within the corresponding cluster plus the corresponding replicates (as identified in the previous analysis step) to finally produce (semi-) quantitative information (number of individual reads representing a taxonomic path). Alpha and Beta diversity were measured using the VEGAN package in R. This analysis was performed by Lifesequencing-ADM.

The diversity of bacteria, fungi and Chlorophyta (green algae) on the N- and W-facing walls of the churches studied was evaluated using the Shannon index [44].

2.3. Statistical Analysis

Analysis of variance was applied using SPSS Statistics v19.0 (IBM) software to statistically determine any significant differences. Differences were considered significant at $p < 0.05$.

3. Results and Discussion

The microbial diversity of granite cultural heritage buildings was analyzed in this work. Both traditional techniques and NGS techniques were used. Traditional culture-dependent techniques allowed the study of phototrophic organisms, such as algae and cyanobacteria, which are easily distinguished and examined by optical microscopy. Other abundant microorganisms in biofilms, such as bacteria and fungi were characterized by NGS techniques, which allow a complete study of those species that are impossible to culture in the laboratory. The combination of both methods made it possible to analyze organism diversity and to study the role of orientation and exposure to climatic factors in the microbiome composition.

3.1. Culture Dependent

Green algae and cyanobacteria were isolated from the three granite churches using a culture-dependent approach. The species/taxon compositions of the sampled biofilms are shown in Table 1 and the most representative species are shown in Figure 2. Four species

(*Trentepohlia* cf. *umbrina*, *Desmococcus olivaceus*, *Tychonema* cf. *bourrellyi*, and *Phormidium autumnale*) predominated in the churches, although differences in the abundance of the species were observed in the three churches and on the N- and W-facing walls of the same church. The N-facing walls of the SMV and BAR churches were dominated by *Trentepohlia* cf. *umbrina*, while the W-facing walls were predominated by *Desmococcus olivaceus* (Figure 3). In the ROC church, the cyanobacteria *Tychonema* cf. *bourrellyi* and *Phormidium autumnale* predominated on the W-facing wall, with a minor presence of *Desmococcus olivaceus*. Green algae (Chlorophyta) and cyanobacteria were similarly represented on the N-facing wall, with *Desmococcus olivaceus and Phormidium autumnale* predominating. Thus, the microbial communities on the SMV and BAR churches were similar to each other, but different from those on the ROC church (Figure 4). The biofilms on the first two churches were mainly composed of green algae, while biofilms on the ROC church contained a higher percentage of cyanobacteria, especially on the W-facing wall.

Table 1. Species composition of the phototrophic communities on the SMV, BAR, and ROC churches, expressed as relative frequencies (%) of the total number of cells observed.

Species		NORTH			WEST		
		SMV	BAR	ROC	SMV	BAR	ROC
Trentepohlia cf. umbrina	Chl	80	75		10	23	
Desmococcus olivaceus	Chl	9	10	53	81	71	15
Phormidium autumnale	Cy		1	40			30
Tychonema cf. bourrellyi	Cy	1					50
Bracteacoccus minor	Chl	1			1		
Chlamydomonas sp. sensulato	Chl		1				
Coccomyxa confluens	Chl	1		1	1	1	
Coenochloris signiensis	Chl	1	1				
Dictyosphaerium chlorelloides	Chl		1				
Diplosphaera chodatii	Chl	1	1	1	1	1	
Elakatothrix gelatinosa	Chl		1				
Cylindrocystis brebissoni	Chl		1				
Elliptochloris sp.	Chl	1	1	1	1	1	1
Gloeocystispoly dermatica	Chl	1	1				
Hantzschia amphyoxis	Ba		1				
Klebsormidium cf. flaccidum	Chl	1	1	1	1	1	1
Mesotaenium cf. caldariorum	Chl		1				
Pseudochlorella sp.	Chl	1	1		1	1	1
Stichococcus bacillaris	Chl	1		1			1
Stichococcus mirabilis	Chl	1	1		1		
Achnantidium cf. minutissimum	Ba		1				
Chlorella vulgaris	Chl				1		
Chroococcus turgidus	Cy				1		1
Nostoc punctiforme	Cy			1			
Oscillatoria tenuis	Cy					1	
Phormidium favosum	Cy		1				

Chl: Chlorophyta; Ba: Bacillariophyta; Cy: Cyanophyta.

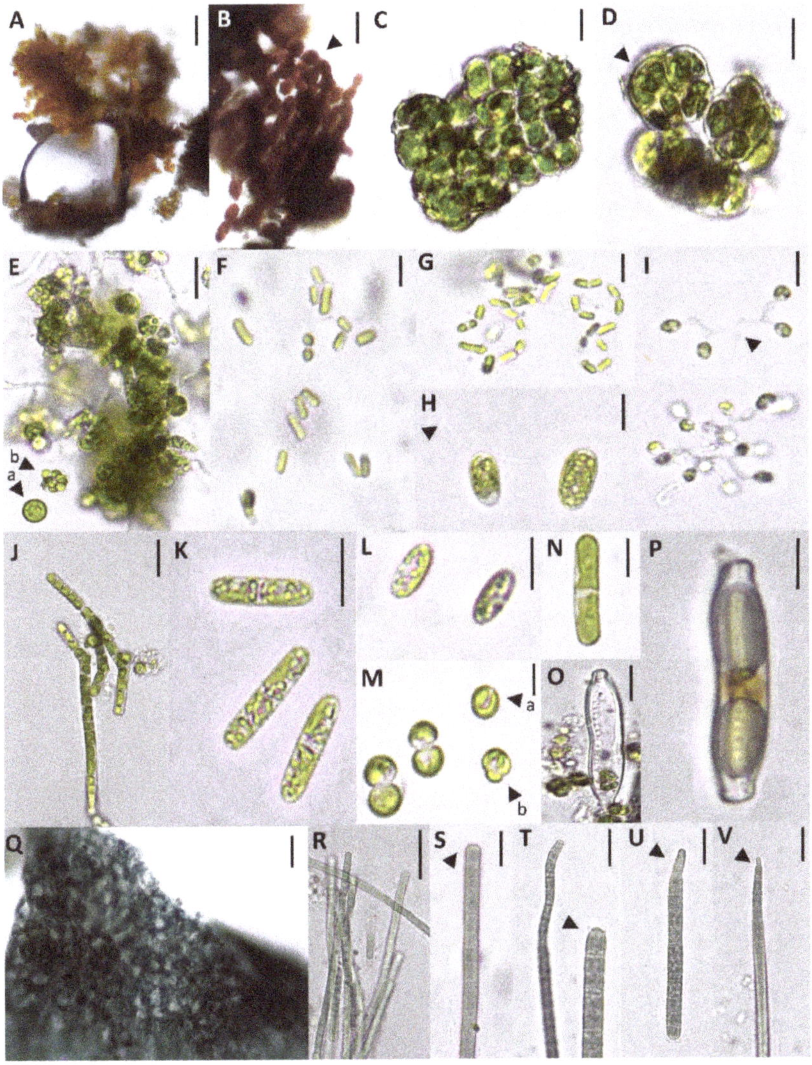

Figure 2. Species/taxa identified in the different samples observed on the surface of the north- and west-facing walls of the SMV, ROC and BAR churches. Chlorophyta or green algae (**A–N**): (**A**,**B**) *Trentepohlia* cf. *umbrina* (the arrow indicates the characteristic constriction of the filament cells); (**C**,**D**) *Desmococcus olivaceus*, genus appearance and cell details (arrow indicates central chloroplast pyrenoid); (**E**) *Bracteacoccus minor* (arrow indicates cell details and autospore); (**F**) *Stichococcus bacillaris*; (**G**) *Stichococcus mirabilis*; (**H**) *Chlamydomonas sp.* (arrow indicates one of the flagella); (**I**) *Dictyosphaerium chlorelloides* (Nauman) (arrow indicates mucilaginous tracts between cells); (**J**) *Klebsormidium* cf. *flaccidum* (Kützing); (**K**) *Mesotaenium* cf. *caldariorum*; (**L**) *Elakatothrix gelatinosa*; (**M**) *Chlorella vulgaris* (the arrows indicate in detail the morphology of the chloroplast (**A**) and the four-celled autospore (**B**)); (**N**) *Cylindrocystis brebissoni*; Bacillariophyta or diatoms: (**O**,**P**) *Hantzschia amphyoxis*, detail of siliceous frustule (**O**) and living cell (**P**); Cyanophyta (**Q–V**) (arrows indicate characteristic terminations of trichomes): (**Q**) Cyanobacterial film found on ROC church walls; (**R**,**S**) *Tychonema* cf. *bourrellyi*; (**T**) *Oscillatoria tenuis*; (**U**) *Phormidium autumnale*; (**V**) *Phormidium favosum* (Scales: **A–C**,**J**,**Q**,**R** = 20 μm; **D–I**,**K–N**,**S–V** = 10 μm; **O–P** = 5 μm).

Figure 3. Example of dominance of *Trentepohlia* cf. *umbrina* on the N-facing wall of the BAR church.

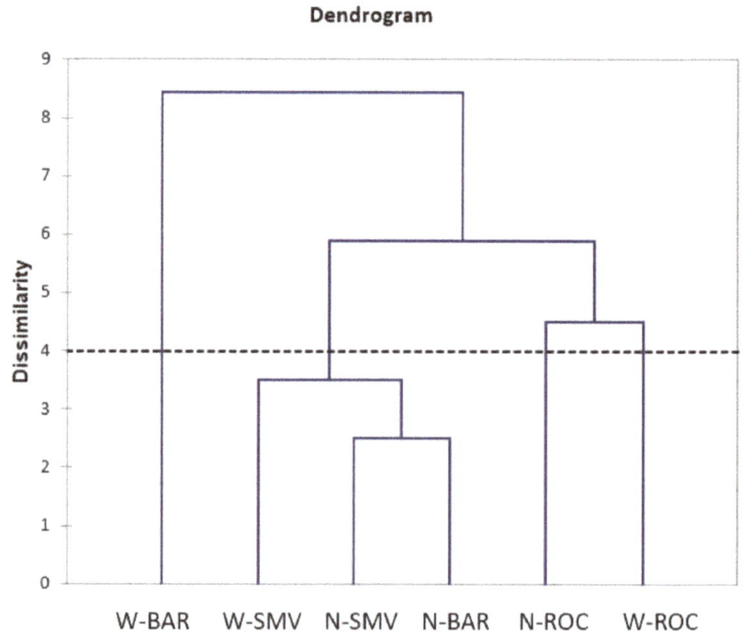

Figure 4. Agglomerative hierarchical analysis of the phototrophic component of the subaerial biofilm samples according to species composition. Dashed line indicates the cut-off point in the clustering of samples.

The species *Trentepohlia* cf. *umbrina*, *Desmococcus olivaceus*, *Stichococcus bacillaris*, and *Klebsormidium* cf. *flaccidum* were previously found to be the most representative dominant species on the exterior granite walls of the Galician Monumental Built Heritage, with *Trentepohlia* spp. characteristically found on N-facing walls [31,38], as in the SMV and BAR churches.

The dominant species were accompanied by others, mainly green algae (Chlorophyta), with a low representation and were identified in in vitro culture (Table 1). The higher species diversity on the N-facing wall of the BAR church was noteworthy. In addition, the presence on that wall of diatom species (*Bacillriophyta*), such as *Hantzschia amphyoxis* and *Achnantidium* cf. *minutissimum*, which lack desiccation resistance mechanisms, and liquid water-loving flagellate motile species, such as *Chlamydomonas sp. sensu lato*, clearly indicates increased water availability at the stone surface.

3.2. Next Generation Sequencing

Microbial community composition (of both bacteria and fungi) of the samples collected from both N- and W-facing walls of the SMV, BAR and ROC churches is represented in Figure 5A (at the phylum level) and Figure 5B (at the class level). The biofilms sampled include a large diversity of species, particularly in the bacterial domain, but also with a strong presence of lichenized fungi. The main species present are shown in Table 2. The value of sampling rural churches to analyze the biodiversity of granite colonizing organisms was previously reported [31,45] and the substrate aspect, inclination, wetness, exposure to rain, degree of disaggregation of the rock, and proximity to mortar were indicated as important factors in determining the composition of the flora.

Bacteria, which are known to discolor, disrupt, weaken and dissolve a wide variety of materials [46], proved the most diverse domain. The main phyla of bacteria that were found, which in several of the churches accounted for up to 50% of the colonization, were Proteobacteria and Bacteroidetes, followed by Acidobacteria, Plantomycetes, and Actinobacteria. The phyla Deinococcus-Thermus, Verrucomicrobia, Armatimonadetes and Cyanobacteria were also present, but were less abundant.

Within the phylum Proteobacteria, the most abundant class was Alphaproteobacteria, with Sphingomonadales and Rhodospirillales being the most abundant orders on most of the walls and with fewer representatives of Caulobacterales, Rhizobiales and Rhodobacteralesorders. Sphingomonads have been reported in a wide diversity of rocky habitats, ranging from a preserved sandstone slab at an exposed site in Oxford, UK [47], to the stone surfaces of monuments on the UNESCO World Heritage List of the Hangzhou West Lake Cultural Landscape in China [48]. The most abundant genus of the order Sphingomonadales found on the church walls was *Sphingomonas* (including the species *Sphingomonas prati, Sphingomonas hengshuiensis* and *Sphingomonas echinoides*). It was identified from biofilms from the N-facing walls of the SMV and BAR churches (31.0 and 15%), as well as from the W-facing walls of SMV, BAR and ROC (4.9, 20.1 and 10.25%, respectively). Species of the genus *Sphingobium* were also present on the W-facing wall of the BAR church (8.3%), and species of the genus *Sphingoaurantiacus*, specifically *Sphingoaurantiacus polygranulatus* was also found in the N- and W-facing wall of BAR (1.3 and 11.3%) and in the N-facing wall of SMV (8.4%). These genera are often regular inhabitants of heritage walls, due to their ability to live in nutrient-poor environments. Their presence causes a visual deterioration of the heritage due to the change of color that their presence produces in the stone, as well as the production of sphingans, a rather viscous exopolysaccharide [49–52].

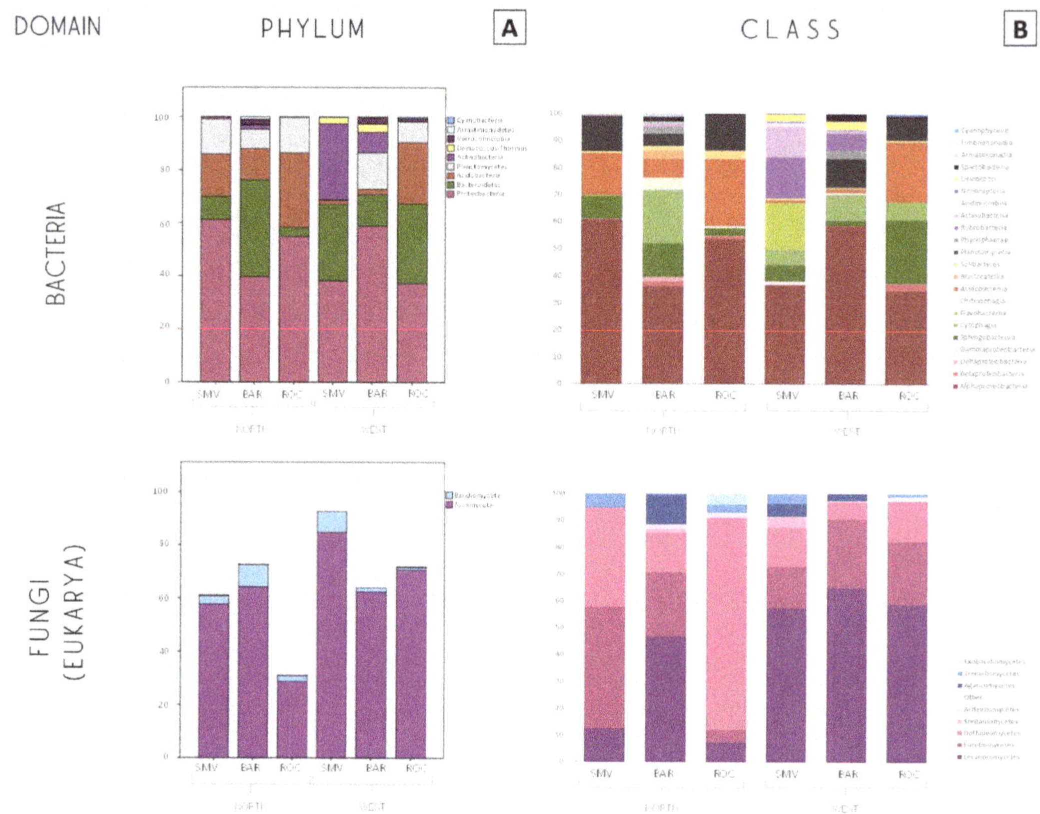

Figure 5. Relative abundances (%) of bacterial and fungi group for each phylum (**A**) and class (**B**). Only species with a frequency greater than 0.001% were considered in the analysis.

All genera of the second most abundant order in the studied biofilms, Rhodospirillales, are part of the so-called acetic acid bacteria. This order was particularly abundant on the N-facing wall of the ROC church, where the genera *Gluconacetobacter* and *Granulibacter* occur (5.7 and 12.4%). The species *Roseomonas arcticisoli* (5.7%) and *Dankookia rubra* (1%) were identified on the N-facing wall of the SMV church.

Other less abundant species of this order, such as *Acidicaldus organivorans*, occurred on the N- and W-facing walls of the BAR church (2.3 and 6.7%), as well as on the N-facing wall of the SMV church (5.2%) and on W-facing wall of the ROC church (1%). Moreover, *Roseomonas riguiloci*, *Acidisphaera rubrifaciens* and *Acidisoma sibiricum* also colonized the N-facing wall of the ROC church (4.1, 1.5 and 3%), but also the W-facing wall of this church (5%), the W-facing wall of the BAR church (1.2%) and the N-facing wall of the SMV church (1%). These species are highly tolerant of low pH conditions, which is advantageous to growth on granitic substrate [53,54] because the substrate acidity is a key factor determining which organisms will develop most successfully. In terms of biodeterioration, some consideration should be given to *Acidicaldus organivorans* as its sulfide-reducing nature contributes to mineral dissolution [55].

Table 2. Species composition and taxonomic classification of the main organism conforming the cultural heritage microbiome identified by NGS. Pink background corresponds to Bacteria phylum, while the brown background corresponds to Fungi kingdom.

DOMAIN	PHYLUM	CLASS	ORDER	NORTH (%)			WEST (%)			GENUS	SPECIES
				SMV	BAR	ROC	SMV	BAR	ROC		
BACTERIA	Proteobacteria	Alphaproteobacteria	Caulobacterales	1.2	2.4	7.9	11.0	1.3	1.9	*Brevundimonas*	*Brevundimonas albigilva*
											Brevundimonas variabilis
										Phenylobacterium	*Phenylobacterium aquaticum*
			Rhizobiales	1.4	6.0	10.3	1.6	2.4	8.7	*Salinarimonas*	*Salinarimonas rosea*
										Bauldia	*Bauldia litoralis*
										Roseiarcus	*Roseiarcus fermentans*
										Aureimonas	*Aureimonas sp.*
			Rhodospirillales	14.4	6.3	29.4	1.0	10.5	10.6	*Acidicaldus*	*Acidicaldus organivorans*
										Gluconacetobacter	*Gluconacetobacter sp.*
										Granulibacter	*Granulibacter sp.*
										Roseomonas	*Roseomonas arcticisoli*
											Roseomonas riguiloci.
										Acidisphaera	*Acidisphaera rubrifaciens*
										Acidisoma	*Acidisoma sibiricum*
										Dankookia	*Dankookia rubra*
			Sphingomonadales	41.9	17.4	3.3	16.5	40.8	12.0	*Sphingobium*	*Sphingobium sp.*
										Sphingomonas	*Sphingomonas prati.*
											Sphingomonas hengshuiensis
											Sphingomonas echinoides
										Sphingoaurantiacus	*Sphingoaurantiacus polygranulatus*
		Betaproteobacteria	Burkholderiales	/	1.8	0.8	0.1	/	2.3	*Massilia*	*Massilia sp.*
		Deltaproteobacteria	Myxococcales	0.3	1.5	0.3	/	/	0.2	*Labilithrix*	*Labilithrix luteola*

Table 2. *Cont.*

DOMAIN	PHYLUM	CLASS	ORDER	NORTH (%)			WEST (%)			GENUS	SPECIES
				SMV	BAR	ROC	SMV	BAR	ROC		
BACTERIA	Bacteroidetes	Sphingobacteriia	Sphingobacteriales	/	11.0	2.6	5.4	1.5	22.5	*Arcticibacter*	*Arcticibacter svalbardensis*
										Pedobacter	*Pedobacter tournemirensis*
		Cytophagia	Cytophagales	/	17.5	/	5.0	9.0	6.3	*Spirosoma*	*Spirosoma rigui*
										Siccationidurans	*Siccationidurans ginsengisoli*
		Flavobacteriia	Flavobacteriales	/	0.1	/	16.6	/	/	*Pricia*	*Pricia antarctica*
										Flavimarina	*Flavimarina pacifica*
	Acidobacteria	Acidobacteriia	Bryobacteraceae	0.8	2.5	3.1	0.4	0.7	0.8	*Paludibaculum*	*Paludibaculum fermentans*
			Acidobacteriales	14.9	6.2	23.5	1.0	1.2	21.3	*Granulicella*	*Granulicella acidiphila*
											Granulicella mallensis
										Silvibacterium	*Silvibacterium bohemicum*
										Edaphobacter	*Edaphobacter modestus*
										Bryocella	*Bryocella elongata*
										Terriglobus	*Terriglobus aquaticus*
											Terriglobus aquaticus
											Terriglobus roseus
	Planctomycetes	Planctomycetia	Planctomycetales	12.7	4.0	4.2	/	2.7	3.2	*Aquisphaera*	*Aquisphaera giovannonii*
										Paludisphaera	*Paludisphaera borealis*
										Singulisphaera	*Singulisphaera rosea*
		Phycisphaerae	Tepidisphaerales	/	/	/	/	2.7	/	*Tepidisphaera*	*Tepidisphaera mucosa*
	Actinobacteria	Actinobacteria	Rubrobacterales	0.1	/	/	14.5	5.8	/	*Rubrobacter*	*Rubrobacter radiotolerans*
			Kineosporiales	/	/	/	2.7	0.1	/	*Pseudokineococcus*	*Pseudokineococcus lusitanus*
			Actinomycetales	0.3	0.3	/	5.6	0.5	/	*Janibacter*	*Janibacter sp.*
										Serinicoccus	*Serinicoccus sp.*

Table 2. *Cont.*

DOMAIN	PHYLUM	CLASS	ORDER	NORTH (%)			WEST (%)			GENUS	SPECIES
				SMV	BAR	ROC	SMV	BAR	ROC		
BACTERIA	Deinococcus-Thermus	Deinococci	Trueperales	0.2	0.2	/	2.4	2.9	/	*Truepera*	*Truepera radiovictrix*
	Verrucomicrobia	Spartobacteria	Cthnoniobacterales	/	1.5	0.2	/	2.3	1.0	*Chthoniobacter*	*Chthoniobacter flavus*
	Cyanobacteria	Cyanophyceae	Nostocales	/	/	/	/	/	0.2	*Loriellopsis*	*Loriellopsis cavernicolae*
			Synechococcales	/	/	/	/	/	0.4	*Trichocoleus*	*Trichocoleus desertorum*
FUNGI (EUKARYA)	Ascomycota	Lecanoromycetes	Lecanorales	6.7	23.0	2.1	52.9	26.1	42.2	*Hypogymnia*	*Hypogymnia metaphysodes*
										Scoliciosporum	*Scoliciosporum umbrinum*
										Lecania	*Lecania cyrtella*
											Lecania erysibe
										Lecanora	*Lecanora dispersa*
											Lecanora horiza
			Pertusariales	0.9	6.6	0.2	/	0.5	/	*Lepra*	*Lepra amara*
			Teloschistales	/	1.0	/	0.2	14.7	/	*Caloplaca*	*Caloplaca maritima*
		Dothideomycetes	Capnodiales	22.3	10.0	24.5	13.4	4.1	10.6	*Catenulostroma*	*Catenulostroma protearum*
										Neodevriesia	*Neodevriesia lagerstroemiae*
										Devriesia	*Devriesia strelitziicola*
		Eurotiomycetes	Chaetothyriales	27.5	17.4	0.8	14.5	16.4	16.7	Unclas. *Herpotrichiellaceae*	*Herpotrichiellaceae sp. MUT 5408*
										Rhinocladiella	*Rhinocladiella sp.*
										Cladophialophora	*Cladophialophora sp.*
										Knufia	*Coniosporium sp. MA 4597*
	Basidiomycota	Agaricomycetes	Cantharellales	/	7.8	/	4.4	1.7	/	*Rhizoctonia*	*Rhizoctonia sp.*
		Tremellomycetes	Tremellales	3.4	0.4	1.1	3.5	/	1.0	*Fellomyces*	*Fellomyces sp.*

Within the order Caulobacterales, the species *Brevundimonas albigilva* was found to colonize the N-facing walls of the BAR and ROC churches (1.4 and 2.8) while *Brevundimonas variabilis* only appeared on the N-facing wall of the ROC church (11%). *Phenylobacterium aquaticum* was found on the N-facing walls of the SMV and ROC churches (1 and 4.6%), as well as on the W-facing wall of the ROC church (1.2). Both genera have previously been found colonizing other monuments, especially on sandstone in polluted areas, as they can survive in such environments due to their capacity to degrade hydrocarbons and use them as a carbon source [56–58].

Members of the order Rhizobiales mainly appeared on the ROC church. More specifically, *Salinarimonas rosea* was found on both the N- and W-facing walls of this church (3.2 and 1%), as well as on the N-facing wall of the BAR church (5.8%). In addition, *Bauldia litoralis* and *Roseiarcus fermentans* were also found on the ROC church, the former on both N- and W-facing walls (4 and 1.5%) and the latter on the N-facing wall (4.1%). This species produces bacteriochlorophyll *a*, and carotenoids, which have been reported to cause an aesthetic impact on the stone where the species occurs [59]. Examples of the genus *Aureimonas* were only found on W-facing walls, in all three churches (approx. 2%). This order is commonly associated with the roots of vascular plants and its occurrence may indicate an advanced level of colonization on the façades (plants rooted in the rock) but may also be due to the influence of endophytes from neighboring forest areas deposited on walls by winds. It has been found to colonize a red sandstone monolithic statue of Buddha [53] in a region of China with a subtropical monsoon climate and a large presence of higher plants.

The other classes of Proteobacteria (Beta and Deltaproteobacteria) were present in much smaller proportions. Species of the order Burkholderiales were found on the N-facing wall of the BAR church and on the W-facing wall of the ROC church, where the genus *Massilia* appeared; members of the order Myxococcales were mainly distributed on the N-facing walls of the churches, along with members of the genus *Labilithrix*.

The most abundant classes within the phylum Bacteroidetes were Sphingobacteriia, Cytophagia and Flavobacteriia. Some widely distributed species, such as *Arcticibacter svalbardensis*, which belongs to the order Sphingobacteriales, were present on the N-facing walls of all the churches (SMV, BAR and ROC: 5, 5.6, 2.4% respectively) as well as on the west-facing wall of the ROC church (9%). Other species within the same order, e.g., *Pedobacter tournemirensis*, were very abundant, but only on the W-facing wall of the ROC church (11%). Within the order Cytophagales, different species of the genus *Spirosoma* were found on both the N-and W-facing walls of the BAR church, specifically *Spirosoma rigui* (3.5%), and *Siccationidurans ginsengisoli* was only found on the W-facing wall of the ROC church (5.6%). Finally, within the order Flavobacteriales, the species *Pricia antarctica* and *Flavimarina pacifica* (6.5 and 8.8%) occurred on the N-facing wall of the SMV church.

The most common orders in the phylum Acidobacteria were Solibacterales and Acidobacteriales. The former is mainly represented by one species, *Paludibaculum fermentans*, which was only found on the N-facing walls of BAR and ROC churches (1.3 and 3%). In the case of the order Acidobacteriales, the species *Granulicella acidiphila* was found on the N- and W-facing walls of the ROC church (11.7 and 11.6%), as well as on the N-facing wall of SMV (14%). This species is clearly acidophilic and is known to produce large amounts of EPS and carotenoids [60] which can be important in terms of biodeterioration. *Silvibacterium bohemicum* was found on both the N- and W-facing walls (8.3 and 1.5%) of the ROC church. Other species of the order, such as *Edaphobacter modestus, Granulicella mallensis* and *Terriglobus aquaticus*, were found in smaller percentages in the N-facing wall of the church of BAR, while *Bryocella elongata, Terriglobus aquaticus* and *Terriglobus roseus* were found on the W-facing wall of the ROC church.

Regarding the phylum Planctomycetes, only species belonging to two classes were found: Planctomycetia and Phycisphaerae. The first class has three different widely distributed species: *Aquisphaera giovannonii* was found at all sites at a percentage abundance of between 1 and 9%, except on the W-facing wall of the ROC church; *Paludisphaera borealis* was found on the N-facing walls of the SMV and ROC churches and on the W-facing wall of the ROC church (between 1 and 3%); and *Singulisphaera rosea*, found on both the N- and W-facing walls of the BAR and ROC churches. The class Phycisphaerae was represented by a single species, *Tepidisphaera mucosa*, which occurred on the BAR church, with a percentage abundance of 2.5%, on both N- and W-facing walls.

The phylum Actinobacteria was mainly found on W-facing walls. The order Rubrobacterales was represented by species such as *Rubrobacter radiotolerans* on the W-facing wall of the SMV and BAR churches (5.7 and 2%). In addition, other species of the Rubrobacter genus were abundant on the SMV and ROC churches (between 1 and 9%). This genus has been widely linked to the biodeterioration of monuments and is known to contribute to the so-called rosy discoloration [61]. The pigment production is related to the high desiccation tolerance of this genus, which confers a selective advantage over other organisms [40]. Other orders such as Kineosporiales and Micrococcales were also abundant on the walls of the SMV church.

In relation to the less abundant phyla, Deinococcus-Thermus, which is associated with tolerance to high radiation and desiccation [50], was only represented by one species, *Truepera radiovictrix*, on the W-facing walls of the SMV and BAR churches (2.3 and 2.9%). The phylum Verrucomicrobia was represented by the species *Chthoniobacter flavus*, on both N- and W-facing walls of the BAR and ROC churches (around 1%). However, the phylum Cyanobacteria was very scarce, and the only species identified were *Loriellopsis cavernicolae* and *Trichocoleus desertorum*, both at abundances of less than 0.5%, on the W-facing wall of the ROC church. In the present study, Cyanobacteria seemed to be rather scarce, in contrast to the results of other studies investigating the diversity of microbial species found in cultural heritage formed by bricks, marbles [48], sandstones [62] and limestones [63], in which cyanobacteria constituted an important fraction of the community. Across Europe, much lower numbers of cyanobacteria are found growing on granite and sandstone than on other types of stone such as marble and limestone [12], which may indicate a preference of these species for alkaline substrates. In this study, the cyanobacteria most frequently detected by NGS was *Loriellopsis cavernicola*, which was previously found in caves in Mediterranean environments [64,65]. The ability of this species to mobilize calcium ions has been previously described [66] and may favor degradation of the substrate.

In relation to the Fungi kingdom (Eukaryota), the phyla most represented on the examined walls are the Ascomycota and Basidiomycota. Within the phylum Ascomycota, the most abundant classes were Lecanoromycetes, Dothideomycetes and Euromycetes, and the least abundant, Sordariomycetes. Lecanoromycetes is the main class conforming lichenized fungi. The most abundant species were *Lecania cyrtella*, which was only found on the W-facing walls of the SMV and BAR churches (51.5 and 19.6), *Scoliciosporum umbrinum*, which was found on the N-facing wall of the BAR church and the W-facing wall of the ROC church (16.5 and 45%), and *Hypogymnia metaphysodes*, which only colonized N-facing walls of the SMV and BAR churches (5.8 and 2.49). Other species of this phylum were also found on different walls of different churches, such as *Lecanora dispersa* (5%), *Lecania erysibe* (1%), on the W-facing wall of the BAR church, and *Lecanora horiza* (2.6%) and *Lepra amara* (5%), on the N-facing wall of the BAR church. Finally, we found a very high percentage of *Caloplaca maritima* (14.3%) on the W-facing wall of the BAR church. *Hypogymnia metaphysodes*, *Lecanora dipersa*, *Schasporom umbirum* and *Lepra amara* have previously been found on stone monuments [45,67,68]: *Lecanora dispersa* and *Lepra amara* are included among the main lichens known to colonize Galician granite [9,45] and have been associated with acidic substrates and cold and humid climates [69]. However, the usual habitat of *Lecania cyrtella*, which is commonly present on the walls of these churches, is wood, and proliferation of this species may be influenced by the presence of surrounding vegetation. Comparison of the effect of the orientation showed that Lecanoromycetes was less abundant on N-facing walls than on W-facing walls. This may be due to the fact that some lichenized fungi have a great capacity to withstand very exposed conditions [70], which may lead to successful colonization of walls that are particularly exposed to solar radiation, unlike the N-facing walls, which are always more humid and shaded. In fact, biofilms samples were taken from N- and W- facing walls of these churches, because the south- and east- facing walls, which are more exposed to solar radiation, were completely covered by lichens.

The main classes of fungi conforming the 'black fungi' identified in this study were the Dothideomycetes and Eurotiomycetes, which is consistent with the findings of previous studies [71–73]. In the Dothideomycetes class, the order Capnoidales occurred on N- and W-facing walls of all churches. *Catenulostroma protearum* occurred on the N-facing walls of all three churches (19, 3, 24% in SMV, BAR and ROC respectively) and on the W-facing wall of the ROC church (7%). *Neodevriesia lagerstroemiae* colonized both N-and W-facing walls of the SMV church (1 and 10.5%) and *Devriesia strelitziicola* colonized the N-facing walls of the SMV and BAR churches (1.8 and 5.6%). Within the class Eurotiomycetes, the most widely represented order is Chaetothyriales. Within this order, species from the family *Herpotrichiellaceae* were found to colonize the N- (12.6%) and W- (1.2%) facing walls of the SMV church and the N-facing wall of the BAR church (11%); species of *Rhinocladiella* colonized the N-facing wall of the SMV church (14.6%) and the W-facing walls of all three churches (3, 15 and 1%); species of *Knufia* colonized the W-facing wall of the SMV (10%) and ROC (2%) churches and species of the genus Cladophialophora colonized the N-facing wall of the BAR church (5.6%). The presence of species of the genera *Catenulostroma, Rhinocladiella,* and *Knufia (Coniosporium sp. MA 4597)* have previously been found to be linked to biodeterioration processes on stone monument façades [74], and its EPS linked with process of corrosion [75]. In fact the vast majority of rock-inhabiting fungi have the ability to penetrate intercrystalline, producing small holes (biopitting) or taking advantage of previous fissures [76–78]. They are also considered to be one of the most stress-tolerant eukaryotes known, in part due to EPS production and their thick cell walls [79].

It is also important to note the absence of ubiquitous genera such as *Aspergillus, Cladosporium, Penicillium,* or *Alternaria,* all of them widely described as members of the microbiome of monuments [80,81]. However, in the case of *Alternaria,* it is a genus more closely linked to Mediterranean environments, and its low presence has previously been reported for the northwestern part of the Iberian Peninsula [82], where the studied churches are located. In addition, *Aspergillus* and *Penicillium* spores have been reported to be more frequent in urban than in rural areas [83], such as the one in the study.

In relation to the phylum Basidiomycota, a low presence was detected on the ROC church. However, species of the *Rhizoctonia* genus (Agaricomycetes) were identified on N-and W-facing walls of the BAR church (7.3 and 1.6%) and on the N-facing wall of the SMV church (4.2%), and species of *Fellomyces* (Tremellomycetes) were found on the N- and W-facing walls of the SMV church (2%).

The presence of lichenized fungi and non-lichenized fungi must be taken into account in evaluating biodeterioration as several lichenized and non-lichenized fungi secrete a variety of acidic primary and secondary metabolites with chelating functions, thus affecting the integrity of the stone [16]. Excreted acids react with different ions in the substrate, leading to biocorrosion and formation of secondary mycogenic minerals [84]. Lichens mainly cause physical weathering of granitic rocks, by inducing disaggregation of surface grains and engulfing them in their thalli [8,9]. The chemical action of the lichens on the granite causes dissolution of the minerals by production of organic acids as well as by an increase in the time of contact between the rock and the attacking solution [7]. On the other hand, black fungi are known to possess thick walls and produce significant amounts of EPS as a method to reduce environmental stress. EPS production allows these organisms to survive extreme conditions, such as temperature fluctuations, water stress, UV radiation, or nutrient deficiency [85]; however, it leads to the deterioration of stone [86].

Thus, taking into consideration the organisms found on the three churches, although the total microbial diversity did not depend on the different locations, it did depend on the orientation of the walls and was highest on the N-facing walls. The values of the Shannon index (Figure 6) indicated that diversity was generally highest among the bacteria, followed by the fungi and green algae. However, regarding Bacteria diversity, no significant differences were found between locations or orientations ($p > 0.05$). In the case of fungi and green algae, a greater diversity of these organisms was found on the N-facing walls ($p < 0.05$). However, the low diversity observed in the green algae group may be due to the use of the culture-dependent technique, which generally reduces the variability detected [22]. In terms of abundance, the orders of bacteria found in highest proportion were the Sphingomonadales and the Acidobacteriales. Most of the fungi present in the microbial consortium are lichenized fungi belonging to the order Lecanorales, although non-lichenized fungi from Capnoidales and Chaetothyriales orders are also present in large proportion. Lichenized fungi were distributed differently in relation to orientation, and they were more abundant on the W-facing walls. In the case of green algae, one of the two most abundant species, *Trentepohlia* cf. *umbrina*, mainly occurred on N-facing walls, while the other, *Desmococcus olivaceus*, was found on W-facing walls. Moreover, diatoms were only present in the microbial consortium on the N-facing wall of the BAR church, which confirms the damper conditions on this wall.

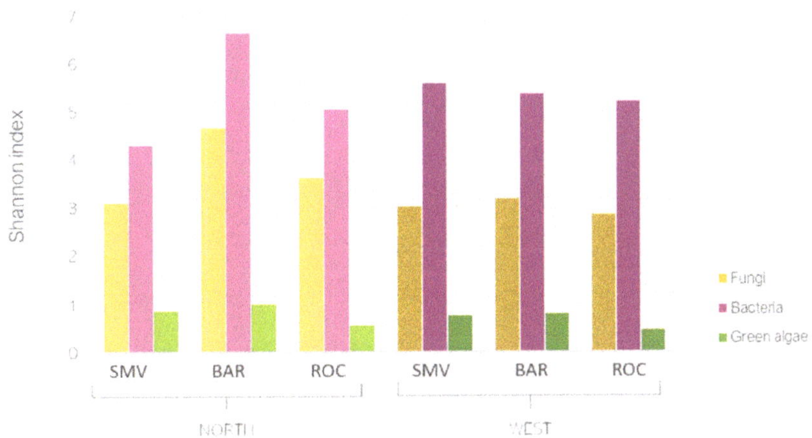

Figure 6. Shannon diversity index for green algae, fungi and bacteria colonizing the three churches under study (SMV, ROC, and BAR).

The low abundance of cyanobacteria on the church walls was noteworthy, especially relative to their presence in biofilms formed on other types of stone. Thus, for instance, cyanobacteria were the most abundant microorganisms in biofilms developed on limestone and marble [12,28]. However, the low abundance of cyanobacteria in the biofilms under study is consistent with the characteristics of biofilms on urban granite buildings in NW Spain [26]. In addition, they did not show any preferential distribution regarding orientation but did show a preferential distribution by location as they were only present on one church (ROC). The presence of cyanobacteria exclusively at ROC church may be due to the absence of close shade-producing structures, which are present in the case of the other churches. The absence of these structures results in a higher solar exposure of the west walls of ROC in relation to the west walls of SMV and BAR, favoring the development of cyanobacteria [87].

On comparing our results with those obtained by Vazquez-Nion et al. [26], who studied the microbial diversity in granitic urban heritage, we found that apart from bacteria (not considered in the aforementioned study), fungi and green algae were still the predominant groups in the biofilms. The groups most commonly present in the microbiota colonizing the churches were green algae (Chlorophyta) and Ascomycota (fungi), with cyanobacteria occurring in lower proportions. The diversity of fungi was very similar in both studies, with average Shannon index values of 3.42 (present study) and 3.43 (previous study).

Vazquez-Nion et al. [26] reported a higher diversity of algae, with Shannon index values between 1.75 and 1.54, in contrast to the values of less than 1 obtained in the present study.

4. Conclusions

The combination of culture-dependent techniques and next-generation sequencing enabled characterization, for the first time, of the microbial consortium present on historical granite buildings in rural areas. In the three buildings examined, the biofilms were mainly composed of bacteria, fungi (lichenized and non-lichenized), green algae and to a lesser extent cyanobacteria. Overall, this composition is very similar to that of biofilms on historical granite buildings in urban settings, especially in terms of fungal diversity. However, some differences were observed regarding the presence of algae, which were more abundant on urban buildings, and of lichenized fungi, which were more abundant in the rural environment.

Orientation also proved to be an important factor regarding both the diversity and abundance of microorganisms on the walls, with environmental factors associated with the north orientation favoring a higher diversity of fungi and green algae, and environmental factors associated with the west orientation determining the abundance of lichenized fungi. Among the two most abundant species of green algae, *Trentepohlia* cf. *umbrina*, mainly occurred on the N-facing wall, while *Desmococcus olivaceus* predominated on the W-facing wall.

These findings support the recommendation to characterize the microbial consortium forming the biofilms developed on the granite cultural heritage. The observed differences in both rural and urban environments, as well as regarding the orientation of different walls and the consequent exposure to different climatic factors, should be considered in selecting techniques to remove the colonizing microorganisms (such as the use of biocides), and when making any decisions related to the conservation of the buildings.

Author Contributions: Conceptualization, B.P.; methodology, B.P. and R.C.; software, R.C.; validation, E.F. and R.C.; formal analysis, E.F.; investigation, R.C., B.P. and E.F.; resources, B.P.; data curation, E.F. and R.C.; writing—original draft preparation, E.F.; writing—review and editing, B.P.; visualization, B.P.; supervision, B.P.; project administration, B.P.; funding acquisition, B.P. All authors have read and agreed to the published version of the manuscript.

Funding: This study was partly financed through project CGL2016-79778-R (AEI/FEDER, UE) and Xunta de Galicia (ED431 2018/32). E. Fuentes was financially supported by a PhD fellowship contract, MICINN-FPI (BES-2017-079927).

Institutional Review Board Statement: Not applicable.

Informed Consent Statement: Not applicable.

Conflicts of Interest: The authors declare no conflict of interest.

References

1. Nemerow, N.L.; Agardy, F.J.; Sullivan, P.; Salvato, J.A. *Environmental Engineering: Environmental Health and Safety for Municipal Infrastructure, Land Use and Planning, and Industry*, 6th ed.; John Wiley & Sons: Hoboken, NJ, USA, 2009.
2. Fassina, V.; Favaro, M.; Crivellari, F.; Naccari, A. The stone decay of monuments in relation to atmospheric environment. *Ann. Chim.* **2001**, *91*, 767–774. [PubMed]
3. Siegesmund, S.; Weiss, T.; Vollbrecht, A. Natural stone, weathering phenomena, conservation strategies and case studies: Introduction. *Geol. Soc. Spec. Publ.* **2002**, *205*, 1–7. [CrossRef]
4. Sabbioni, C. Contribution of atmospheric deposition to the formation of damage layers. *Sci. Total Environ.* **1995**, *167*, 49–55. [CrossRef]
5. Baptista-Neto, J.A.; Smith, B.J.; McAllister, J.J.; Silva, M.A.M.; Castanheira, F.S. Surface modification of a granite building stone in central Rio de Janeiro. *An. Acad. Bras. Cienc.* **2006**, *78*, 317–330. [CrossRef]
6. Silva, B.; Rivas, T. Relation Between Type of Soluble Salt and Decay Forms in Granitic Coastal Churches in Galicia (NW Spain). In *European Commission Research Workshop Origin, Mechanisms and Effects of Salts on Degradation of Monuments in Marine and Continental Environments*; Zezza, Ed.; DGXII, Protection and Conservation of the European Cultural Heritage: Bari, Italy, 1996.
7. Silva, B.; Prieto, B.; Rivas, T.; Sanchez-Biezma, M.J.; Paz, G.; Carballal, R. Rapid biological colonization of a granitic building by lichens. *Int. Biodeterior. Biodegrad.* **1997**, *40*, 263–267. [CrossRef]

8. Silva, B.; Rivas, T.; Prieto, B. Effects of lichens on the geochemical weathering of granitic rocks. *Chemosphere* **1999**, *39*, 379–388. [CrossRef]
9. Prieto, B.; Seaward, M.R.D.; Edwards, H.G.M.; Rivas, T.; Silva, B. An Fourier transform-Raman spectroscopic study of gypsum neoformation by lichens growing on granitic rocks. *Spectrochim. Acta Part A Mol. Biomol. Spectrosc.* **1998**, *55*, 211–217. [CrossRef]
10. Miller, A.Z.; Sanmartín, P.; Pereira-Pardo, L.; Dionísio, A.; Saiz-Jimenez, C.; Macedo, M.F.; Prieto, B. Bioreceptivity of building stones: A review. *Sci. Total Environ.* **2012**, *426*, 1–12. [CrossRef]
11. Crispim, C.A.; Gaylarde, C.C. Cyanobacteria and biodeterioration of cultural heritage: A review. *Microb. Ecol.* **2005**, *49*, 1–9. [CrossRef]
12. Macedo, M.F.; Miller, A.Z.; Dionísio, A.; Saiz-Jimenez, C. Biodiversity of cyanobacteria and green algae on monuments in the Mediterranean Basin: An overview. *Microbiology* **2009**, *155*, 3476–3490. [CrossRef]
13. Piñar, G.; Ripka, K.; Weber, J.; Sterflinger, K. The micro-biota of a sub-surface monument the medieval chapel of St. Virgil (Vienna, Austria). *Int. Biodeterior. Biodegrad.* **2009**, *63*, 851–859. [CrossRef]
14. Sterflinger, K.; Ettenauer, J.; Piñar, G. Microbes, Science, Art and Conservation, Who Wins the Game. In *Science, Technology and Cultural Heritage*; Rogerio-Candelera: London, UK, 2014; pp. 191–204.
15. Piñar, G.; Ettenauer, J.; Sterflinger, K. "La vie en ros": A review of the rosy discoloration of subsurface monuments. *Conserv. Subterr. Cult. Herit.* **2014**, 113–124.
16. Gadd, G.M. Geomicrobiology of the built environment. *Nat. Microbiol.* **2017**, *2*, 1–9. [CrossRef]
17. Carter, N.E.A.; Viles, H.A. Bioprotection explored: The story of a little known earth surface process. *Geomorphology* **2005**, *67*, 273–281. [CrossRef]
18. Favero-Longo, S.E.; Viles, H.A. A review of the nature, role and control of lithobionts on stone cultural heritage: Weighing-up and managing biodeterioration and bioprotection. *World J. Microbiol. Biotechnol.* **2020**, *36*, 100. [CrossRef]
19. Fuentes, E.; Prieto, B. Recovery Capacity of Subaerial Biofilms Grown on Granite Buildings Subjected to Simulated Drought in a Climate Change Context. *Microb. Ecol.* **2021**. [CrossRef]
20. Prieto, B.; Vázquez-Nion, D.; Fuentes, E.; Durán-Román, A.G. Response of subaerial biofilms growing on stone-built cultural heritage to changing water regime and CO2 conditions. *Int. Biodeterior. Biodegrad.* **2020**, *148*, 104882. [CrossRef]
21. Viles, H.A.; Cutler, N.A. Global environmental change and the biology of heritage structures. *Glob. Chang. Biol.* **2012**, *18*, 2406–2418. [CrossRef]
22. Ward, D.M.; Weller, R.; Bateson, M.M. 16S rRNA sequences reveal numerous uncultured microorganisms in a natural community. *Nature* **1990**. [CrossRef]
23. Marvasi, M.; Cavalieri, D.; Mastromei, G.; Casaccia, A.; Perito, B. Omics technologies for an in-depth investigation of biodeterioration of cultural heritage. *Int. Biodeterior. Biodegrad.* **2019**, *144*, 104736. [CrossRef]
24. Hugenholtz, P.; Goebel, B.M.; Pace, N.R. Impact of culture-independent studies on the emerging phylogenetic view of bacterial diversity. *J. Bacteriol.* **1998**, *180*, 4765–4774. [CrossRef]
25. Sterflinger, K.; Little, B.; Pinar, G.; Pinzari, F.; de los Rios, A.; Gu, J.D. Future directions and challenges in biodeterioration research on historic materials and cultural properties. *Int. Biodeterior. Biodegrad.* **2018**, *129*, 10–12. [CrossRef]
26. Vázquez-Nion, D.; Rodríguez-Castro, J.; López-Rodríguez, M.C.; Fernández-Silva, I.; Prieto, B. Subaerial biofilms on granitic historic buildings: Microbial diversity and development of phototrophic multi-species cultures. *Biofouling* **2016**, *32*, 657–669. [CrossRef]
27. Gaylarde, C.; Baptista-Neto, J.A.; Ogawa, A.; Kowalski, M.; Celikkol-Aydin, S.; Beech, I. Epilithic and endolithic microorganisms and deterioration on stone church facades subject to urban pollution in a sub-tropical climate. *Biofouling* **2017**, *33*, 113–127. [CrossRef]
28. Brewer, T.E.; Fierer, N. Tales from the tomb: The microbial ecology of exposed rock surfaces. *Environ. Microbiol.* **2018**, *20*, 958–970. [CrossRef]
29. Carballal, R.; Paz-Bermúdez, G.; Sánchez-Biezma, M.J.; Prieto, B. Lichen colonization of coastal churches in Galicia: Biodeterioration implications. *Int. Biodeterior. Biodegrad.* **2001**, *47*, 157–163. [CrossRef]
30. Prieto, B.; Silva, B.; Carballal, R.; de Silanes, M.E.L. Colonization by lichens of granite dolmens in Galicia (NW Spain). *Int. Biodeterior. Biodegrad.* **1994**, *34*, 47–60. [CrossRef]
31. Rifón-Lastra, A.; Noguerol-Seoane, Á. Green algae associated with the granite walls of monuments in Galicia (NW Spain). *Cryptogam. Algol.* **2001**, *22*, 305–326. [CrossRef]
32. Barberousse, H.; Lombardo, R.J.; Tell, G.; Couté, A. Factors involved in the colonisation of building façades by algae and cyanobacteria in France. *Biofouling* **2006**, *22*, 69–77. [CrossRef]
33. Ortega-morales, B.O.; Gaylarde, C.; Anaya-hernandez, A.; Chan-bacab, M.J. International Biodeterioration & Biodegradation Orientation affects Trentepohlia -dominated bio fi lms on Mayan monuments of the Rio Bec style. *Int. Biodeterior. Biodegrad.* **2013**, *84*, 351–356. [CrossRef]
34. Gauch, H.G.; Whittaker, R.H. Hierarchical Classification of Community Data. *J. Ecol.* **1981**. [CrossRef]
35. Rippka, R.; Deruelles, J.; Waterbury, J.B. Generic assignments, strain histories and properties of pure cultures of cyanobacteria. *J. Gen. Microbiol.* **1979**. [CrossRef]
36. Bold, H.C. The Morphology of Chlamydomonas chlamydogama, Sp. Nov. *Bull. Torrey Bot. Club* **1949**, *76*, 101–108. [CrossRef]
37. Bischoff, H.W.; Bold, H.C. Some soil algae from enchanted rock and related algae species. *Phycol. Stud.* **1963**, *6318*, 1–95.

38. Rifón Lastra, A.; Noguerol Seoane, Á. Algunas Chaetophorales (Chlorophyta) subaéreas interesantes para la Península Ibérica. *Port. Acta Biológica* **2000**, *19*, 81–90.
39. Škaloud, P.; Rindi, F.; Boedeker, C.; Leliaert, F. *Freshwater Flora of Central Europe, Vol 13: Chlorophyta: Ulvophyceae (Süßwasserflora von Mitteleuropa, Bd. 13: Chlorophyta: Ulvophyceae)*; Springer: Berlin/Heidelberg, Germany, 2018.
40. Komárek, J. *Freshwater Flora of Central Europe, Vol 13. Cyanophyta (Nostocales) (Süßwasserflora von Mitteleuropa, Bd. 13: Cyanophyta (Nostocales)*; Springer Spektrum: Berlin/Heidelberg, Germany, 2016.
41. Lange-Bertalot, H.; Hoffman, G. *Diatomeen im Süsswasser-Benthos von Mitteleuropa: Bestimmungsflora Kiesenalgen für die Ökologische Praxis. Über 700 der Häufigsten Arten und ihre Ökologie Gebundene Ausgabe*; Gantner Verlag and location: Berlin, Germany, 2011.
42. Klindworth, A.; Pruesse, E.; Schweer, T.; Peplies, J.; Quast, C.; Horn, M.; Glöckner, F.O. Evaluation of general 16S ribosomal RNA gene PCR primers for classical and next-generation sequencing-based diversity studies. *Nucleic Acids Res.* **2013**, *41*, e1. [CrossRef]
43. Edgar, R.C.; Haas, B.J.; Clemente, J.C.; Quince, C.; Knight, R. UCHIME improves sensitivity and speed of chimera detection. *Bioinformatics* **2011**, *27*, 2194–2200. [CrossRef]
44. Shannon, C.E.; Weaver, W. *The Mathematical Theory of Communication*; Univ of Illinois Press: Urbana, IL, USA, 1949.
45. Prieto, B.; Rivas, M.T.; Silva, B.M. Colonization by lichens of granite churches in Galicia (northwest Spain). *Sci. Total Environ.* **1995**, *167*, 343–351. [CrossRef]
46. Crispim, C.A.; Gaylarde, P.M.; Gaylarde, C.C. Algal and cyanobacterial biofilms on calcareous historic buildings. *Curr. Microbiol.* **2003**, *46*, 79–82. [CrossRef]
47. Cutler, N.A.; Chaput, D.L.; Oliver, A.E.; Viles, H.A. The spatial organization and microbial community structure of an epilithic biofilm. *FEMS Microbiol. Ecol.* **2015**, *91*. [CrossRef]
48. Li, Q.; Zhang, B.; He, Z.; Yang, X. Distribution and diversity of bacteria and fungi colonization in stone monuments analyzed by high-throughput sequencing. *PLoS ONE* **2016**, *11*, 1–17. [CrossRef]
49. Berdoulay, M.; Salvado, J.C. Genetic characterization of microbial communities living at the surface of building stones. *Lett. Appl. Microbiol.* **2009**, *49*, 311–316. [CrossRef]
50. Lan, W.; Li, H.; Wang, W.D.; Katayama, Y.; Gu, J.D. Microbial community analysis of fresh and old microbial biofilms on Bayon Temple sandstone of Angkor Thom, Cambodia. *Microb. Ecol.* **2010**, *60*, 105–115. [CrossRef]
51. de Leo, F.; Iero, A.; Zammit, G.; Urzì, C.E. Chemoorganotrophic bacteria isolated from biodeteriorated surfaces in cave and catacombs. *Int. J. Speleol.* **2012**. [CrossRef]
52. Kim, Y.J.; Lim, J.; Sukweenadhi, J.; Seok, J.W.; Lee, S.W.; Park, J.C.; Taizhanova, A.; Kim, D.; Yang, D.C. Genomic Characterization of a Newly Isolated Rhizobacteria Sphingomonas panacis Reveals Plant Growth Promoting Effect to Rice. *Biotechnol. Bioprocess Eng.* **2019**, *24*, 119–125. [CrossRef]
53. Bai, F.Y.; Chen, X.P.; Huang, J.Z.; Lu, Y.S.; Dong, H.Y.; Wu, Y.H.; Song, S.L.; Yu, J.; Bai, S.; Chen, Z.; et al. Microbial biofilms on a giant monolithic statue of Buddha: The symbiosis of microorganisms and mosses and implications for bioweathering. *Int. Biodeterior. Biodegrad.* **2021**, *156*, 105106. [CrossRef]
54. Mihajlovski, A.; Gabarre, A.; Seyer, D.; Bousta, F.; Di Martino, P. Bacterial diversity on rock surface of the ruined part of a French historic monument: The Chaalis abbey. *Int. Biodeterior. Biodegrad.* **2017**, *120*, 161–169. [CrossRef]
55. Johnson, D.B.; Quatrini, R. Acidophile microbiology in space and time. *Curr. Issues Mol. Biol.* **2020**, *39*, 63–76. [CrossRef]
56. Meng, H.; Zhang, X.; Katayama, Y.; Ge, Q.; Gu, J.D. Microbial diversity and composition of the Preah Vihear temple in Cambodia by high-throughput sequencing based on genomic DNA and RNA. *Int. Biodeterior. Biodegrad.* **2020**, *149*, 104936. [CrossRef]
57. Jroundi, F.; Fernández-Vivas, A.; Rodriguez-Navarro, C.; Bedmar, E.J.; González-Muñoz, M.T. Bioconservation of deteriorated monumental calcarenite stone and identification of bacteria with carbonatogenic activity. *Microb. Ecol.* **2010**, *60*, 39–54. [CrossRef]
58. Tahon, G.; Willems, A. Isolation and characterization of aerobic anoxygenic phototrophs from exposed soils from the Sør Rondane Mountains, East Antarctica. *Syst. Appl. Microbiol.* **2017**, *40*, 357–369. [CrossRef]
59. Kulichevskaya, I.S.; Danilova, O.V.; Tereshina, V.M.; Kevbrin, V.V.; Dedysh, S.N. Descriptions of Roseiarcus fermentans gen. nov., sp. nov., a bacteriochlorophyll a-containing fermentative bacterium related phylogenetically to alphaproteobacterial methanotrophs, and of the family Roseiarcaceae fam. nov. *Int. J. Syst. Evol. Microbiol.* **2014**, *64*, 2558–2565. [CrossRef]
60. Falagán, C.; Foesel, B.; Johnson, B. ranulicella acidiphila sp. nov: Novel acidobacteria isolated from metal-rich acidic waters. *Extremophiles* **2017**, *21*, 459–469. [CrossRef]
61. Laiz, L.; Miller, A.Z.; Jurado, V.; Akatova, E.; Sanchez-Moral, S.; Gonzalez, J.M.; Dionísio, A.; MacEdo, M.F.; Saiz-Jimenez, C. Isolation of five Rubrobacter strains from biodeteriorated monuments. *Naturwissenschaften* **2009**, *96*, 71–79. [CrossRef]
62. Zhang, X.; Ge, Q.; Zhu, Z.; Deng, Y.; Gu, J.D. Microbiological community of the Royal Palace in Angkor Thom and Beng Mealea of Cambodia by Illumina sequencing based on 16S rRNA gene. *Int. Biodeterior. Biodegrad.* **2018**, *134*, 127–135. [CrossRef]
63. Li, Q.; Zhang, B.; Wang, L.; Ge, Q. Distribution and diversity of bacteria and fungi colonizing ancient Buddhist statues analyzed by high-throughput sequencing. *Int. Biodeterior. Biodegrad.* **2017**, *117*, 245–254. [CrossRef]
64. Dominguez-Moñino, I.; Diaz-Herraiz, M.; Jurado, V.; Laiz, L.; Miller, A.Z.; Saiz-Jimenez, C.; Santos, J.L.; Alonso, E. Nature and origin of the violet stains on the walls of a Roman tomb. *Sci. Total Environ.* **2017**, *598*, 889–899. [CrossRef]
65. Ramirez, M.; Hernandez-Marine, M.; Novelo, E.; Roldan, M. Cyanobacteria-containing biofilms from a Mayan monument in Palenque, Mexico. *Biofouling* **2010**, *26*, 399–409. [CrossRef]
66. Popović, S.; Krizmanić, J.; Vidaković, D.; Karadžić, V.; Milovanović, Ž.; Pećić, M.; Subakov Simić, G. Biofilms in caves: Easy method for the assessment of dominant phototrophic groups/taxa in situ. *Environ. Monit. Assess.* **2020**, *192*. [CrossRef]

67. McIlroy de la Rosa, J.P.; Casares Porcel, M.; Warke, P.A. Mapping stone surface temperature fluctuations: Implications for lichen distribution and biomodification on historic stone surfaces. *J. Cult. Herit.* **2013**, *14*, 346–353. [CrossRef]
68. Halda, J.P.; Janeček, V.P.; Horák, J. Important part of urban biodiversity: Lichens in cemeteries are influenced by the settlement hierarchy and substrate quality. *Urban For. Urban Green.* **2020**, *53*, 1–6. [CrossRef]
69. Giordani, P. Variables influencing the distribution of epiphytic lichens in heterogeneous areas: A case study for Liguria, NW Italy. *J. Veg. Sci.* **2006**, *17*, 195. [CrossRef]
70. Kranner, I.; Beckett, R.; Hochman, A.; Nash, T.H. Desiccation-tolerance in lichens: A review. *Bryologist* **2008**, *111*, 576–593. [CrossRef]
71. Ruibal, C.; Gueidan, C.; Selbmann, L.; Gorbushina, A.A.; Crous, P.W.; Groenewald, J.Z.; Muggia, L.; Grube, M.; Isola, D.; Schoch, C.L.; et al. Phylogeny of rock-inhabiting fungi related to Dothideomycetes. *Stud. Mycol.* **2009**, *64*, 123–133. [CrossRef]
72. Egidi, E.; De Hoog, G.S.; Isola, D.; Onofri, S.; Quaedvlieg, W.; De Vries, M.; Verkley, G.J.M.; Stielow, J.B.; Zucconi, L.; Selbmann, L. Phylogeny and taxonomy of meristematic rock-inhabiting black fungi in the Dothideomycetes based on multi-locus phylogenies. *Fungal Divers.* **2014**, *65*, 127–165. [CrossRef]
73. Selbmann, L.; Zucconi, L.; Isola, D.; Onofri, S. Rock black fungi: Excellence in the extremes, from the Antarctic to space. *Curr. Genet.* **2015**, *61*, 335–345. [CrossRef]
74. Kirchhoff, N.; Hoppert, M.; Hallmann, C. Algal and fungal diversity on various dimension stone substrata in the Saale/Unstrut region. *Environ. Earth Sci.* **2018**, *77*, 1–10. [CrossRef]
75. Breitenbach, R.; Silbernagl, D.; Toepel, J.; Sturm, H.; Broughton, W.J.; Sassaki, G.L.; Gorbushina, A.A. Corrosive extracellular polysaccharides of the rock-inhabiting model fungus Knufia petricola. *Extremophiles* **2018**, *22*, 165–175. [CrossRef]
76. De Leo, F.; Antonelli, F.; Pietrini, A.M.; Ricci, S.; Urzì, C. Study of the euendolithic activity of black meristematic fungi isolated from a marble statue in the Quirinale Palace's Gardens in Rome, Italy. *Facies* **2019**, *65*, 1–10. [CrossRef]
77. Urzì, C.; De Leo, F.; De Hoog, S.; Sterflinger, K. Recent Advances in the Molecular Biology and Ecophysiology of Meristematic Stone-Inhabiting Fungi. *Microbes Art* **2000**, 3–19. [CrossRef]
78. Knabe, N.; Gorbushina, A.A. Territories of Rock-Inhabiting Fungi: Survival on and Alteration of Solid Air-Exposed Surfaces. *Methods Microbiol.* **2018**, *45*, 145–169. [CrossRef]
79. Gorbushina, A. Microcolonial fungi: Survival potential of terrestrial vegetative structures. *Astrobiology* **2003**, *3*, 543–554. [CrossRef]
80. Jurado, V.; Sanchez-Moral, S.; Saiz-Jimenez, C. Entomogenous fungi and the conservation of the cultural heritage: A review. *Int. Biodeterior. Biodegrad.* **2008**, *62*, 325–330. [CrossRef]
81. Sterflinger, K.; Piñar, G. Microbial deterioration of cultural heritage and works of art-Tilting at windmills? *Appl. Microbiol. Biotechnol.* **2013**, *97*, 9637–9646. [CrossRef]
82. Aira, M.J.; Rodríguez-Rajo, F.J.; Fernández-González, M.; Seijo, C.; Elvira-Rendueles, B.; Abreu, I.; Gutiérrez-Bustillo, M.; Pérez-Sánchez, E.; Oliveira, M.; Recio, M.; et al. Spatial and temporal distribution of Alternaria spores in the Iberian Peninsula atmosphere, and meteorological relationships: 1993–2009. *Int. J. Biometeorol.* **2013**, *57*, 265–274. [CrossRef]
83. Guinea, J.; Peláez, T.; Alcalá, L.; Bouza, E. Outdoor environmental levels of *Aspergillus* spp. conidia over a wide geographical area. *Med. Mycol.* **2006**, *44*, 349–356. [CrossRef]
84. Gadd, G.M. Metals, minerals and microbes: Geomicrobiology and bioremediation. *Microbiology* **2010**, *156*, 609–643. [CrossRef]
85. Selbmann, L.; De Hoog, G.S.; Mazzaglia, A.; Friedmann, E.I.; Onofri, S. Fungi at the edge of life: Cryptoendolithic black fungi from Antarctic desert. *Stud. Mycol.* **2005**, *51*, 1–32.
86. Gorbushina, A.A.; Broughton, W.J. Microbiology of the Atmosphere-Rock Interface: How Biological Interactions and Physical Stresses Modulate a Sophisticated Microbial Ecosystem. *Annu. Rev. Microbiol.* **2009**, *63*, 431–450. [CrossRef]
87. Shang, J.L.; Zhang, Z.C.; Yin, X.Y.; Chen, M.; Hao, F.H.; Wang, K.; Feng, J.L.; Xu, H.F.; Yin, Y.C.; Tang, H.R.; et al. UV-B induced biosynthesis of a novel sunscreen compound in solar radiation and desiccation tolerant cyanobacteria. *Environ. Microbiol.* **2018**, *20*, 200–213. [CrossRef]

applied sciences

MDPI

Article

Diversity of Biodeteriorative Bacterial and Fungal Consortia in Winter and Summer on Historical Sandstone of the Northern Pergola, Museum of King John III's Palace at Wilanow, Poland

Magdalena Dyda [1,2,*], Agnieszka Laudy [3], Przemyslaw Decewicz [4], Krzysztof Romaniuk [4], Martyna Ciezkowska [4], Anna Szajewska [5], Danuta Solecka [6], Lukasz Dziewit [4], Lukasz Drewniak [4] and Aleksandra Skłodowska [1]

1 Department of Geomicrobiology, Institute of Microbiology, Faculty of Biology, University of Warsaw, Miecznikowa 1, 02-096 Warsaw, Poland; asklodowska@biol.uw.edu.pl
2 Research and Development for Life Sciences Ltd. (RDLS Ltd.), Miecznikowa 1/5a, 02-096 Warsaw, Poland
3 Laboratory of Environmental Analysis, Museum of King John III's Palace at Wilanow, Stanislawa Kostki Potockiego 10/16, 02-958 Warsaw, Poland; alaudy@muzeum-wilanow.pl
4 Department of Environmental Microbiology and Biotechnology, Institute of Microbiology, Faculty of Biology, University of Warsaw, Miecznikowa 1, 02-096 Warsaw, Poland; decewicz@biol.uw.edu.pl (P.D.); romaniuk@biol.uw.edu.pl (K.R.); mciezkowska@biol.uw.edu.pl (M.C.); ldziewit@biol.uw.edu.pl (L.D.); ldrewniak@biol.uw.edu.pl (L.D.)
5 The Main School of Fire Service, Slowackiego 52/54, 01-629 Warsaw, Poland; aszajewska@sgsp.edu.pl
6 Department of Plant Molecular Ecophysiology, Institute of Experimental Plant Biology and Biotechnology, Faculty of Biology, University of Warsaw, Miecznikowa 1, 02-096 Warsaw, Poland; solecka@biol.uw.edu.pl
* Correspondence: magdalena.dyda@biol.uw.edu.pl or magdalena.dyda@rdls.pl; Tel.: +48-786-28-44-96

Citation: Dyda, M.; Laudy, A.; Decewicz, P.; Romaniuk, K.; Ciezkowska, M.; Szajewska, A.; Solecka, D.; Dziewit, L.; Drewniak, L.; Skłodowska, A.; et al. Diversity of Biodeteriorative Bacterial and Fungal Consortia in Winter and Summer on Historical Sandstone of the Northern Pergola, Museum of King John III's Palace at Wilanow, Poland. *Appl. Sci.* **2021**, *11*, 620. https://doi.org/10.3390/app11020620

Received: 1 December 2020
Accepted: 4 January 2021
Published: 10 January 2021

Publisher's Note: MDPI stays neutral with regard to jurisdictional clai-ms in published maps and institutio-nal affiliations.

Abstract: The aim of the presented investigation was to describe seasonal changes of microbial community composition in situ in different biocenoses on historical sandstone of the Northern Pergola in the Museum of King John III's Palace at Wilanow (Poland). The microbial biodiversity was analyzed by the application of Illumina-based next-generation sequencing methods. The metabarcoding analysis allowed for detecting lichenized fungi taxa with the clear domination of two genera: *Lecania* and *Rhinocladiella*. It was also observed that, during winter, the richness of fungal communities increased in the biocenoses dominated by lichens and mosses. The metabarcoding analysis showed 34 bacterial genera, with a clear domination of *Sphingomonas* spp. across almost all biocenoses. Acidophilic bacteria from Acidobacteriaceae and Acetobacteraceae families were also identified, and the results showed that a significant number of bacterial strains isolated during the summer displayed the ability to acidification in contrast to strains isolated in winter, when a large number of isolates displayed alkalizing activity. Other bacteria capable of nitrogen fixation and hydrocarbon utilization (including aromatic hydrocarbons) as well as halophilic microorganisms were also found. The diversity of organisms in the biofilm ensures its stability throughout the year despite the differences recorded between winter and summer.

Keywords: cultural heritage; biodeterioration; biodiversity of microorganisms; stone surfaces; historical sandstone; next-generation sequencing

1. Introduction

Microorganisms have the ability to colonize different organic and inorganic materials used for the construction of monuments [1], surfaces of historical glasses [2], or historical sandstone and limestone objects [3–5]. Microbial colonization of cultural heritage objects determines deterioration processes, which are also enhanced by a wide range of abiotic (physical and chemical) factors, such as humidity, insolation, temperature, exposure condition, and chemical composition of the air [6,7]. Microorganisms commonly found in the environment can result in damage to various materials including sandstone [8,9]. These undesirable degradative processes are called biodeterioration [10] and may affect valuable objects of cultural heritage [11].

Biodeterioration of stone surfaces mainly depends on their porosity, roughness, and mineral composition [12]. The porosity and roughness affect the rate of microbial colonization for physical reasons, while the mineral composition of stones determines the availability of nutrients for different groups of colonizing microorganisms. Generally, stones pioneering microorganisms mainly belong to chemolithotrophs, chemoorganotrophs, and phototrophs [13,14] and their activity may lead to the chemical dissolution of the stone material by production and secretion of dissolution agents such as organic and inorganic acids and ligands [15].

Besides the above-listed abiotic factors affecting the presence and activity of various bacterial and fungal groups inhabiting the historical objects, the presence of more complex organisms, including lichens, mosses, and algae, is also important. These lithobiontic, epilithic and endolithic organisms are considered to be the main factors involved in the deterioration of historical stone objects [5,16–18]. A wide variety of lichens, mosses and algae have been found on rocks, historical mural paintings, as well as stone monuments and buildings. However, bacteria and fungi still play a key role in the primary colonization and initial destruction of such facilities and objects [19–23].

Generally, data regarding the biodiversity and the biological activity of biofilms participating in stone deteriorating processes are scarce and incomplete. Molecular genetic analyses have been performed over the last ten years and they have proved to be valuable. Good examples are the biodiversity studies performed on biofilms occurring on different stone surfaces located inside and outside historical buildings, caves, and grottoes [7,9,23,24]. Further studies were performed using next-generation sequencing analyses of whole microbial communities to establish microbial biodiversity in spatial structure (2D) of the lithobiontic biofilm on sandstone surfaces [9]. These analyses enhance our knowledge regarding the microbial diversity on historical stone surfaces. Yet, there is still a considerable lack of systemic studies describing the seasonal changes of microbial community composition in situ in relation with biodeterioration patterns (BPs) or with anthropogenic factors, e.g., air pollution. The results of such studies will be an indication for conservators which biocide to use depending on the season of work or BPs type.

In the presented study, we described the changes in the composition of the microbial community in situ in biocenosis of lithobiontic communities in subsequent winter and summer seasons on historical sandstone of Pergola in Museum of King John III's Palace at Wilanow (Poland). The specific objective of this study was to identify microorganisms and show the relationship between microbiological depending on the of BPs.

2. Materials and Methods

2.1. Characteristic of Samples and Sampling Location

All analyses were conducted for the historical Northern Pergola in Wilanow gardens in Warsaw (Poland). Pergola was designed by Franciszek Maria Lanci and built in 1852 on the extension of the northern wing the Museum of King John III's Palace, located between courtyard and garden. Pergola was made of sandstone with iron gates and fencing spans. The sandstone slabs in Pergola were not cleaned and underwent maintenance in recent years. Furthermore, visible biodeterioration patterns (BPs) were monitored on Northern Pergola surfaces for the last four years. As a result of these observations, it was found that the pattern of biodeterioration is constant and does not change from years. Sampling for quantitative analysis of microbial colonization of surface, microbial biodiversity analysis, identification of visible organisms on Pergola surfaces and SEM images were performed in winter (December) and summer seasons (June). Analyses were carried out for eight samples that display different morphology and BPs. Characteristic biocenoses areas were documented by color photography and by the scanning electron microscopy (SEM) images. Documentation was repeated over time to demonstrate seasonal changes in the studied area. All samples were taken from the north-facing side of the Northern Pergola and between 30 to 70 cm above the ground. Samples were scraped with sterile scalpels to a depth of up to 5 mm, and immediately placed in sterile plastic vessels. During the months in which the analyses were carried out, the physicochemical parameters of the air were as follows: the average daily value of relative humidity in June was 56% (ranged between 25 to 96%) and temperature ranged between 9 to 29 °C (average daily value 19 °C). In December, the relative humidity mean value was higher: 83% (ranged between 60% and 96%) and temperature ranged from −9 to 11 °C (average daily value 1 °C) (data from https://sggw.meteo.waw.pl/hist.pl). These parameters are characteristic for moderate transient climate with seasons.

2.2. Mosses and Lichens Identification Methodology

The identification of lichens and mosses was based on the diagnostic characters of the different species [25,26], concerning form of growth, color and surface structures. Identification was generally performed in situ. When laboratory observations were necessary, adequate fragments of lichen or mosses were carefully removed with a scalpel without damaging the rock surface. Fragments were placed on sterile Petri dishes covered with cellulose filters, moisturized with diluted (1:20) Murashige Skoog solution, sealed with Parafilm, and processed in the laboratory within 12 h. Samples of lichens and mosses were observed (if necessary) under fluorescent microscope in visible or ultraviolet light.

2.3. Microscopic Observations of the Biofilm Surfaces through Scanning Electron Microscopy (SEM)

Biofilm samples (of approx. $2cm^2$) for microscopic imaging were gently collected in winter and summer seasons from different sampling sites on surfaces of the Northern Pergola. The preparations were fixed in formaldehyde vapor in desiccator for 3 weeks in the presence of silica gel desiccant. Prior to observation, the samples were sputtered with gold. Preparations were viewed in a scanning electron microscope LEO 1430VP (LEO Electron Microscopy).

2.4. Quantitative Analysis of Particulate Matter (PM) Concentrations in Air

The DustTrak Monitor 8533 (TSI) was located 1.5 m above the ground surface inside of the Northern Pergola. The device measures the following dust fractions: PM1, PM2.5, PM4, PM10 and TOTAL; where PM means particulate matter with an aerodynamic diameter grain size expressed in µm. The term "TOTAL" means the total dust, which is all dust with an aerodynamic diameter larger than 10 µm. Each dust measurement lasted for 5 min and was done in triplicate in each season. Results are shown as average values. The Student's *t*-test was used to evaluate differences in the PM concentration of the air between the two seasons. Results with $p \leq 0.05$ were considered statistically significant.

2.5. Measurements of the Chemical Composition of Air

At least 14 measurements of the air in the gardens surrounding the Northern Pergola were done using GASMET DX–4000 per season in order to study the chemical characteristics of air. In each series of measurements readings concentration were performed for 37 of selected gases by determining the infrared spectrum using Fourier transformation (FT-IR). Each reading was collected for 24 s. Results are shown as average values with standard deviations and were analyzed statistically using the two-tailed unpaired Student's *t*-tests to determine differences. Results with $p \leq 0.05$ were considered statistically significant.

2.6. DNA Isolation and Purification

Total DNA isolation from environmental probes was performed using Power Soil® DNA Isolation Kit (Mo Bio Laboratories Inc., Carlsbad, CA, USA). Then, obtained DNA was additionally purified using Wizard® DNA Clean-Up System (Promega, Madison, WI, USA).

2.7. Amplicon Preparation

For the amplicon preparation in PCR reaction the following primer pairs were used: 16S_V3-F: 5′ TCGTCGGCAGCGTCAGATGTGTATAAGAGACAGCCTACGGGNGGCWG CAG 3′ and 16S_V4-R: 5′ GTCTCGTGGGCTCGGAGATGTGTATAAGAGACAGGAC-TACHVGGGTATCTAATCC 3′ targeting the variable region V3 and V4 of a bacterial 16S rRNA gene, and ITS_F: 5′ TCGTCGGCAGCGTCAGATGTGTATAAGAGACAGCTTGGT-CAT 3′ and ITS_R: 5′ GTCTCGTGGGCTCGGAGATGTGTATAAGAGACAGGCTGCGTTC TTCATCGATGC 3′ targeting the ITS-1 region of fungal ribosomal RNA unit. Each reaction was prepared using KAPA HiFi polymerase (KAPA Biosystems, Wilmington, MA, USA) in a Mastercycler Nexus GX2 termocycler (Eppendorf, Hamburg, Germany). After 3 min of denaturation of DNA in 95 °C, 30 cycles including: denaturation (95 °C, 30 s), primer annealing (60–65 °C, 30 s) and DNA synthesis (72 °C, 30 s) were set. The last cycle was finished by 5 min of the final synthesis (72 °C). Each PCR reaction was repeated in triplicate and then each three probes were mixed and used for the sequencing.

2.8. DNA Sequencing

Illumina Nextera XT adapter overhang nucleotide sequences were included to the gene-specific sequences to allow further sample indexing. An amplicon library was sequenced on Illumina MiSeq instrument in the DNA Sequencing and Oligonucleotide Synthesis Laboratory-oligo.pl IBB PAS using the v3 600 cycle chemistry kit in paired-end mode, which allowed generation of long paired reads.

2.9. Processing of Raw Amplicon Reads

Raw reads were subjected to quality filtering using Prinseq-lite (v0.20.4) [27] and QIIME 2 package (v08.2018) [28] prior to analysis. During filtering reads shorter than 100 bp were excluded from further analysis. The remaining sequences were trimmed at 3′ end, using a window size of 20 bp and Phred quality score ≥Q30. Processed reads were imported into QIIME 2 and run through Dada2 library for denoising, merging, removing chimeras and dereplication, in order to obtain amplicon sequence variants (ASV). Taxonomy was assigned for each of the obtained ASVs using Naive Bayes classifier (QIIME 2 package) with Silva 132 database and UNITE v7.2 reference datasets [29,30]. The numbers of sequences after each step are shown in Table S1.

Rarefaction curves of observed amplicon sequence variants (ASV) are presented in the Figure S1, and indicate that sufficient readings have been obtained to perform reliable analyses.

QIIME 2 was also used to remove the order Chlorophyta and mitochondria family derived from the analyzed 16S rDNA amplicons set to avoid influence of eukaryotic sequences on biodiversity composition [31,32]. Differences of the biodiversity between samples collected in winter and summer were analyzed for two taxonomic levels: family and genus. Presence of all families, where abundance in samples was higher than 3%, are shown in stacked-bar plots for bacterial and fungal amplicons, respectively. The two additional groups "other" and "unspecified" were created to collect families with an abundance lower than 3% (the first one) and those that had their family unspecified or uncertain (i.e., 'f__' or 'Incertae sedis') or couldn't be assigned to any reference sequence (the latter one). In the second analysis, to show the dominating genera in each sample, only known genera were used. Analyses of bacterial and fungal diversity at lower taxonomic levels (at the phylum and class levels) are presented in the Supplementary Figures S2 and S3.

Alpha diversity indices (Shannon, Simpson), and the richness estimator (Chao1) of bacterial and fungal amplicons were performed in R with the use of Phyloseq's v1.28 (http://dx.plos.org/10.1371/journal.pone.0061217) plot_richness function and adjusted with ggplot2 v3.2.1 package (https://doi.org/10.1007/978-3-319-24277-4). Principal Coordinate Analyses (PCoA) were generated to visualize the distribution of the microbial diversity across samples.

Statistica 13.1 (Statsoft) was used to determine the correlation between the presence of mosses or lichens and the number of sequences of dominant bacterial and fungal taxa on the sandstone surfaces.

2.10. Accession Numbers of Nucleotide Sequences

Raw sequences obtained in this study have been deposited in the Sequence Read Archive (SRA) as BioProject (accession number: SRP100727, https://www.ncbi.nlm.nih.gov/sra/SRP100727). In particular, samples with the following accession numbers: SAMN06447814-SAMN06447821, SAMN06447826-SAMN06447833, SAMN06447839-SAMN06447846 and SAMN06447847-SAMN06447854 correspond to the raw 16S rDNA amplicons samples 1–8 collected in winter and summer, as well as raw ITS1 amplicons for samples 1–8 collected in winter and summer, respectively.

2.11. Isolation and Analysis of Acidification Properties of Bacterial Strains

Bacterial strains from the Northern Pergola were isolated and examined using the plate method. Swabs were taken from different sampling sites on the Pergola surfaces from areas of 25 cm^2 each. The sampling swabs were immersed in sterile saline solution (2 mL) in laboratory tubes, then shaken and spread (0.1 mL) onto agar plates dedicated to determining acidifying microorganisms (Blickfeldt medium (BTL): yeast extract 2.5 g/L, peptone 10 g/L, glucose 10 g/L, lactose 10 g/L, $CaCO_3$ 5 g/L, agar 20 g/L). The agar plates were incubated at 25 $\pm$ 0.5 °C for up to 7 days. All tests were done in triplicate. Different isolated bacteria strains obtained from the Blickfeldt plates in winter and summer seasons respectively were passaged twice on fresh plates to obtain pure cultures before experiment. The acidifying ability of different bacterial isolates were assayed in Blickfeldt liquid medium (yeast extract 2.5 g/L, peptone 10 g/L, glucose 10 g/L, lactose 10 g/L, pH 6.7). The test was performed on 96-well plates in 200 μL medium. Medium was inoculated with overnight bacteria cultures to the optical density 0.1. Cultures were grown with shaking (160 rpm) at 26 °C. After 24, 48 and 72 h of cultivation pH was measured using microelectrodes. Optical density of cultures was also measured at wavelength 600 nm in TECAN microplate reader (data not shown). All tests were done in triplicate and reported results are shows as average value.

3. Results

3.1. Biodeterioration Patterns

The macroscopic observations of the Northern Pergola revealed that the biological colonization was more mature and complex on the north-facing side compared to the southern exposure. Apart from the visible discoloration, various multicolored, powdery, crusty and shrubby biocenoses were observed on the sandstone surfaces (Figure 1).

Figure 1. General view of the Northern Pergola in winter from the southern (**a**) and the north-facing side (**b**).

Samples that display different biodeterioration patterns (BPs) were taken (except sample no 2) from vertical surfaces and from a similar height from the ground (Figure 2). According to the scale used by Caneva et al. [33] both insolation and water availability of the analyzed samples were medium high. Based on these macroscopic examinations, eight different sampling sites from the northern exposition were selected for the diversity analyses.

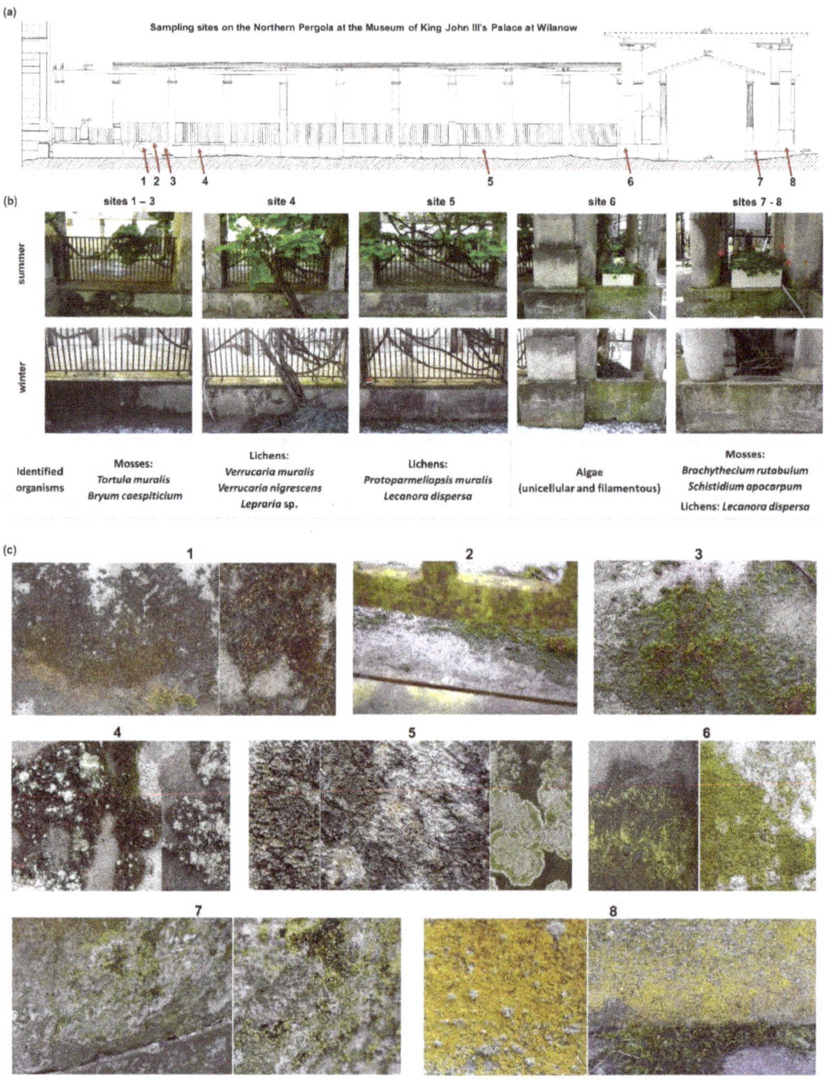

Figure 2. Architectural chart of the Northern Pergola at the Museum of King John III's Palace at Wilanow with sampling locations (**a**), most important identified characteristic species of organisms inhabiting sandstone surfaces (**b**) and biodeterioration patterns of the sandstone surfaces (**c**).

We observed various BPs [34] of the sandstone surfaces (Figure 2):

- green biofilms (GB) in sites 6 and 7.
- yellow biofilms (YB) with lichen crusts (LC) and mosses (M) in site no. 8.
- mosses (M) in sites 1, 2, and 3.

- mosses (M) with lichen crusts (LC) in site no. 4.
- lichen crusts (LC) in site no. 5.

The most shadowed sites (sites 1, 2, and 3) were each covered with a continuous thick layer of mosses, mainly *Tortula muralis* and *Bryum caespiticium* (Figure 2). The next two sites (sites 4 and 5) were mostly colonized by lichens *Protoparmeliopsis muralis* and *Lecanora dispersa* (Figure 2) with traces of *Candelariella aurella* and *Caloplaca citrina*. The other sites were dominated by unicellular and filamentous algae (site 6), mosses (sites 7 and 8)-*Brachythecium rutabulum*, *Schistidium apocarpum*, *Bryum argenteum*, and *Homalothecium sericeum* and lichens *Lecanora dispersa* (Figure 2).

The scanning electron microscopy (SEM) technique was used to visualize porosity and damages of the sandstone surfaces caused by microbial communities (Figure 3a). Detailed analyses of SEM images (Figure 3b) confirmed that organisms, including bacterial structures, are present on sandstone surfaces regardless of season. This is due to the adaptation of microorganisms to low temperatures, as well as the effect of climate warming.

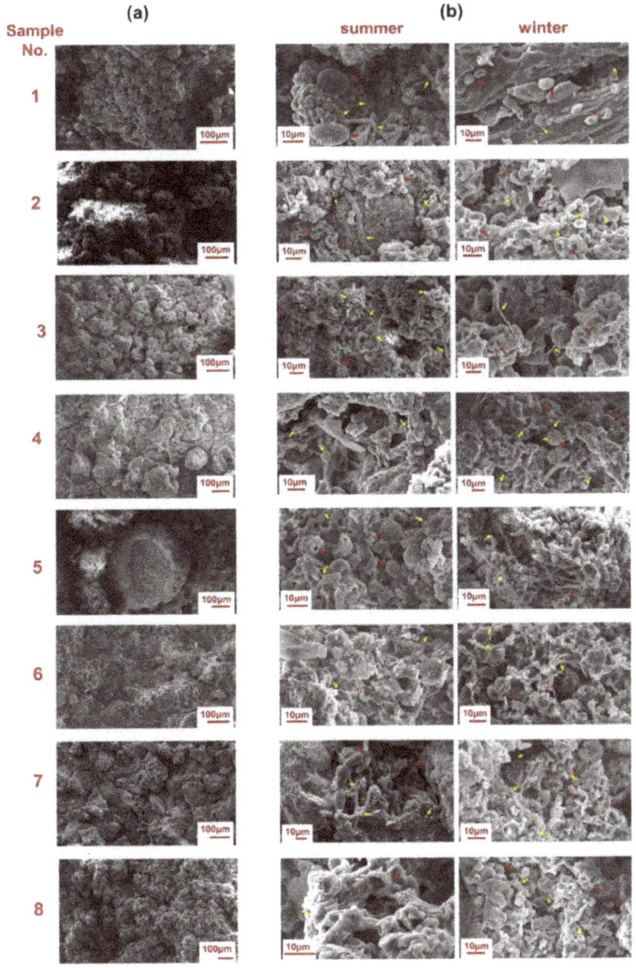

Figure 3. Scanning Electron Microscopy (SEM) images of lithobiontic organisms (red arrows) and of microorganisms (yellow arrows) inhabiting historical Pergola. SEM images of larger biofilm areas (**a**) and SEM images at higher magnification for biofilms samples analyzed in summer and winter (**b**).

Nevertheless, microscopic observations (SEM images) of samples taken in winter as compared to summer samples allowed to visualize finer sediment structures in the analyzed material in samples 4–8. These differences can result from night frosts, which can affect both the weathering processes and the biodeteriogen consortia themselves, such as cell shrinkage.

The visible biocenoses showed different appearances during winter and summer seasons not only in sites 4 and 5 inhabited by *Caloplaca citrina*–species known from their ability to produce yellow-red pigments [35]. They were distinctly green during summer while brownish and greyish colors were more common in winter (Figures 1 and 2). On the contrary, there were no pronounced changes in look of lichens. The composition of lichens also did not change within subsequent seasons.

The visible color changes of BPs depending on the seasons became the starting point to determine their cause. Since Warsaw is one of the cities with the highest air pollution indexes in Poland, in the first stage we measured concentration of air pollutants in winter and summer. In the second stage, a metabarcoding analysis was performed to obtain data regarding the seasonal differences in diversity of microbial consortia inhabiting the mature biofilms on historical sandstone surfaces.

3.2. Abiotic Factors Affecting Biodeterioration

Our results demonstrated relevant differences of concentrations in particulate matter (PM) fractions of dust in the palace garden's air (e.g., 32 μg/m^3 PM 10 in winter versus 20 μg/m^3 PM 10 in summer) (Figure 4). The higher concentration of PM in December confirms the presence of London smog in Warsaw during the heating period.

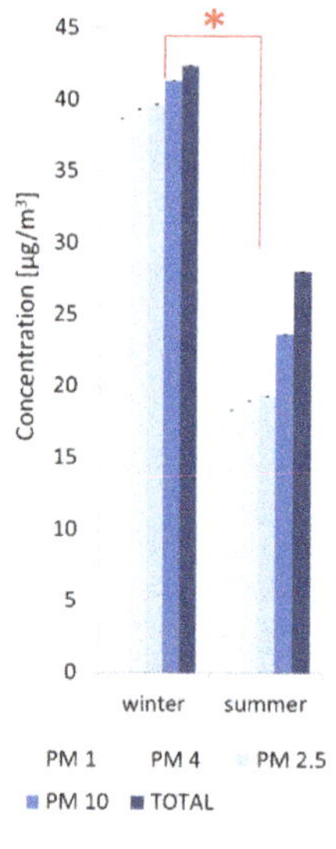

Figure 4. Particulate matter (PM) concentrations in the air in summer and winter in the gardens surrounding the Northern Pergola at the Museum of King John III's Palace at Wilanow. The PM abbreviation means particulate matter with an aerodynamic diameter grain size expressed in μm. The data represent averages with standard deviation. * -Indicates significant differences according to the Student's *t*-test ($p \leq 0.05$) when comparing PM concentrations between winter and summer.

The results were consistent with the data from the measuring station located in Ursynów, near the Museum of King John III's Palace at Wilanow: mean daily value of PM10: 33.77 μg/m^3 in December and 22.09 μg/m^3 in June; (data from: http://powietrze.gios.gov.pl/pjp/archives?lang=en). In addition, PM measurements were performed with the same measuring device as those installed in Polish measuring stations. Differences in numerical data were due to the distance of the measuring station from the Northern Pergola and due to different measurement methodology-at the stations data are collected continuously while we made only temporary measurements. Nevertheless, the same trends of seasonal changes were observed.

In addition, in June we noticed a significant increase in concentrations of nitrogen dioxide, formaldehyde and benzene-compounds characteristic for photochemical smog. The results were consistent with the data from the measuring station located in Ursynów-33 $\mu g/m^3$ NO_x in June and 26 $\mu g/m^3$ NO_x in December (data from: http://powietrze.gios.gov.pl/pjp/archives?lang=en). A measuring station on Ursynow district collected data only for nitrogen oxide, but we observed the same seasonal dependencies.

Furthermore, chemical analysis of air (Figure 5) in the garden show that the highest concentration had: methane 2 part per million (ppm) (~1.31 mg/m^3) in both seasons, carbon monoxide and propane 1–1.5 ppm (~1.15–1.72 mg/m^3 and 1.80–2.70 mg/m^3 respectively), ethylene and 1,3-butadiene 0.5–1.0 ppm in winter (~1.15–1.72 mg/m^3 and 2.19–3.31 mg/m^3 respectively) (Figure 5). Concentration of carbon dioxide was high in both seasons (416 and 417 ppm; about 749.5 mg/m^3, data not shown) and other chemicals, except sulfur dioxide, were below 0.5 ppm. Correlation between concentration of carbon dioxide and methane (two hundred times higher concentration of CO_2 than CH_4) indicated that performed measurements were correct.

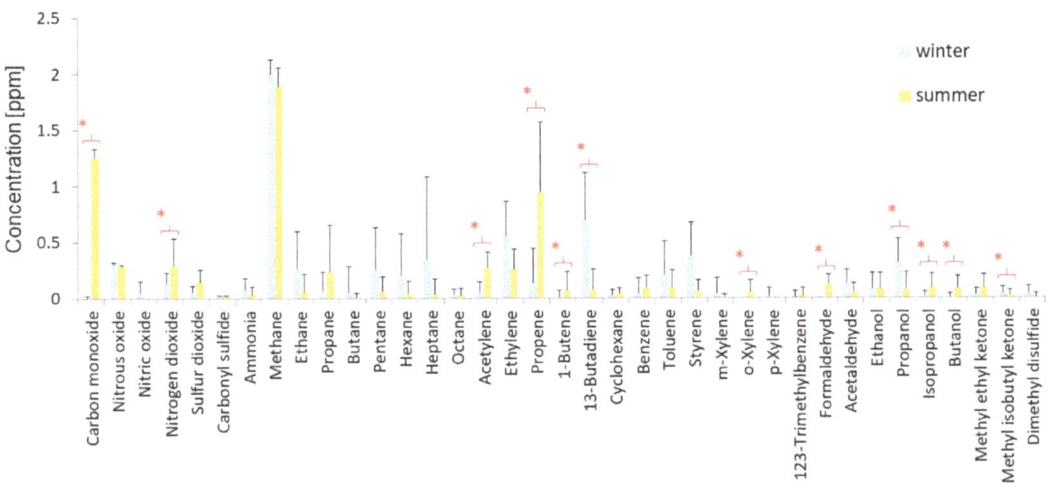

Figure 5. Concentrations in ppm of selected chemical compounds in the air in the Northern Pergola at the Museum of King John III's Palace at Wilanow. The data represent averages with standard deviation. * -Indicates significant differences according to the Student's *t*-test ($p \leq 0.05$) when comparing the concentration of selected chemical compounds between winter (blue bars) and summer (yellow bars).

Summarizing, the obtained results of seasonal changes in concentrations of airborne pollutants as well as seasonal climate changes (described in the Section 2.1) may affect visible color changes of BPs on historical sandstone surfaces. The second important factors may be related with seasonal changes in the diversity of microorganisms inhabiting biofilms, so the taxonomic diversity of the bacteria and fungi have been analyzed using the new generation sequencing methodology (matebarcoding).

3.3. Seasonal Changes of Microbial Biodiversity

3.3.1. Alpha Diversity of Bacterial and Fungal Communities

To determine significant changes in microbial diversity associated with subsequent seasons a high-throughput sequencing (Illumina technology) of PCR amplicons (of marker genes, i.e., regions V3/V4 of 16S rRNA genes for bacteria and ITS1 region for fungi) was applied. Analysis of preliminary reads showed a high diversity of the bacterial communities, with a total of 1,551,839 reads belonging to 15,310 amplicon sequence variants (ASVs), and 697,259 sequences finally analyzed. The number of reads for fungal biodiversity analysis was higher–4,389,046 reads with 2,640,212 finally analyzed sequences that clustered into 1706 ASVs. The numbers of sequences excluded from the samples are shown in Supplementary Table S1. DNA sequencing data of analyzed samples revealed the presence of 177 different bacterial genera (without uncultured or unidentified genera), belonging to 32 different classes, and 118 fungal genera (excluding unidentified genera), belonging to 29 classes (Supplementary Figure S2). Furthermore, the percentage of identified bacterial taxa was higher compared to fungal taxa, which is associated with a still intensively verified and supplemented classification within the fungi kingdom [36,37].

We measured alpha diversity using the Chao1, Shannon, and Simpson indices to evaluate differences between the microbial richness and biodiversity of the samples collected in winter and summer seasons (Figure 6). The Chao1 index, used to calculate the microbiological richness, ranged from 100 to 1000 in summer and from 100 to 1600 in winter for bacterial amplicons, and from 80 to 400 in summer and from 120 to 820 in winter for fungal amplicons. Our results also showed that the bacterial communities are more diverse than the fungal consortia (higher Shannon and Simpson indices for bacterial amplicons). In addition, both the diversity and richness of bacteria and fungi show a greater range of indices values for samples taken in winter compared to those analyzed in summer.

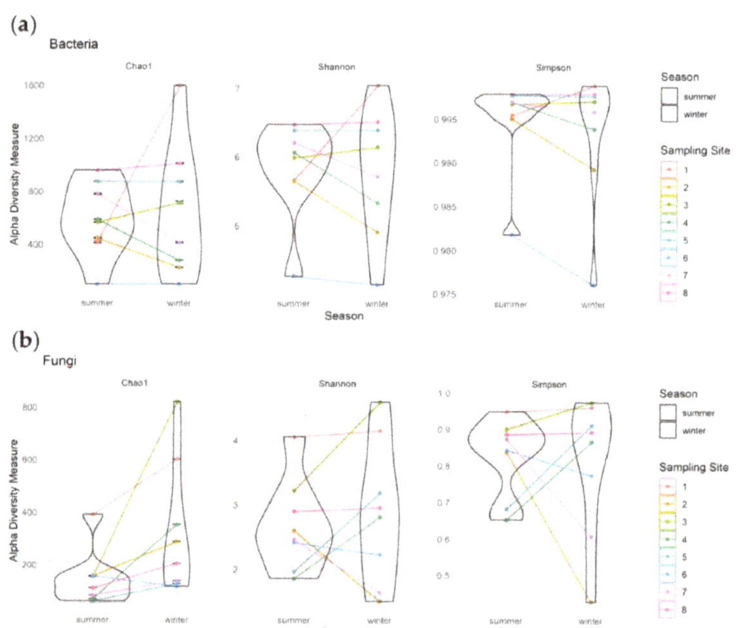

Figure 6. Alpha diversity of bacterial (**a**) and fungal (**b**) communities. For all analyzed samples, Chao1, Shannon and Simpson indices were calculated. The samples were colored and connected by sampling site, as well as grouped by season as indicated in legend.

The comparison of alpha diversity in samples collected in summer and winter in a pairwise manner allowed to identify microbiologically stable biocenoses. In case of bacterial communities (Figure 6a), these seem to be stable in sites 5 and 8 (with lichen crusts). Moreover, these stable biocenoses had also the highest values of analyzed indices in both seasons. Our results showed that other biocenoses were affected by seasons. Sites

2, 4, and 7 had lower bacterial community richness and diversity in winter comparing with summer, while the opposite effect was observed in the site no 1. Higher bacterial community richness in winter and similar diversity in both seasons was observed in site no 3, while in site no 6 with green biofilms, the richness (Chao1 indices) was stable and diversity (Simpson indices) decreased in winter.

In the case of fungi, richness (Chao1 indices) increased during winter in all sites except no 6 (Figure 6b). Despite a decrease in richness in that site in winter, as well as the increase of richness in sites no 1 and 8, the mycological diversity (Shannon and Simpson indices) in these samples was stable. The evenness (Simpson indices) of fungal communities in sites 3, 4, and 5 increased in winter, while the opposite effect was observed in sites 2 and 7.

3.3.2. Seasonal Biodiversity of Bacterial and Fungal Communities

Dominating bacterial families (Figure 7a) identified in both seasons in stable biocenoses in sites 5 and 8 (according to alpha diversity analysis) belonged to: Hymenobacteraceae (both sites), Sphingomonadaceae (both sites), Geodermatophilaceae (site 8), Microbacteriaceae (site 8), Nocardioidaceae (site 8), Rhizobiaceae (site 5), Sphingobacteriaceae (site 8) and Trueperaceae (site 5).

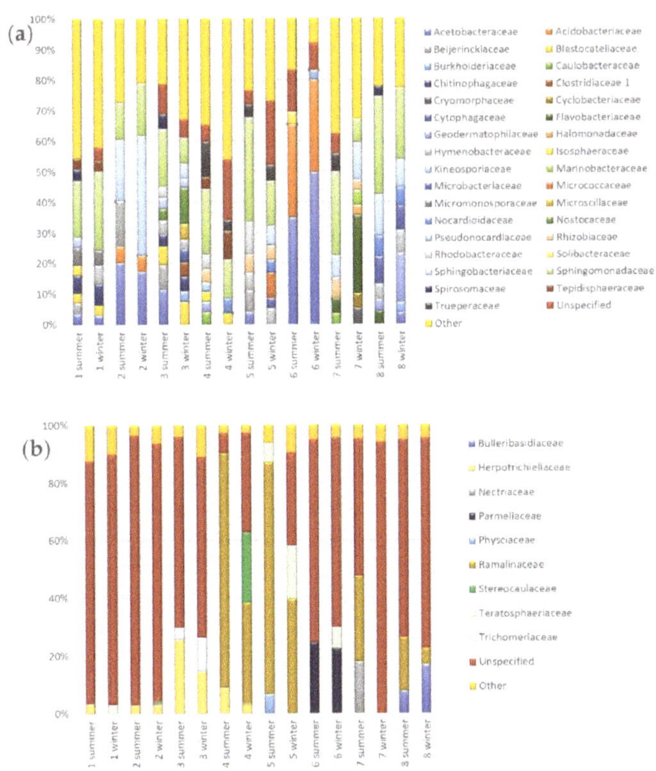

Figure 7. Seasonal biodiversity based on bacterial and fungal family composition. Two stacked-bar plots show percentage composition of bacterial (**a**) and fungal (**b**) families identified across samples in summer and winter. Taxonomy was assigned using Naive Bayes classifier (QIIME 2 package) with Silva 132 database and UNITE v7.2 reference datasets. The total number of 33 bacterial and 9 fungal families constituted more than 3% in overall compositions. Two additional and artificial groups were created to include the families with abundance lower than the threshold (*Other* below 3%) and the ones with unspecified (i.e., '*f__*') or uncertain (i.e., '*Incertae_sedis*') taxonomic position assigned (*Unspecified*).

Biocenosis in site 6 was strictly dominated in both seasons by Acetobacteraceae and Acidobacteriaceae while in site 2, in addition to these two families, also Sphingobacteriaceae and Sphingomonadaceae were presented in both seasons with high abundance. In site no 1, bacterial sequences that presented with high abundance in both seasons belonged to Acetobacteraceae, Blastocatellaceae, Chitinophagaceae, Micromonosporaceae, Sphingomonadaceae, and Tepidisphaeraceae families, while in site no 3 the following families dominated in both seasons: Blastocatellaceae, Chitinophagaceae, Hymenobacteraceae, Nostocaceae, Pseudonocardiaceae, and Sphingomonadaceae. In the last two sites, i.e., 4 and 7, there were only three and four families identified in both seasons respectively. These were Sphingomonadaceae, Tepidisphaeraceae and Trueperaceae in site 4 and Flavobacteriaceae, Rhizobiaceae, Sphingobacteriaceae and Sphingomonadaceae in site 7. In analyzed biocenoses we also identified families that were dominated in samples collected only in winter or summer. To "summer-specific" bacterial families belonged: Beijerinckiaceae (sites 1–3 and 5), Caulobacteraceae (sites 4 and 7), Isosphaeraceae (sites 1 and 4), Kineosporiaceae (site 4), Rhodobacteraceae (sites 3–5), Solibacteraceae (site 6) and Spirosomaceae (sites 1, 3, and 8). The analysis of diversity at the family level showed that there are more families specific to winter. These "winter-specific" families on Pergola's surfaces were: Burkholderiaceae (sites 3, 6 and 8), Clostridiaceae 1 (site 3), Cryomorphaceae (site 7), Cyclobacteriaceae (site 7), Cytophagaceae (site 3), Halomonadaceae (site 7), Marinobacteraceae (site 7), Micrococcaceae (site 5), and Microscillaceae (site 3).

In sites 1 and 7, there were no fungal families identified in both seasons (Figure 7b). The Herpotrichiellaceae family was identified in sites from 2 to 4 in both seasons, Ramalinaceae dominated in winter and summer in sites 4, 5, and 8. Parmeliaceae family was identified in both seasons only in site 6, Bulleribasidiaceae in site 8, Teratosphaeriaceae in site 5 while Trichomeriaceae in site 3. Nectriaceae and Physciaceae families were identified with high abundance only in samples collected in summer (from sites 7 and 5 respectively), while Stereocaulaceae only in sample collected in winter (site 4).

Moreover, a positive correlation between the presence of lichen crusts on the sandstone surface and the presence of *Rhinocladiella* spp. and *Lecania* spp. fungi, bacterial genus *Truepera* spp., as well as bacteria from Nocardioidaceae family were found. For Acidobacteriaceae and Acetobacteraceae families and the presence of lichen crusts, a negative correlation was observed. A positive correlation between the presence of mosses and the number of sequences of *Methylobacterium* spp. and *Roseomonas* spp., as well as those from the families Hymenobacteraceae and Pseudonocardiaceae, was found (Supplementary Table S2).

The composition of the bacterial and fungal communities in samples from subsequent winter and summer seasons was also visualized using Principal Coordinates Analysis (PCoA). Our data indicate that the highest differences between seasons were documented in samples from sites 3, 4, 5, 7, and 8 for bacterial diversity, and from sites 4 and 5 for fungal diversity (Figure 8).

Family proportion analysis underlines microbial biodiversity dependence from the BPs. Furthermore, this analysis establishes microbial taxa that are seasonally stable and where domination is strictly correlated to seasonal changes (Figure 8). It is worth noting that the seasonal variability of individual biocenoses resulted from differences in bacterial diversity. In the case of fungi, seasonal variability was observed mostly in samples from sites 4 and 5 (Figure 8). For stable biocenoses (sites 1, 2, and 6), the domination of the following bacterial families was observed: Sphingomonadaceae (sites 1 and 2), Chitinophagaceae, Micromonosporaceae, Tepidisphaeraceae, and Blastocatellaceae (site 1), Acetobacteraceae, Acidobacteriaceae (sites 2 and 6), and Sphingobacteriaceae (site 2) (Figure 7a).

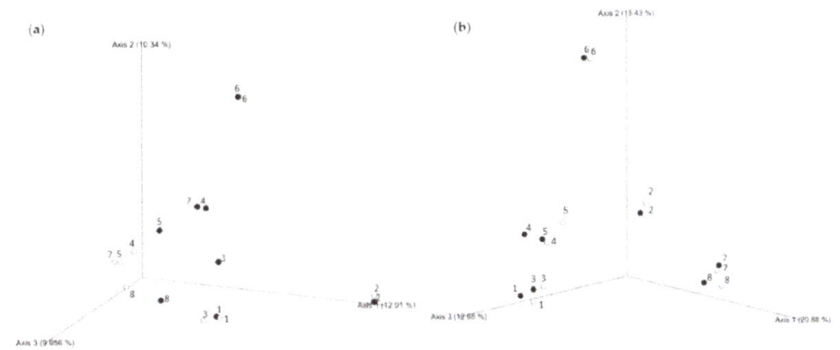

Figure 8. Principal Coordinates Analysis (PCoA) of seasonal microbial biodiversity changes using Bray-Curtis distance. Seasonal relationships and distribution of bacteria (**a**) and fungi (**b**) identified across samples collected from Pergola surfaces in summer (○–white spheres) and winter (•–black spheres).

Interestingly, dominating taxa characteristic were observed only in summer or winter seasons despite the seasonal stability of bacterial families in the biocenoses listed above. In the sample from site 1, the following families were identified only during the summer period (Figure 7a): Beijerinckiaceae, Isosphaeraceae, Pseudonocardiaceae, and Spirosomaceae; and during winter the Hymenobacteraceae family was identified in addition. For site 2 the Beijerinckiaceae family was identified only during the summer period. For site 6 the Solibacteracaceae family was identified only during the summer and Burkholderiaceae only during the winter season.

Our data indicate that in the remaining five biocenoses (sites 3, 4, 5, 7 and 8) the microbial communities manifested significant seasonal changes (Figure 7). Bacterial families dominating only during the winter season included (Figure 7a): Burkholderiaceae, Clostridiaceae 1, Cytophagaceae, Microscillaceae and Sphingomonadaceae (site 3) and Blastocatellaceae (site 4); Sphingobacteriaceae, Micrococcaceae, Nocardioidaceae and Microbacteriaceae (site 5); Marinobacteraceae, Cyclobacteriaceae, Halomonadaceae, and Cryomorphaceae (site 7); and Acetobacteraceae, and Burkholderiaceae (site 8).

Specific to summer, dominating bacteria in biocenoses belonged to the families (Figure 7a): Trueperaceae and Caulobacteraceae (site 7); Rhodobacteraceae (sites 3, 4, and 5); Geodermatophilaceae, Rhizobiaceae, Kineosporiaceae, and Isosphaeraceae (site 4); Acetobacteraceae (site 5); Beijerinckiaceae (sites 3 and 5); Spirosomaceae (sites 3 and 8); and Flavobacteriaceae (site 8).

The bacterial families occurring independently to the seasonal changes were (Figure 7a): Sphingomonadaceae (sites 1–5 and 7–8); Rhizobiaceae (sites 5 and 7); Caulobacteraceae (site 4); Flavobacteriaceae (site 7); Hymenobacteraceae (sites 5 and 8); Sphingobacteriaceae (sites 7 and 8); Spirosomaceae and Tepidisphaeraceae (site 4); Microbacteriaceae, Nocardioidaceae and Geodermatophilaceae (site 8); and Chitinophagaceae, Blastocatellaceae and Nostocaceae (site 3).

Analysis of the distribution of fungal families demonstrated that their seasonal stability was associated with the dominance of (Figure 7b): Herpotrichiellaceae (sites 2, 3 and 4); Bulleribasidiaceae (site 8); Parmeliaceae (site 6); Ramalinaceae (sites 4, 5 and 8); Teratosphaeriaceae (site 5); and Trichocomaceae (site 3). There were also fungal families which dominated in the summer season, i.e., Herpotrichiellaceae (site 1); Physciaceae (site 5); and Ramalinaceae and Nectriaceae (site 7); or in the winter season: Teratosphaeriaceae (sites 1 and 6); and Stereocaulaceae (site 4).

Our data established a total of 34 identified bacterial genera (Figure 9a) that appeared to be the most represented (abundance of more than 3%) with a clear domination of *Sphingomonas* spp. in seven out of eight biocenoses studied (sites 1–6 and 8 with the exception of site 6). One third of the identified genera were also present in both seasons analyzed, i.e., *Acidiphilium* spp. (site 2); *Aureimonas* spp. (site 5); *Blastocatella* spp. and *Pseudonocardia* spp. (site 3); *Bryocella* spp., *Endobacter* spp. and *Granulicella* spp. (site 6); *Hymenobacter* spp. (sites 3 and 8); *Marmoricola* spp. (site 8); *Pedobacter* spp. (sites 7 and 8); *Polymorphobacter* spp. (site 1); and *Truepera* spp. (sites 4 and 5).

(a)

Genus	S1	W1	S2	W2	S3	W3	S4	W4	S5	W5	S6	W6	S7	W7	S8	W8
Acidiphilium			16.6%	15.1%							5.1%					
Arthrobacter										8.4%						
Aureimonas							3.9%		6.0%	4.4%			6.9%			
Blastocatella	3.0%				4.8%	3.7%										
Brevundimonas													3.5%			
Bryobacter											4.1%					
Bryocella											23.8%	23.4%				
Clostridium sensu stricto 13						4.8%										
Craurococcus					3.3%											
Cytophaga						4.0%										
Endobacter											3.8%	9.6%				
Flavimarina															3.3%	
Flavobacterium													4.4%			
Geodermatophilus																15.0%
Granulicella			3.9%								6.0%	5.8%				
Halomonas														3.3%		
Hymenobacter		6.0%			4.3%	3.4%			5.2%						4.6%	4.5%
Marinobacter														4.8%		
Marmoricola															3.3%	3.5%
Methylobacterium	3.6%				7.5%				5.7%							
Nocardioides										3.1%					3.6%	
Pedobacter													6.1%	4.2%	11.9%	5.3%
Polymorphobacter	4.3%	5.1%			3.5%											
Pontibacter										5.3%						
Pseudonocardia					3.2%	3.0%										
Ralstonia												3.2%				
Roseomonas									3.8%							3.8%
Rubellimicrobium					3.7%		3.6%		9.5%							
Scytonema UTEX 2349						10.2%										
Singulisphaera	3.4%						3.0%									
Sphingomonas	13.8%	19.4%	11.6%	15.9%	12.1%	4.9%	10.2%	5.8%	19.1%	13.2%			17.1%	4.0%	30.6%	22.0%
Spirosoma	3.1%				4.8%											
Tepidisphaera	3.1%							5.6%								
Truepera							11.6%	3.6%	4.3%	5.1%			5.8%			

(b)

Genus	S1	W1	S2	W2	S3	W3	S4	W4	S5	W5	S6	W6	S7	W7	S8	W8
Bacidina								23.2%								
Knufia										2.5%						
Kondoa													1.1%		1.4%	1.7%
Lecania							81.0%	12.2%	80.1%	39.7%			29.5%	1.2%	18.4%	5.5%
Lepraria								24.3%								
Phaeophyscia									6.9%							
Rhinocladiella	3.5%	1.5%	1.2%	1.9%	25.7%	14.7%	8.8%	3.0%	2.7%							
Rhizoplaca											1.6%	1.3%				
Teratosphaeria									1.2%	18.4%			1.4%			
Vishniacozyma															8.1%	17.2%

Figure 9. Bacterial (**a**) and fungal (**b**) genera dominating across samples. On two heat map-like figures, genera that represent at least 3% (in case of bacteria) or 1% (in case of fungi) relative abundance in each sample are shown. Overall, there are 34 bacterial genera (**a**) that seem to be the most represented. In case of fungal analysis (**b**), 10 different genera are shown. The abundance of each genus above the threshold was colored from green (the smallest percentage) to red (the highest percentage) while the empty cells were colored according to the season, i.e., white–summer (S prefix), grey–winter (W prefix).

For the analyzed fungi, 10 different genera were shown (Figure 9b), among which the following dominants (more than 1%) were present regardless of the season: *Kondoa* (site 8); *Lecania* (sites 4, 5, 7 and 8); *Rhinocladiella* (sites 1–4); *Rhizoplaca* (site 6); *Teratosphaeria* (site 5); and *Vishniacozyma* (site 8). The greatest diversity of identified dominating genera was observed in sites 4 and 5 for fungi (Figure 9b).

3.4. Acidifying Properties of Bacterial Strains Isolated within Subsequent Seasons

All samples showed a high percentage of bacteria from the families Acidobacteriaceae and Acetobacteraceae, families with known acidifying properties. Therefore, to establish if bacteria identified in samples from Pergola's surfaces can demonstrate a biodeteriorative potential, acidifying properties of isolated strains were analyzed.

Tests were performed for 32 morphologically different, randomly selected bacterial strains isolated (on a medium with calcium carbonate) from biocenoses on the Pergola's surfaces in each season (Figure 10). Most bacterial strains isolated during summer demonstrated ability to acidify the environment (pH of culture medium raised above 7 only for 5 isolates), while strains isolated in winter showed alkalizing properties (only four isolates reduced pH below 6 during 72 h cultivation).

(a)	pH during cultivation of bacterial strains isolated in summer			(b)	pH during cultivation of bacterial strains isolated in winter		
	24h	48h	72h		24h	48h	72h
1	6.05	5.72	5.49	1	6.21	7.17	7.14
2	5.81	5.44	5.34	2	7.88	8.17	8.05
3	6.45	4.69	4.61	3	5.64	5.50	5.42
4	6.79	6.67	6.41	4	6.85	6.79	6.50
5	6.70	6.27	6.36	5	6.92	6.89	6.48
6	5.41	5.39	5.40	6	6.03	6.51	6.20
7	6.06	5.39	5.87	7	7.33	8.30	8.16
8	5.39	5.95	5.83	8	5.90	5.74	6.61
9	6.46	6.89	7.18	9	6.77	6.91	7.36
10	6.29	6.50	6.45	10	5.90	5.68	6.68
11	5.47	5.34	6.32	11	5.90	5.81	6.78
12	6.70	6.70	7.20	12	6.09	6.46	6.44
13	5.51	5.06	5.16	13	6.11	6.56	6.89
14	6.80	6.69	7.07	14	8.07	6.95	7.77
15	5.77	5.43	5.52	15	6.95	6.84	7.13
16	6.56	6.88	6.85	16	6.86	6.97	7.65
17	5.56	6.09	6.93	17	7.67	7.75	8.19
18	5.90	6.57	7.03	18	5.78	5.66	6.00
19	6.55	6.84	6.68	19	7.21	7.41	7.59
20	6.63	6.85	6.57	20	6.88	6.28	5.93
21	6.57	6.94	6.50	21	5.89	6.81	7.04
22	5.29	5.41	5.83	22	5.92	5.67	5.13
23	6.43	6.78	6.84	23	6.28	6.32	6.32
24	6.78	6.96	7.29	24	5.83	5.62	6.23
25	6.89	6.62	6.43	25	5.12	4.91	5.16
26	5.10	5.17	5.52	26	5.81	6.83	6.90
27	5.12	5.27	6.40	27	6.17	6.44	6.96
28	5.99	5.38	5.22	28	5.65	5.39	7.44
29	6.29	6.84	5.63	29	8.22	8.15	8.20
30	5.22	5.13	6.20	30	8.02	7.85	7.84
31	5.54	6.37	5.40	31	7.31	7.00	6.41
32	5.53	5.45	5.54	32	6.81	7.18	7.34
Control	6.75	6.59	6.41	Control	6.77	6.72	6.66

Figure 10. Acidifying activity of the selected bacteria strains (1–32) isolated in summer (**a**) and winter (**b**) seasons from biofilm surfaces located on the Northern Pergola. Heat maps present the pH after 24, 48 and 72 h of different bacterial isolates incubation in Blickfeldt liquid medium. The data represent averages of three independent experiments.

4. Discussion

In this study, we sought to characterize the biocenoses on historical sandstone surfaces with different biodeterioration patterns (BPs) of the Northern Pergola in subsequent winter and summer seasons. Macroscopic analysis of the Pergola's slab surface revealed the presence of mosses (M), lichen crusts (LC), and green (GP) and yellow patinas (YP). The degradation potential of mosses is based primarily on the physical interaction with the stone surface, especially in places where stone slabs are joined. Additionally, mosses accumulate water, which helps to retain the humidity necessary for the growth of lichens and microorganisms. Lichens are symbiotic organisms composed of the cells typical for algae or cyanobacteria and fungi. All lichen communities show biodeteriorating abilities [38]. They cause mechanical damage due to the penetration of their thallus as the result of their expansion and contraction. They have the unique ability to produce chemical compounds that are characteristic to them. The number of identified substances produced by lichens as secondary metabolites is estimated at about 800. Some of them are strong chelating compounds, which have the ability to selectively remove certain metal ions, such as calcium or magnesium ions, from the rock/stones and these features are correlated to polar groups such as –OH and –COOH (e.g., lecanoric acid produced by *Lecanora* sp.) [39,40]. These substances play a significant role in the biodeterioration and bioweathering processes. Moreover, these compounds show antibacterial and antifungal activity, and an ability to absorb harmful UVA and UVB rays, and thus fulfill the role of a photoprotector [39,41,42].

All these features affect the formation of the entire biocenosis including microbiome. Unfortunately, most of the literature data concern calcium carbonate minerals, especially different types of limestones, which are more reactive to acid solutions than sandstone [43]. Most sandstone is composed of quartz or feldspar (both silicates) and is thus a mineral more resistant to weathering processes at the Earth's surface, but cementing materials may be either silicate minerals or non-silicate minerals, such as calcite [43]. Thus, mosses and lichens may be an important factor of sandstone bioweathering processes. Most of the data concerning the presence of mosses and lichens describe the biocenoses occurring on sandstone surfaces in tropical zones and in Great Britain [8,9,22,38,44,45]. Therefore, the results of microbial diversity obtained for the Pergola surfaces were compared with results obtained for microbiota inhabiting sandstones in other climatic zones or other types of stones.

For the Angkor temples complex in Cambodia in tropical monsoon climate, with distinctive rainy and dry seasons, visible biocenoses on monuments' sandstone surfaces consisted mainly of different filamentous and coccoid cyanobacteria, along with green algae, other bacteria, and, in some cases, fungi. Some biocenoses were dominated by the lichen community (*Leprarietum, Cryptothecietum, Pyxinetum*), and some moss species were identified [22,38]. Generally, their composition is different from that observed on the Pergola surfaces. Mosses and lichens species dominating Pergola surfaces were also observed on other historical buildings located in Poland built of other materials than sandstone. The same species (excluding the *Homalothecium sericeum*) were also identified as predominant taxa on bricks and wooden historical buildings in the Former Auschwitz II–Birkenau State Museum. The barracks' exteriors were colonized by bryophytes, lichens, and, to a lesser extent, algae [46].

Since the Pergola has not been cleaned in the last decade, it is legitimate to state that the biofilm under study is already stabilized and the pioneer organisms (algae, autotrophic bacteria, etc.) have been replaced by successors. The maturity and stability of the biofilm present on stone surfaces of the Pergola may be the reason for the relatively low abundance of Cyanobacteria phylum representatives in the studied biocenoses. Our results are contrary to the results obtained by Zhang et al. [47] for consortia responsible for the biodeterioration of Angkor temples in Cambodia. They collected two types of samples from two monuments made from sandstone: exfoliated/degraded sandstone materials loosely on sandstone block and visible biofilms on the sandstone wall. In sandstone exfoliation sediments they observed domination of heterotrophic bacteria (Actinobacteria, Proteobateria and Acidobaceria), while Cyanobacteria and Chloroflexi were more abundant in biofilms samples. The differences in microbial composition described above may result from the highly different climate zones in Poland and Cambodia. Similar conclusions were derived on the basis of analyses made for tombstones located in different continents [48]. Structure of bacterial and eukaryotic communities presented on tombstones were significantly affected not only by climate, but also by rock type (limestone or granite). In the case of Pergola, the stone surface is eutrophicated and other microorganisms, such as microfungi, heterotrophic bacteria, and nitrophilous lichens (e.g., some species of *Caloplaca* or *Candelariella*) can develop. Therefore, in this situation, the chemical composition of the stone may become less important [8].

The alpha diversity analyses showed that observed bacterial communities on Pergola's surfaces are more diverse than the fungal ones. Similar results were obtained even for interior sandstone structures of the Rotunda of Sts. Felix and Adauctus, which is a part of the Wawel Royal Castle located in Krakow, Poland [49] or for limestone and granite tombstones [48]. Zhang et al. [47] observed that the alpha diversity of the bacterial community is also due to the depth from which the sample was taken for analysis (the lowest diversity was observed in the bottom depth layers of biofilms collected from Angkor temples). Noteworthy, richness of microbial communities on Pergola's surfaces as well as their structures are more similar in summer than in winter when the diversity across sampling sites is more versatile.

The bacteria and fungi diversity in biocenoses on the historical Pergola at the phylum level (Supplementary Figure S3) showed similarities with the biodiversity observed in samples of biofilms on sandstone slabs in Belfast and Oxford [9,45], and in the West Lake Cultural Landscape of Hangzhou in China [50]. Moreover, in the case of Pergola, taxonomic composition at the phylum level remained constant regardless of the season of the study. Differences were observed at lower taxonomic levels (families, genus). Independently from the seasons, biocenoses on the Northern Pergola were dominated by sequences representing the genera *Sphingomonas* spp. Sphingomonadales were also most abundant in biofilms on sandstone in Oxford [9] and on stone surfaces of UNESCO's World Heritage listed monuments and exteriors of the temples located in the West Lake Cultural Landscape of Hangzhou in China [50]. They were also identified on carbonate stones of a medieval church in Italy [51], as well as on granite and limestone tombstones in different climatic zones [48]. On the surface of Pergola sandstone slabs, bacteria belonging to the phylum Acidobacteria and Actinobacteria were notably abundant as in Oxford [9], Hangzhou (China) [50], and in Angkor temples complex in Cambodia [47]. Moreover, in samples from Pergola's biocenoses, bacteria belonging to the phylum Firmicutes and Gemmatimonadetes were also identified. Zhang et al. [47] observed high abundance of these phyla only in the samples of sandstone materials loosely on sandstone block from two Angkor temples. Li et al. [50] identified bacteria at the level of genera and, as in the case of Pergola, they noted the presence of *Hymenobacter, Methylobacterium Pseudonocardia, Roseomonas, Scytonema, Spirosoma,* and *Truepera* with abundances similar to those found in our study.

Furthermore, the results of metagenomic analyses allowed for the identification of bacteria that have the ability to produce natural pigmented compounds. The genus *Sphingomonas* identified with high abundance in almost all analyzed biocenoses and *Ralstonia* genus are known for its properties to produce red pigments [35]. Moreover, in the analyzed biofilms there were numerous bacteria, especially during winter, from the genera *Arthrobacter* which produce blue and dark green pigments and *Cytophaga* which produce yellow pigments. On the other hand, *Flavobacterium* genus, which also produce yellow pigments [35] were identified with high abundance in summer. The above observations indicate that the main factor influencing the visible color changes of BPs depending on the seasons may be related with seasonal variation in the number of microorganisms capable of producing pigments.

The fungal composition of all samples was dominated mainly by lichenized fungi (e.g., *Lecania* spp., *Lepraria* spp.), while other taxa were also observed. Domination of Ascomycota is clearly visible and similar to the observations of Cutler et al. [9,45] and Li et al. [50,52]. Our data underlines the domination mostly by two genera: *Lecania* and *Rhinocladiella,* presence of which was correlated with the presence of lichens on the sandstone surface. Many microbial taxa and eukaryotes that are lichen symbionts have also been identified on granite and limestone surfaces of tombstones in tropical, subtropical as well as more temperate regions [48].

Moreover, microbial biodiversity analysis indicates high abundance of acidifying bacteria taxa in biocenoses structures, i.e., Acetobacteraceae and Acidobacteriaceae. During summer, ASVs of Acetobacteraceae family (producing acetic acid during fermentation) were most frequent compared to winter. High percentage of Acetobacteraceae in different biocenoses can affect the diversity of bacterial strains isolated on the Blickfeldt medium, and, in consequence, distinct acidifying properties were shown for summer isolates. Our data appears to correlate with the observation that different acidifying bacteria dominate the biofilm structures with seasonal dependence. The data quoted here refers only to cultured bacterial strains. Unfortunately, cultivation methods do not allow for the isolation of microorganisms in the proportions in which they are actually present in a given sample. Also in vitro studies of the activity of the culturable strains do not give us real insight into the actual metabolic changes that can be demonstrated by entire complex consortia of microorganisms in situ, in the environment. Therefore, in the next stage, we would like to verify which metabolic pathways dominate in a given season and which pathways are characteristic for given BPs using metabolomic analyses for sites with different BPs in dependent of seasons. Direct metabolic analyses of microorganisms colonizing the surfaces of historical sandstone would undoubtedly broaden the knowledge of geomicrobiology of building materials [53] in temperate climates, especially since sandstone is also a popular material nowadays. Such research would allow for the development of dedicated conservation schemes, e.g., for the selection of a biodeteriogens removal agent depending on the season or BPs.

One of the possible explanation of differences in biochemical properties of isolated strains is that during seasons bacteria can change their metabolisms. These theses are consistent with the results of the metabolomic analysis of historical brick and wood samples activated by incubation in atmosphere with high humidity. Gutarowska and colleagues established that humidity can activate numerous metabolic pathways, including those regulating the production of primary and secondary metabolites [54]. Microbiological metabolites responsible for the biodeterioration of stone surfaces include siderophores, inorganic acids (e.g., sulfuric, carbonic) and organic acids (e.g., citric, oxalic) [53,55]. Therefore, it is not excluded that an increase in humidity or seasonality may affect the secretion of specific metabolites, including organic acids. This is important because organic acids have a different mode of action than carbonic acid (large number of carbonates are sparingly soluble in water). Many organic acids exhibit chelating activity in contrast to carbonic acid, which is a well-known feature described in chemistry handbooks.

Also, the rock type determines the type of metabolic pathways of microorganisms. Shotgun metagenomics analyses of bacterial communities on granite revealed a high number of genes associated with acid tolerance and chemotaxis, while the bacteria inhabiting limestone had a high proportion of genes involved in photosynthesis and radiation resistance which seems to be related to the occurrence of lichens [48]. In addition, the domination of acid tolerant microbial taxa (Acidobacteriaceae, Beijerinckiaceae and Methylocystaceae) was found on granite tombstones, while alkaliphilic (*Spirosoma*, *Rubellimicrobium* and *Truepera*) constituted most abundant microbial taxa on limestone tombstones surfaces. The above observations allowed the authors to hypothesize that microorganisms that inhabit different types of stones may be predisposed to different (symbiotic versus free-living lifestyles) ecological strategies and that the main factor influencing the diversity of bacterial taxa are pH preferences of these microorganisms [48].

Our observations, in turn, may indicate the adaptation of microorganisms to the changing seasons, which results in the intensification of sandstone weathering in summer (strongly acidifying properties of bacterial strains), while in winter the alkalization of the substrate may be related with the biomineralization process, especially precipitation of amorphous or microcrystalline forms of some minerals by hydrolysis [56]. Unfortunately, there are only a few studies concerning sandstone surfaces. Zhang et al. [47] observed that both sandstone pH, as well as nitrate concentration depends on sample location. The above may be due to the fact that sandstone is not a homogeneous rock and, in addition, weathered sandstone materials may have contained acidifying and alkalizing compounds secreted by microorganisms during their growth. The authors also confirmed that alkalizing bacteria (Nitrospirae) were presented in exfoliated sandstone materials loosely on sandstone block of two Angkor temples [47].

The accumulation of polluting particles on the stone surfaces, especially as an effect of London smog during heating period and photochemical smog during summer, results in the formation of crust enriching the substratum for colonization of microorganisms and thereby could influence the microbial biodiversity pattern [57]. One of the most important microbial factors influencing colonization of exposed stone surfaces is the presence of chemoorganotrophs that trap minerals and organic substances from the air or nutrients from the stone [57]. Microorganisms utilizing organic substances from the air or nutrients from the stone can often be found in subaerial biofilms [58] and our data corroborate these findings. The presence of *Methylobacterium* in biocenoses located in sites 1-3 was observed, which corresponds well with the presence of methane in atmospheric air. Also, Nocardioidaceae, which can use hydrocarbons compounds from air pollution, were detected on the Pergola surfaces.

5. Conclusions

Taken together, analyses of bacterial and fungal communities' biodiversity in biocenoses from historic sandstone Pergolas revealed that the observed changes were mostly associated with the biodeterioration patterns (BPs) (presence of mosses or lichen crusts). However, dominating microbial groups that correlate with seasonal changes were also identified. What is more, visible color changes in BPs may be caused by seasonal changes in the number of microorganisms capable of producing pigments. The above observation may be a starting point for the implementation of dedicated strategies for the protection of historical stone objects, taking into account the season of the year in which renovation works are carried out and depending on the type of BPs.

The diversity of organisms in the biofilm ensures its stability throughout the year despite the differences recorded between winter and summer. Additionally, the maturity and stability of the biofilm present on stone surfaces of the Pergola may be the reason for the relatively low abundance of Cyanobacteria phylum representatives in the studied biocenoses because of the pioneer organisms (e.g., algae, autotrophic bacteria, etc.) have been replaced by successors. Moreover, a positive correlation between the presence of lichen crusts on the sandstone surface and the presence of lichenized fungi *Rhinocladiella* and *Lecania* as well as the presence of alkaliphilic bacterial genus *Truepera* was found, whereas a negative correlation was observed for Acidobacteriaceae and Acetobacteraceae acidophilic bacterial families.

Identified correlations give the evidence to the complex relationships between the organisms that build described biocenoses, but the functional analysis of the processes taking place on the surface of the historical sandstone in different seasons requires further research. In the next stage we would like to verify, which metabolic pathways dominate in a given season and which pathways are characteristic for a given BPs. Thus, metabarcoding analyses in combination with metabolic analyses will form the basis for the implementation of dedicated and more effective methods of removing biodeteriogens from the surfaces of historic buildings made of sandstone. This is an important issue in the context of the protection of sandstone objects in a moderate climate zone and taking into account the number of buildings made of sandstone or with elements made of this material.

An additional conclusion is that metabarcoding researches seem to be an excellent and convenient method for analyzing the biodeterioration processes of stone monuments. It allows for quick and relatively accurate determination of the composition of biocenoses and, as a consequence, to predict its biodeteriorative potential. Metabarcoding provides a much larger amount of data in a short time compared to more conventional methods of identifying microorganisms. Metagenomic methods in combination with the metabolomics studies should allow conservators to determine the least destructive ways of cleaning historic buildings and ensuring optimal methods for their protection/preservation.

Supplementary Materials: The following are available online at https://www.mdpi.com/2076-3417/11/2/620/s1, Figure S1: Rarefaction curves for observed Amplicon Sequence Variants (ASVs) in each sample collected in winter (suffix W) and summer (suffix S) from the sandstone surfaces of the Northern Pergola at the Museum of King John III's Palace at Wilanow. Saturation curves are presented for bacterial (a) and fungal (b) ASVs, respectively; Figure S2. Identified bacteria (a) and fungi (b) at the class level in samples taken from the surface of the of the Northern Pergola at the Museum of King John III's Palace at Wilanow. One additional group was created to include the class with unspecified (i.e., '*p__*') or uncertain (i.e., '*Unassigned*') taxonomic position assigned (*Unspecified*); Figure S3. Identified bacteria (a) and fungi (b) at the phylum level in winter and summer at each sampling sites on the sandstone surfaces of the of the Northern Pergola at the Museum of King John III's Palace at Wilanow. Two additional and artificial groups were created to include the phyla with abundance lower than the threshold (*Other*, below 0.1%) and the ones with unspecified (i.e., '*p__*') or uncertain (i.e., '*Unassigned*') taxonomic position assigned (*Unspecified*); Table S1. Summary of reads number changes after each step of processing of V3/V4 16S bacterial rDNA and fungal ITS1 amplicons in each sample collected in winter and summer from the sandstone surfaces of the Northern Pergola at the Museum of King John III's Palace at Wilanow; Table S2: Correlation factors between the presence of mosses and lichen crusts and the dominant sequences of identified bacterial and fungal taxa in the analyzed samples.

Author Contributions: Conceptualization, L.D. (Lukasz Drewniak), L.D. (Lukasz Dziewit) and A.L.; methodology, L.D. (Lukasz Drewniak). and L.D. (Lukasz Dziewit); software, P.D.; validation, M.D., P.D., A.S. (Anna Szajewska), D.S., M.C. and K.R.; formal analysis M.D. and P.D.; investigation, M.D., P.D., A.S. (Anna Szajewska), D.S., M.C. and K.R.; resources, M.D., A.S. (Anna Szajewska), D.S., M.C., K.R. and A.L.; data curation, M.D., P.D., A.S. (Anna Szajewska), D.S., M.C., and A.L.; writing—original draft preparation, M.D.; P.D and A.S. (Aleksandra Skłodowska); writing—review and editing, A.S. (Aleksandra Skłodowska) and L.D. (Lukasz Drewniak); visualization, M.D. and P.D.; supervision, A.S. (Aleksandra Skłodowska); project administration, A.L.; funding acquisition, A.L. All authors have read and agreed to the published version of the manuscript.

Funding: This work was supported in the framework of the project "Revitalization and digitalization of Wilanow, the only Baroque royal residence in Poland"; POIS.11.01.00-00.068/14; co-financed by the European Regional Development Fund as part of the Operational Program Infrastructure and Environment 2007–2013, Priority 11 "Culture and cultural heritage", Activity 11.1 Protecting and maintaining cultural heritage of over-regional importance.

Institutional Review Board Statement: Not applicable for studies not involving humans or animals.

Informed Consent Statement: Not applicable for studies not involving humans.

Data Availability Statement: Data is contained within the article or supplementary material.

Acknowledgments: Next generation sequencing was carried out at the DNA Sequencing and Oligonucleotide Synthesis Laboratory IBB PAS using the CePT infrastructure financed by the European Union–European Regional Development Fund (Innovative economy 2007–13, Agreement POIG.02.02.00-14-024/08-00).

Conflicts of Interest: Author Magdalena Dyda was employed by the company Research and Development for Life Sciences Ltd. (RDLS Ltd.). The remaining authors declare that the research was conducted in the absence of any commercial or financial relationships that could be construed as a potential conflict of interest. The funders had no role in the design of the study; in the collection, analyses, or interpretation of data; in the writing of the manuscript, or in the decision to publish the results.

References

1. Pangallo, D.; Bučková, M.; Kraková, L.; Puškárová, A.; Šaková, N.; Grivalský, T.; Chovanová, K.; Zemánková, M. Biodeterioration of epoxy resin: A microbial survey through culture-independent and culture-dependent approaches. *Environ. Microbiol.* **2015**, *17*, 462–479. [CrossRef] [PubMed]
2. Piñar, G.; Garcia-Valles, M.; Gimeno-Torrente, D.; Fernandez-Turiel, J.L.; Ettenauer, J.; Sterflinger, K. Microscopic, chemical, and molecular-biological investigation of the decayed medieval stained window glasses of two Catalonian churches. *Int. Biodeter. Biodegr.* **2013**, *84*, 388–400. [CrossRef] [PubMed]

3. Sterflinger, K.; Piñar, G. Microbial deterioration of cultural heritage and works of art—Tilting at windmills? *Appl. Microbiol. Biotechnol.* **2013**, *97*, 963–9646. [CrossRef] [PubMed]
4. Zanardini, E.; May, E.; Inkpen, R.; Cappitelli, F.; Murrell, J.C.; Purdy, K.J. Diversity of archaeal and bacterial communities on exfoliated sandstone from Portchester Castle (UK). *Int. Biodeter. Biodegr.* **2016**, *109*, 78–87. [CrossRef]
5. Pinheiro, A.C.; Mesquita, N.; Trovão, J.; Soaresa, F.; Tiagoa, I.; Coelhoa, C.; Paiva de Carvalhoa, H.; Gil, F.; Catarino, L.; Piñar, G.; et al. Limestone biodeterioration: A review on the Portuguese cultural heritage scenario. *J. Cult. Herit.* **2019**, *36*, 275–285. [CrossRef]
6. Gadd, G.M. Geomycology: Biogeochemical transformations of rocks, minerals, metals and radionuclides by fungi, bioweathering and bioremediation. *Mycol. Res.* **2007**, *111*, 3–49. [CrossRef]
7. Gaylarde, C.C.; Rodríguez, C.H.; Navarro-Noya, Y.E.; Ortega-Morales, B.O. Microbial Biofilms on the Sandstone Monuments of the Angkor Wat Complex, Cambodia. *Curr. Microbiol.* **2012**, *64*, 85–92. [CrossRef]
8. Cuzman, O.A.; Tiano, P.; Ventura, S.; Frediani, P. Biodiversity on Stone Artifacts. In *The Importance of Biological Interactions in the Study of Biodiversity*, 1st ed.; Lõpez-Pujol, J., Ed.; InTech: Rijeka, Croatia, 2011; pp. 367–390. Available online: http://www.intechopen.com/books/the-importance-ofbiological-interactions-in-the-study-of-biodiversity/biodiversity-on-stone-artifacts (accessed on 8 September 2020). [CrossRef]
9. Cutler, N.A.; Chaput, D.L.; Oliver, A.E.; Viles, H.A. The spatial organization and microbial community structure of an epilithic biofilm. *FEMS Microbiol. Ecol.* **2015**, *91*, fiu027. [CrossRef]
10. Hueck, H.J. The biodeterioration of materials—An appraisal. *Int. Biodeter. Biodegr.* **2001**, *48*, 5–11. [CrossRef]
11. Perito, B.; Cavalieri, D. Innovative metagenomic approaches for detection of microbial communities involved in biodeteriorattion of cultural heritage. In Proceedings of the Florence Heri-Tech—The Future of Heritage Science and Technologies, Florence, Italy, 16–18 May 2018. *IOP Conf. Ser. Mater. Sci. Eng.* **2018**, *364*, 012074. [CrossRef]
12. Korkanç, A.; Savran, A. Impact of the surface roughness of stones used in historical buildings on biodeterioration. *Construct. Build. Mater.* **2015**, *80*, 279–294. [CrossRef]
13. Warscheid, T.; Braams, J. Biodeterioration of stone: A review. *Int. Biodeter. Biodegr.* **2000**, *46*, 343–368. [CrossRef]
14. Kusumi, A.; Li, X.S.; Katayama, Y. Mycobacteria isolated from Angkor monument sandstones grow chemolithoautotrophically by oxidizing elemental sulfur. *Front. Microbiol.* **2011**, *2*, 104. [CrossRef] [PubMed]
15. Gadd, G.M.; Bahri-Esfahani, J.; Li, Q.; Rhee, Y.J.; Wei, Z.; Fomina, M.; Liang, X. Oxalate production by fungi: Significance in geomycology, biodeterioration and bioremediation. *Fungal Biol. Rev.* **2014**, *28*, 36–55. [CrossRef]
16. Lisci, M.; Monte, M.; Pacini, E. Lichens and higher plants on stone: A review. *Int. Biodeter. Biodegr.* **2003**, *51*, 1–17. [CrossRef]
17. Hoppert, M.; Flies, C.; Pohl, W.; Gunzl, B.; Schneider, J. Colonization strategies of lithobiontic microorganisms on carbonate rocks. *Environ. Geol.* **2004**, *46*, 421–428. [CrossRef]
18. McNamara, C.J.; Perry, T.D., 4th; Bearce, K.A.; Hernandez-Duque, G.; Mitchell, R. Epilithic and Endolithic Bacterial Communities in Limestone from a Maya Archaeological Site. *Microb. Ecol.* **2006**, *51*, 51–64. [CrossRef] [PubMed]
19. Gorbushina, A.A.; Heyrman, J.; Dornieden, T.; Gonzalez-Delvalle, M.; Krumbein, W.E.; Laiz, L.; Petersen, K.; Saiz-Jimenez, C.; Swings, J. Bacterial and fungal diversity and biodeterioration problems in mural painting environments of St. Martins church (Greene-Kreiensen, Germany). *Int. Biodeter. Biodegr.* **2004**, *53*, 13–24. [CrossRef]
20. Gorbushina, A.A. Life on the rocks. *Environ. Microbiol.* **2007**, *9*, 1613–1631. [CrossRef]
21. Mohammadi, P.; Maghboli-Balasjin, N. Isolation and molecular identification of deteriorating fungi from Cyrus the Great tomb Stones. *Iran. J. Microbiol.* **2014**, *6*, 361–370. Available online: https://www.ncbi.nlm.nih.gov/pmc/articles/PMC4385579/ (accessed on 11 September 2020).
22. Caneva, G.; Bartoli, F.; Ceschin, S.; Salvadori, O.; Futagami, Y.; Salvati, L. Exploring ecological relationships in the biodeterioration patterns of Angkor temples (Cambodia) along a forest canopy gradient. *J. Cult. Herit.* **2015**, *16*, 728–735. [CrossRef]
23. Cennamo, P.; Montuori, N.; Trojsi, G.; Fatigati, G.; Moretti, A. Biofilms in churches built in grottoes. *Sci. Total Environ.* **2016**, *543*, 727–738. [CrossRef] [PubMed]
24. Kusumi, A.; Li, X.S.; Osuga, Y.; Kawashima, A.; Gu, J.-D.; Nasu, M.; Katayama, Y. Bacterial Communities in Pigmented Biofilms Formed on the Sandstone Bas-Relief Walls of the Bayon Temple, Angkor Thom, Cambodia. *Microbes Environ.* **2013**, *28*, 422–431. [CrossRef] [PubMed]
25. Kremer, B.P.; Muhle, H. *Porosty, Mszaki, Paprotniki (Polish Edition)*; Swiat Ksiazki Press: Warsaw, Poland, 1998; ISBN 8372270619.
26. Wójciak, H. *Flora Polski. Porosty, Mszaki, Paprotniki*, 2nd ed.; MULTICO Oficyna Wydawnicza: Warsaw, Poland, 2007; ISBN 9788370735524. (In Polish)
27. Schmieder, R.; Edwards, R. Quality control and preprocessing of metagenomic datasets. *Bioinformatics* **2011**, *27*, 863–864. [CrossRef]
28. BBolyen, E.; Rideout, J.R.; Dillon, M.R.; Bokulich, N.A.; Abnet, C.C.; Al-Ghalith, G.A.; Alexander, H.; Alm, E.J.; Arumugam, M.; Asnicar, F.; et al. Reproducible, interactive, scalable and extensible microbiome data science using QIIME 2. *Nat. Biotechnol.* **2019**, *37*, 852–857. [CrossRef]
29. Quast, C.; Pruesse, E.; Yilmaz, P.; Gerken, J.; Schweer, T.; Yarza, P.; Peplies, J.; Glöckner, F.O. The SILVA ribosomal RNA gene database project: Improved data processing and web-based tools. *Nucleic Acids Res.* **2013**, *41*, D590–D596. [CrossRef]

30. Kõljalg, U.; Nilsson, R.H.; Abarenkov, K.; Tedersoo, L.; Taylor, A.F.S.; Bahram, M.; Bates, S.T.; Bruns, T.D.; Bengtsson-Palme, J.; Callaghan, T.M.; et al. Towards a unified paradigm for sequence-based identification of fungi. *Mol. Ecol.* **2013**, *22*, 5271–5277. [CrossRef]
31. Herrero, A.; Flores, E. (Eds.) *The Cyanobacteria: Molecular Biology, Genomics and Evolution*, 1st ed.; Caister Academic Press: Sevilla, Spain, 2008; ISBN 978-1-904455-15-8.
32. Shi, X.L.; Lepère, C.; Scanlan, D.J.; Vaulot, D. Plastid 16S rRNA Gene Diversity among Eukaryotic Picophytoplankton Sorted by Flow Cytometry from the South Pacific Ocean. *PLoS ONE* **2011**, *6*, e18979. [CrossRef]
33. Caneva, G.; Bartoli, F.; Savo, V.; Futagami, Y.; Strona, G. Combining Statistical Tools and Ecological Assessments in the Study of Biodeterioration Patterns of Stone Temples in Angkor (Cambodia). *Sci. Rep.* **2016**, *6*, 32601. [CrossRef]
34. Caneva, G.; Fidanza, M.R.; Tonon, C.; Favero-Longo, S.E. Biodeterioration Patterns and Their Interpretation for Potential Applications to Stone Conservation: A Hypothesis from Allelopathic Inhibitory Effects of Lichens on the Caestia Pyramid (Rome). *Sustainability* **2020**, *12*, 1132. [CrossRef]
35. Ramesh, C.; Vinithkumar, N.V.; Kirubagaran, R.; Venil, C.K.; Dufossé, L. Multifaceted Applications of Microbial Pigments: Current Knowledge, Challenges and Future Directions for Public Health Implications. *Microorganisms* **2019**, *7*, 186. [CrossRef] [PubMed]
36. Panzer, K.; Yilmaz, P.; Weiß, M.; Reich, L.; Richter, M.; Wiese, J.; Schmaljohann, R.; Labes, A.; Imhoff, J.F.; Glöckner, F.O.; et al. Identification of Habitat-Specific Biomes of Aquatic Fungal Communities Using a Comprehensive Nearly Full-Length 18S rRNA Dataset Enriched with Contextual Data. *PLoS ONE* **2015**, *10*, e0134377. [CrossRef] [PubMed]
37. Stielow, J.B.; Lévesque, C.A.; Seifert, K.A.; Meyer, W.; Iriny, L.; Smits, D.; Renfurm, R.; Verkley, G.J.M.; Groenewald, M.; Chaduli, D.; et al. One fungus, which genes? Development and assessment of universal primers for potential secondary fungal DNA barcodes. *Persoonia* **2015**, *35*, 242–263. [CrossRef] [PubMed]
38. Bartoli, F.; Casanova Municchia, A.; Futagami, Y.; Kashiwadani, H.; Moon, K.H.; Caneva, G. Biological colonization patterns on the ruins of Angkor temples (Cambodia) in the biodeterioration vs bioprotection debate. *Int. Biodeter. Biodegr.* **2014**, *96*, 157–165. [CrossRef]
39. Matwiejuk, A. Matuzalem Genus of Lichens—*Rhizocarpon geographiucm* (L.) dc.—Its Properties and Application. *Kosmos* **2007**, *56*, 175–180. Available online: https://kosmos.ptpk.org/index.php/Kosmos/article/view/1322/1301 (accessed on 15 October 2020). (In Polish).
40. Gehrmann, C.; Krumbein, W.E.; Petersen, K. Lichen weathering activities on mineral and rock surfaces. *Stud. Geobot.* **1988**, *8*, 33–45.
41. Seaward, M.R.D. Lichens, agents of monumental destruction. *Microbiol. Today* **2003**, *30*, 110–112. Available online: https://socgenmicrobiol.org.uk/pubs/micro_today/pdf/080303.pdf (accessed on 15 October 2020).
42. Studzińska, E.; Witkowska-Banaszczak, E.; Bylka, W. Bioactive compounds of lichen. *Herb. Polon.* **2008**, *54*, 79–88. Available online: http://herbapolonica.pl/magazines-files/9811528-Zwi%C4%85zki%20biologicznie.pdf (accessed on 16 October 2020). (In Polish)
43. Boggs, J.R. *Principles of Sedimentology and Stratigraphy*, 3rd ed.; Pentice Hall: New York, NY, USA, 2000; ISBN 0-13-099696-3.
44. Scheerer, S.; Ortega-Morales, O.; Gaylarde, C. Microbial deterioration of stone monuments-an updated overview. *Adv. Appl. Microbiol.* **2009**, *66*, 97–139. [CrossRef]
45. Cutler, N.A.; Oliver, A.E.; Viles, H.A.; Ahmad, S.; Whiteley, A.S. The characterisation of eukaryotic microbial communities on sandstone buildings in Belfast, UK, using TRFLP and 454 pyrosequencing. *Int. Biodeter. Biodegr.* **2013**, *82*, 124–133. [CrossRef]
46. Rajkowska, K.; Otlewska, A.; Koziróg, A.; Piotrowska, M.; Nowicka-Krawczyk, P.; Hachułka, M.; Wolski, G.J.; Kunicka-Styczyńska, A.; Gutarowska, B.; Żydzik-Białek, A. Assessment of biological colonization of historic buildings in the former Auschwitz II-Birkenau concentration camp. *Ann. Microbiol.* **2014**, *64*, 799–808. [CrossRef] [PubMed]
47. Zhang, X.; Ge, Q.; Zhu, Z.; Deng, Y.; Gu, J.-D. Microbiological community of the Royal Palace in Angkor Thom and Beng Mealea of Cambodia by Illumina sequencing based on 16S rRNA gene. *Int. Biodeter. Biodegr.* **2018**, *134*, 127–135. [CrossRef]
48. Brewer, T.A.; Fierer, N. Tales from the tomb: The microbial ecology of exposed rock surfaces. *Environ. Microbiol.* **2018**, *20*, 958–970. [CrossRef] [PubMed]
49. Dyda, M.; Pyzik, A.; Wilkojc, E.; Kwiatkowska-Kopka, B.; Sklodowska, A. Bacterial and Fungal Diversity Inside the Medieval Building Constructed with Sandstone Plates and Lime Mortar as an Example of the Microbial Colonization of a Nutrient-Limited Extreme Environment (Wawel Royal Castle, Krakow, Poland). *Microorganisms* **2019**, *7*, 416. [CrossRef] [PubMed]
50. Li, Q.; Zhang, B.; He, Z.; Yang, X. Distribution and Diversity of Bacteria and Fungi Colonization in Stone Monuments Analyzed by High-Throughput Sequencing. *PLoS ONE* **2016**, *11*, e0163287. [CrossRef] [PubMed]
51. Chimienti, G.; Piredda, R.; Pepe, G.; van der Werf, I.D.; Sabbatini, L.; Crecchio, C.; Ricciuti, P.; D'Erchia, A.M.; Manzari, C.; Pesole, G. Profile of microbial communities on carbonate stones of the medieval church of San Leonardo di Siponto (Italy) by Illumina-based deep sequencing. *Appl. Microbiol. Biotechnol.* **2016**, *100*, 8537–8548. [CrossRef]
52. Li, Q.; Zhang, B.; Wang, L.; Ge, Q. Distribution and diversity of bacteria and fungi colonizing ancient Buddhist statues analyzed by high-throughput sequencing. *Int. Biodeter. Biodegr.* **2017**, *117*, 245–254. [CrossRef]
53. Gadd, G.M. Geomicrobiology of the built environment. *Nat. Microbiol.* **2017**, *2*, 16275. [CrossRef]

54. Gutarowska, B.; Celikkol-Aydin, S.; Bonifay, V.; Otlewska, A.; Aydin, E.; Oldham, A.L.; Brauer, J.I.; Duncan, K.E.; Adamiak, J.; Sunner, J.A.; et al. Metabolomic and high-throughput sequencing analysis—modern approach for the assessment of biodeterioration of materials from historic buildings. *Front. Microbiol.* **2015**, *6*, 979. [CrossRef]
55. Gadd, G.M. Metals, minerals and microbes: Geomicrobiology and bioremediation. *Microbiology* **2010**, *156*, 609–643. [CrossRef]
56. Otlewska, A.; Gutarowska, B. Environmental parameters conditioning microbially induced mineralization under the experimental model conditions. *Acta Biochim. Pol.* **2016**, *63*, 343–351. [CrossRef] [PubMed]
57. Nuhoglu, Y.; Oguz, E.; Uslu, H.; Ozbek, A.; Ipekoglu, B.; Ocak, I.; Hasenekoglu, I. The accelerating effects of the microorganisms on biodeterioration of stone monuments under air pollution and continental-cold climatic conditions in Erzurum, Turkey. *Sci. Total Environ.* **2006**, *364*, 272–283. [CrossRef] [PubMed]
58. Viles, H.A.; Gorbushina, A.A. Soiling and microbial colonisation on urban roadside limestone: A three year study in Oxford, England. *Build. Environ.* **2003**, *38*, 1217–1224. [CrossRef]

applied sciences

MDPI

Review

Biodeterioration of Glass-Based Historical Building Materials: An Overview of the Heritage Literature from the 21st Century

Maria Filomena Macedo [1,2,*], Márcia Gomes Vilarigues [1,2] and Mathilda L. Coutinho [3]

1 Departamento de Conservação e Restauro, Faculdade de Ciências e Tecnologia, Universidade NOVA de Lisboa, Campus Caparica, 2829-516 Caparica, Portugal; mgv@fct.unl.pt
2 VICARTE, Research Unit Vidro e Cerâmica para As Artes, Faculdade de Ciências e Tecnologia, Universidade NOVA de Lisboa, Campus Caparica, 2829-516 Caparica, Portugal
3 Laboratório HERCULES-IFAA, Universidade de Évora, 7000-809 Évora, Portugal; mathildal@gmail.com
* Correspondence: mfmd@fct.unl.pt

Abstract: The main goal of this work was to review the 21st century literature (2000 to 2021) regarding the biological colonisation and biodeterioration of glass-based historical building materials, particularly stained glass and glazed tiles. One of the main objectives of this work was to list and systematize the glass-colonising microorganisms identified on stained glass and glazed tiles. Biodiversity data indicate that fungi and bacteria are the main colonisers of stained-glass windows. Glazed tiles are mainly colonised by microalgae and cyanobacteria. Several studies have identified microorganisms on stained glass, but fewer studies have been published concerning glazed tiles. The analysis of colonised samples is a vital mechanism to understand biodeterioration, particularly for identifying the colonising organisms and deterioration patterns on real samples. However, the complexity of the analysis of materials with high biodiversity makes it very hard to determine which microorganism is responsible for the biodeteriogenic action. The authors compared deterioration patterns described in case studies with laboratory-based colonisation experiments, showing that many deterioration patterns and corrosion products are similar. A working group should develop guidelines or standards for laboratory experiments on fungi, bacteria, cyanobacteria, and algae on stained glass and glazed tiles.

Keywords: stained glass; glazed tiles; biodiversity; biodeterioration; cultural heritage; laboratory experiments

Citation: Macedo, M.F.; Vilarigues, M.G.; Coutinho, M.L. Biodeterioration of Glass-Based Historical Building Materials: An Overview of the Heritage Literature from the 21st Century. *Appl. Sci.* **2021**, *11*, 9552. https://doi.org/10.3390/app11209552

Academic Editor: Sergio Montelpare

Received: 30 July 2021
Accepted: 9 October 2021
Published: 14 October 2021

1. Introduction

Glass-based building materials have a long tradition in the construction and ornamentation of buildings. They have been applied in two forms: glass coatings applied over ceramics, or plate glass. The unique properties of glass, namely its transparency and wide variety of colours, have allowed it to become a vehicle for visual arts, contributing to the complex ornamentation of buildings and thus their valuable cultural heritage [1–5].

Glass is produced by fusing a mixture of raw materials—vitrifiers, fluxes, or network modifiers and stabilisers—each with different functions, to obtain a solid amorphous inorganic material [6]. In silicate glasses, which are the focus of the present work, silica (dioxide of silicon, SiO_2) is the main vitrifier (or network former) building up the glass network. Fluxes (or network modifiers) are added to the composition to lower the melting temperature of silica. To increase its durability, stabilisers are added to replace the highly mobile alkaline ions (network modifiers) in the glass network. To achieve specific properties, minor elements can be added, such as colorants, bleaches, and opacifiers. In ancient glass production, sand and quartz were the main sources of silica [7]. The other compounds varied through time and location. Therefore, ancient glasses were produced with a wide variety of chemical compositions [7], which influence their durability.

Stained glass, mosaics, and glazed tiles are an integral part of the structure of a historic building, and were designed to be a part of the architectural ensemble. The removal of these architectural elements from the original location signifies a considerable value loss; consequently, their preservation on site is of the utmost importance [5]. However, their preservation on-site raises concerns regarding their conservation, since glasses are particularly sensitive to environmental degradation, and are also very vulnerable to biodeterioration [8–10]. Microbial colonisation is a major problem of these building materials due to the damaging consequences of biodeterioration [10,11]. For decades, research has focused on the characterisation of biological communities based on the taxonomic identification of colonising organisms [12]. The interactions between the substrate and organisms to unveil deterioration processes and, more recently, the influence of environmental conditions on biological colonisation and biodeterioration are now being analysed due to advances in genomic and analytical techniques. In general, studies focus on the relationship between three main factors—organisms, the material, and the environment [13]. Therefore, the research on biodeterioration has been based on the analysis of the colonised substrate of artworks [14,15], on laboratory experiments [16,17], or combining laboratory experiments with case studies [18]. This work aimed to review the 21st century literature regarding glass-based historical building materials, particularly stained glass and glazed tiles. Although glass has been used in mosaics tesserae, and some studies have mentioned biodeterioration, no consistent literature was found [2,19]. Therefore, this type of application was not considered in the present work due to the lack of data. The 21st century literature was analysed and discussed, focusing on the biodiversity and the biodeterioration mechanisms, and comparing case studies and laboratory experiments.

Another goal of this work was to perform a critical review based on the 21st century literature regarding stained glass and glazed tiles, that can be used to assess cultural heritage biodiversity and biodeterioration. This knowledge is crucial to understand biodeterioration mechanisms, and is also very important for planning laboratorial experiments. The analysis of colonised samples is a vital mechanism to understand biodeterioration, particularly for identifying the colonising organisms and deterioration patterns on real samples. In this work, the objective was to analyse the materials and complexity of colonised samples in order to determine which microorganism was responsible for the biodeteriogenic action. The authors compared the deterioration patterns described in case studies with laboratory-based colonisation experiments, showing that many deterioration patterns and corrosion products are similar. A working group should develop guidelines or standards for laboratory-based experiments on fungi, bacteria, cyanobacteria, and algae on stained glass and glazed tiles.

2. Materials and Methods

The peer-reviewed literature concerning stained glass and glazed tiles included in the category of cultural heritage, dating from the 21st century (since 2000 to 2021), was dispersed and sometimes difficult to access. Two databases, WoS and Scopus, were consulted for this review, using the specific terms "stained glass" or "glazed tiles", which were combined with the keywords "biodeterioration", "bioreceptivity", "microorganisms", "microbial", "fungi", "algae" and "bacteria". All three authors undertook distinct data searches. From the total number of obtained results, the final results were manually selected. The total amount of biodiversity data collected from the gathered literature can be seen in the supplementary word document (Tables S1–S3). These data were organized in graphs and tables to allow a better understanding of the subject. The data allowed the study of biological colonisation in cultural heritage buildings, and should provide a basis for future laboratory experiments.

3. Results

3.1. Type of Silicate Glasses

Research on the biodeterioration of glass-based historical building materials has focused on several different glass types (Table 1). The classification of silicate glasses depends on the raw materials used for their production, and their final composition [4,7,20]. For this work, we based the classification of glass on the main flux: K-rich (main flux is K), mixed alkali (main fluxes are Na + K), Na-rich (main flux is Na), and Pb-rich (main flux is Pb). Glass types may have additional classifications within subgroups of the categories mentioned above [4,20]. Studies on stained glass were the most diverse regarding the type of glass. Three distinct compositional groups of silicate glasses were investigated: K-rich, mixed-alkali, and Na-rich (Table 1).

Table 1. Type of silicate glass used in the glass-based building materials described in the 21th century literature.

Type of Silicate Glass	Vitrifier	Flux	Stabilizer	Type of Building Material	Nr. of Studies	References
K-rich	SiO_2	K_2O	CaO	Stained glass	5	[18,21–23]
Mixed-alkali	SiO_2	$Na_2O + K_2O$	CaO	Stained glass	1	[18]
Na-rich	SiO_2	Na_2O	CaO	Stained glass	3	[24,25]
Pb-rich	SiO_2	PbO	Al_2O_3/CaO	Glazed tiles	6	[26–31]

3.2. Provenance

The provenance of cultural heritage stain glass and glazed tiles can be seen in Figure 1.

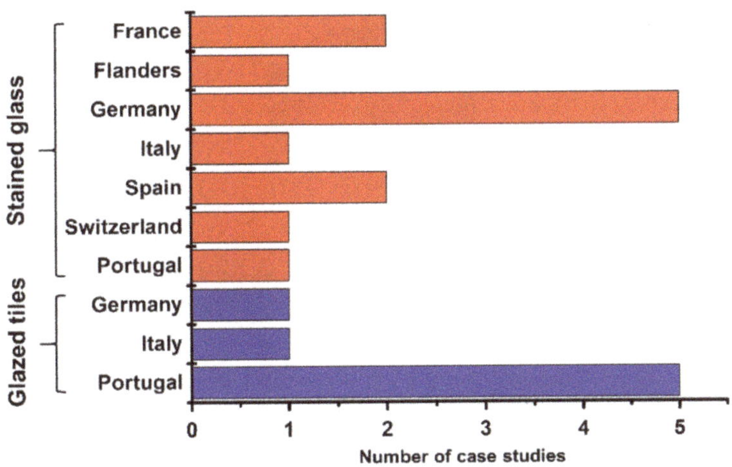

Figure 1. Provenance of the stained glass and glazed tiles with biological colonization from the 21st century literature.

Stained glass studies were made in more locations than glazed tiles. Moreover, most of the stained glass works of art were from Germany, while the glazed tiles are mostly Portuguese. The reason could be the restoration of several cathedrals and churches in Germany, during the 21st century. These glass windows date from the 12th to 20th century, but the majority belong to the 15th century. The provenance is not necessarily related to the place where these works of art were applied. For example, some studies investigated German tiles located in Brazil, and some Portuguese tiles were also studied in Brazil.

3.3. *Biodiversity*

The strategy of many microorganisms for settling and developing in hostile environments, such as stained glasses and glazed tiles, is through the production of biofilms (i.e., cells embedded in a matrix of extracellular polymeric substances (EPSs)) that can act as a protective layer [32]. The biofilm can limit solute diffusion and restrict the movement of water and nutrients into or out of its embedded microorganisms [10]. Biofilm constituents include EPSs, water, organic acids, lipids, enzymes, DNA, and organophosphates. Biofilms also have dust, pollen, spores, oil, and coal-fired carbonaceous particles from the atmosphere adhered to them or incorporated as part of the biofilm [33–35].

3.3.1. Fungi

The various fungi that have been identified on stained glass and glazed tiles in the literature of the 21st century can be seen in Table S1 and Figure 2. A total of 86 fungal specimens have been identified on stained glass and glazed tiles. The majority belong to the Ascomycota division, and a few to the Basidiomycete (Table S1). On stained glass, 24 different genera were identified: *Alternaria, Aspergillus, Aureobasidium, Capnobotryella, Chaetomium, Cladosporium, Coniosporum, Didymella, Engyodontium, Fusarium, Geomyces, Hortaea, Kirschsteiniothelia, Leptosphaeria, Myrothecium, Penicillium, Penidiella, Phoma, Rhodotorula, Sistotrema, Stanjemonium, Trichoderma, Ustilago,* and *Verticillium*. The most diverse genera on stained glass were *Aspergillus* [21,36,37], *Cladosporium* [18,23], and *Penicillium* [18] (Figure 2).

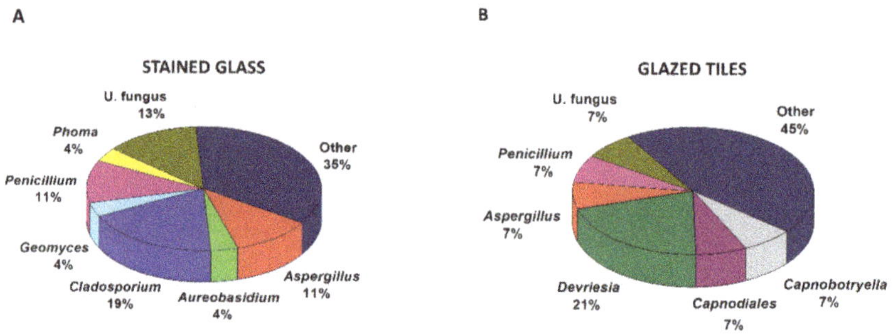

Figure 2. Relative % of fungal genera reported from the 21st literature regarding stained glass and glazed tiles. The category "other" refers to a compilation of several genera that occurs less than 3.4% and "U. fungus" are unidentified fungi. (**A**) Stained glass. (**B**) Glazed tiles.

A total of 26 fungal specimens were identified on glazed tiles, which is a much lower number than the fungal biodiversity reported for stained glass (Table S1). These fungi belong to 19 different genera, with the most diverse identified genera on glazed tiles being *Devriesia*, followed by *Aspergillus, Capnobotryella, Capnodiales,* and *Penicilium* (Figure 2). Four *Devriesia* species, namely *Devriesia imbrexigena, Devriesia modesta, Deveriesia* neodevriesiaceae, and *Devriesia xanthorrheae,* were identified on majolica glazed tiles belonging to outdoor wall coverings of two Portuguese monuments [28,38]. In fact, a novel species of this genera, *Devriesia imbrexigena,* was isolated and described for the first time after being collected from the majolica glazed tiles of the Pena National Palace (Sintra, Portugal) [38] (Table S1).

Figure 3 presents a SEM image of different fungi inoculated over the glass. The glass was partially clean allowing us to see hyphae fingerprints and hyphae over the glass surface.

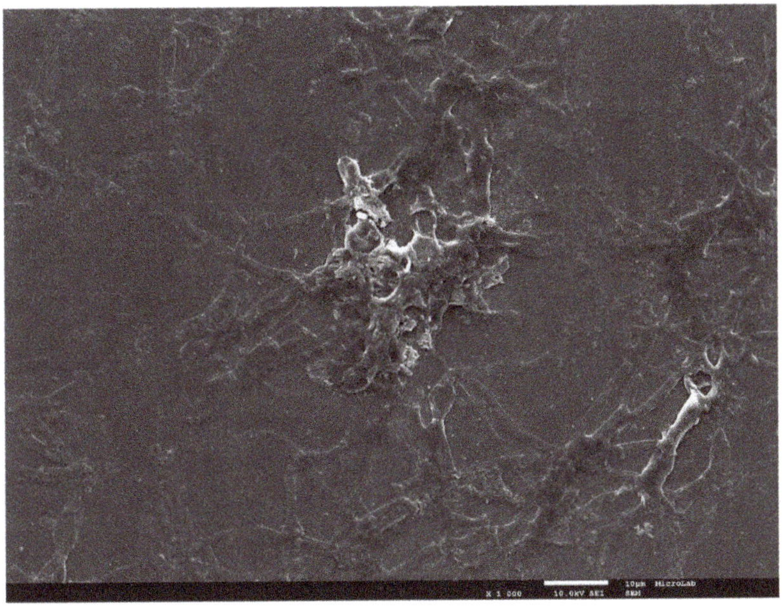

Figure 3. SEM micrography with fungi hyphae on top of glass and hyphae fingerprints (image taken from Ref. [39] with permission of the author).

A small number of fungal genera were simultaneously identified on both types of substrate: *Aspergillus*, *Aureobasidium*, *Capnobotryella*, *Fusarium*, *Penicillum*, and *Phoma* (Table S1). Among the most diverse genera, only *Aspergillus* and *Penicillum* occurred on both stained glass and glazed tiles (Figure 2). *Aspergillus* and *Penicillium* produce spores which are easily released and dispersed into the air, are omnipresent saprophytes, and dominate in temperate soils where they are frequently identified in both indoor and outdoor environments [40].

3.3.2. Bacteria

The bacteria that have been identified on stained glass and glazed tiles from the 21st century literature are summarized in Figure 4 (Table S2). A total of 87 specimens of bacteria have been identified on stained glass and glazed tiles. On stained glass windows, bacteria belonged to five different phyla have been found: Actinobacteria, Bacteroidetes, Firmicutes, Nitrospirae, and Proteobacteria (Figure 4). The most diverse genera were *Bacilus* [25], *Paenibacillus* [25], and *Kocuria* [23]. The microbial populations of the glass biofilms consisted mostly of fungi and bacteria, and fungi were often the dominant group.

A total of 26 bacterial specimens have been identified on glazed tiles, belonging to five different phyla: Actinobacteria, Bacteoidetes, Chloroflexi, Firmicutes, and Proteobacteria. Among these, 13 different genera of bacteria were identified, although few genera had a representative biodiversity (Table S2). Except *Methylibium* and *Microcella*, all other genera were represented by only one specimen. This is due to the characterisation of the bacterial communities of glazed tiles being provided from just two case studies [26,29]. Both studies applied molecular biology for the characterisation of the microbial community. The fact that many uncultured species appeared in these results highlights one of the drawbacks of molecular biology (Table S2). *Methylibium* was the only genera identified simultaneously on stained glass and glazed tiles (Table S2).

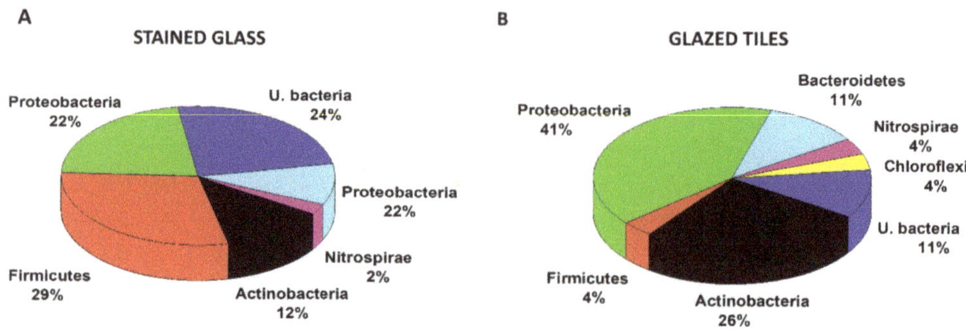

Figure 4. Relative % of Bacteria Phyla of stained glass and glazed tiles, reported in the literature for the 21st century. The category "U. Bacteria" are unidentified bacteria. (**A**) Stained glass. (**B**) Glazed tiles.

3.3.3. Microalgae, Cyanobacteria and Diatoms

The phototrophs, including green microalgae, cyanobacteria and diatoms that have been identified in stained glass and glazed tiles literature of the 21st century and were we summarized this knowledge Figure 5 (Table S3). The presence of phototrophs on stained, was limited to 2 cyanobacteria (one *Gloeocapsa* sp. and one *Oscillatoria* sp.), both were reported on stained glass in Mausoleum of the Assis Chermont in Brazil (Table S3) [24]. In the same stained glass window a rotifera was also identified. Nevertheless, other authors, not from the 21st century literature refer green biofilms on glass windows [41].

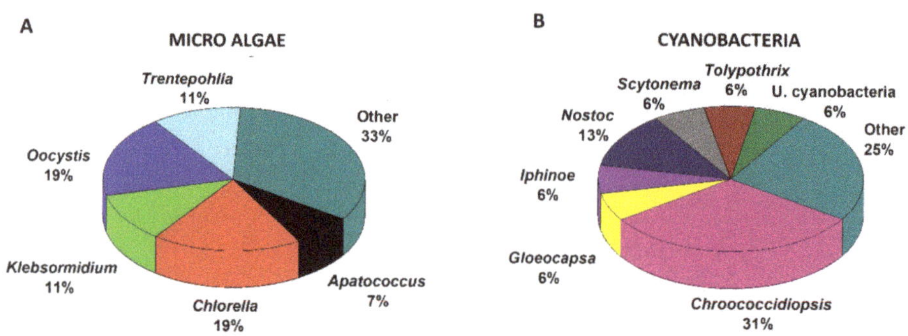

Figure 5. Relative percentage of genera of phototrophic microorganisms reported in the literature for the 21st century on glazed tiles. (**A**) Micro algae and (**B**) Cyanobacteria. The category "Other" refers to a compilation of several genera that occurs less than 3.4% and "U. cyanobacterium" are unidentified cyanobacterium.

On glazed tiles, a total of 27 specimens of green algae (Table S3), belonging to 14 different genera, were identified. The majority of the identified genera belonged to the Chlorophyta phylum, although a few Charophyta were also identified. The genera of green algae in which more than one specimen was reported were *Apatococcus, Chlorella, Klebsormidium, Oocystis,* and *Trentepohlia* (Figure 5). All these genera are common colonisers of stone monuments located in the Mediterranean basin [42].

Regarding the bacterial phototrophs, a total of 14 genera of cyanobacteria were identified on glazed tiles. The most diverse cyanobacterial genera were *Chroococidiopsis, Gleocapsa, Iphinoe, Nostoc, Scytomena,* and *Tolypotrix*. Except for *Iphinoe* and *Tolypotrix*, all the have been found colonising several monuments of the Mediterranean basin [42]. *Tolytrix* has been identified on several building façades in India [43].

Most of the identified microorganisms were detected by traditional molecular biology methods. However, nowadays, the characterisation of the colonising organisms involves understanding the structure and function of the microbial communities, detecting the metabolically active fraction, and identifying the functional diversity. This information can only be obtained by multiproxy approaches combining: (i) cultivation and classical taxonomy [44]; (ii) classical molecular biology techniques, based on Sanger-sequencing and widely applied in the field of cultural heritage [45,46]; and (iii) next generation sequencing (NGS), which is still poorly explored in the field cultural heritage for functional gene analysis to identify metabolic processes related to biodeterioration mechanisms [46]. Several studies have been undertaken to identify the microbial communities colonising stained glass and glazed tiles (e.g., [18,24,27,29,47]). In these studies, the identification of microorganisms was by direct observation [24,27]. Some works presented a more detailed characterisation of the stained glass biodeteriogens by using molecular techniques that allowed the identification of bacterial communities [25], or both fungal and bacterial communities [23]. On glazed tiles, some used exclusively molecular techniques [26] and others also combined both culture or direct observation [28,29]. However, among studies on stained glass and glazed tiles, no NGS study has yet been conducted (Supplemental Tables: Tables S1–S3).

Lichens have been studied on stained glass [48], but not recently. On both glazed roofing tiles and glazed tiles, lichens have been reported [11]. However, only glazed roofing tiles have been more extensively studied [49,50]. Bryophyta has been reported on glazed tiles, although no specific identification was made. This type of organism may not be able to grow on stained glass, since they need a substrate with a high porosity.

The data reported from the 21st century literature indicate that fungi and bacteria dominate the stained glass window colonisers. Glazed tiles can also be colonized by macroalgae and cyanobacteria.

3.4. Biodeterioration Studies

The analysis of colonised glass-based historical materials is the first approach to tackle biodeterioration through the identification of biophysical and biochemical alterations that appear on those materials. Many studies have focused on the analysis of deteriorated stained glass windows or glazed tiles with biological colonisation. The microbe–substrate interactions can be investigated through microscopic techniques, such as optical microscopes [21,24], scanning electron microscopes [21,24,28,29], or confocal microscopes [29]. Aesthetical and morphological alterations of the glass surface can be detected through these techniques. The main surface alterations reported on the colonised glasses were iridescence, etching, cracks, and corrosion pits (Table 2). Often, these deterioration forms were observed close to the colonized areas, and were consequently attributed to biodeterioration [21,23,29]. Besides these alterations, shifts in the chemical composition of the surface, due to leaching or enrichment of certain elements, are often mentioned. Moreover, the identification of crystals and other deposits on the surface is frequent on colonised glasses or glazes (Table 2). Often, the chemical or mineralogical characterisation of these compounds finds that Ca-rich minerals are the most common compounds (e.g., calcite, oxalates, and syngenite) [21,24] (Table 2). Piñar et al. [23] performed X-ray diffraction and identified several types of compounds on the stained glass windows of two churches located in Spain, specifically sulfates (Ca, K, and Na), carbonates (Ca), oxides (Si and Mn), and oxalates (Ca). Mn compounds, identified as bixbyite, were also identified on these medieval stained glass windows [23]. Dark stains rich in Mn were also reported on the stained glass windows from the Les Noes-Pres-Troyes Church (Aube, France) [51]. On glazed tiles, the presence of surface deposits has also been described. The CMSL analysis of the surface of glazed tiles from the Pena National Palace revealed the presence of inorganic deposits close to microorganisms, and a matrix of extracellular polymeric substances (EPSs) [29]. On glazed tiles, microbial colonisation does not always occur on the surface, and some studies have analysed chasmolithic colonisation under the glaze [26,27,30,52]. Only one study identified a green chemical compound, chromium oxide, under the surface, although the authors did not consider it to have a biological origin [30]. In other studies, the identification of precipitation or chemical alteration was not performed (Table 2). Research of the glaze–ceramic interface could be a crucial step for understanding whether microorganisms are indeed capable of causing the detachment of the glaze, which is major concern in the preservation of glazed ceramic building materials.

Table 2. Main biodeterioration patterns and alteration products identified on case studies on stained glass and glaze tiles from the 21st century bibliography.

Material	Type of Silicate Glass	Organisms	Location	Deterioration	Alteration Products [1]	Ref.
Stained glass	K-rich and Na-rich	Bacteria and fungi	Outdoors	Pitting, micro-cracks and some interconnected micro-cracks	gypsum ($CaSO_4{\cdot}2H_2O$) bixbyite (Mn_2O_3) syngenite ($K_2Ca(SO_4){\cdot}H_2O$) thenardite (a-Na_2SO_4) calcite ($CaCO_3$) quartz (SiO_2) weddellite, ($CaC_2O_4{\cdot}2H_2O$) whewellite ($CaC_2O_4{\cdot}H_2O$)	[23]
	K-rich	Bacteria and fungi	Outdoors	Crusts, interconnected craters and pits, whitish deposits	White Ca-rich	[21]
	K-rich glass	Bacteria and Fungi	Outdoors	Dark crusts, pits and craters, white deposits	Brown-black Mn-rich deposits	[51]
	Na-rich	Cyanobacteria rotifer, fungi and green algae	Indoors	Flaking-off of the enamels, interconnected pits, Iridescence	Soot and calcium carbonate	[24]
Glazed tiles	Pb-rich	Bacteria, cyanobacteria, algae and fungi	Outdoors	Stains under glaze		[26]
	Pb-rich	Bacteria, Algae, cyanobacteria and fungi	Outdoors	Pitting	n.i. inorganic crystals	[29]
	Pb-rich	Algae, cyanobacteria and fungi	Outdoors	Staining and pitting	n.i. crystals	[28]
	Pb-rich	Bacteria and Fungi	Indoors	Underglaze staining	n.i. crystals	[30]
	Pb-rich	Cyanobacteria and diatomacae	Outdoors	Underglaze staining	-	[27]

[1] "n.i." non-identified crystals.

The attribution of these forms of alteration to biodeterioration is based on the known ability of some microorganisms to excrete metabolic products, such as organic or inorganic acids, that can react with the substrate and solubilise or precipitate compounds [10,53–56]. Except for oxalates and Mn-compounds, all the compounds described in Table 2 have previously been identified on glass exposed to outdoor weathering [9,57,58]. Minerals such as syngenite ($K_2Ca(SO_4)_2{\cdot}H_2O$), gypsum ($CaSO_4$), and potassium carbonate (K_2CO_3) were detected on K-rich silicate glasses exposed to sheltered outdoor conditions [57]. On soda–lime glass, sodium carbonate (Na_2CO_3) and calcium carbonate ($CaCO_3$) crystals have been described as the first signs of deterioration caused by atmospheric corrosion [58]. Analysis of the compounds identified on glass with biological colonisation seem to follow the same trend of abiotic deterioration, in which the cationic part of these salts depends on the composition of the glass: Ca and K-minerals occur in K-rich glass, Na- and K-minerals on Na-K glass samples, and Na and Ca minerals on Na-rich glasses [58].

The analysis of colonized samples is vital for understanding biodeterioration, particularly identifying the colonizing organisms and deterioration patterns on real samples. However, the complexity of deterioration resulting from many synergetic and antagonistic factors, such as the environmental factors, makes drawing conclusions based solely on the investigation of colonized samples very difficult. In addition, the analysis of materials with high biodiversity makes it impossible to determine which microorganism was responsible for the biodeteriogenic action.

3.5. Laboratory Assays with Glass-Based Historical Building Materials

Laboratory experiments have the advantage of testing the damaging potential of a specific type of organism or group of organisms under controlled conditions. In the gathered literature these experiments use glass models with a chemical composition similar to historical samples. Not only chemical composition has been taken into account for the production of models, but some studies produced samples according to ancient technology. Rodrigues et al. [18] used blown glass since the effect of surface finishing can influence deterioration of glass. Studies regarding stained glass biodeterioration have been performed using K-rich, Mixed alkalis and Na-rich glasses (Table 3), the same glass compositions identified on the case studies (Table 2). K-rich glass was the most studied type of glass. Besides being widely used in stained glass production, it is also the most unstable type of glass so its conservation is a major concern [59].

Table 3. Laboratory experiments performed on glass or glaze models to simulate stained glass and glazed tiles biodeterioration. Time in *w* means weeks, *m* means months and *d* stands for days. The alteration products and biodeterioration patterns are also on this table.

Material	Organisms	Time	Type of Silicate Glass		Alt. Products	Biodeterioration	Ref.
Stained glass	Bacteria	6 w.	K-rich	Pristine	$CaCO_3$ $Ca_3(PO_4)_2$	Micro-cracking Depletion of K, Ca, P and Na on surface layer Alteration Mn oxidation-state	[51]
	Fungi	6 m.	Colourless Mixed alkali	Pristine	$CaCO_3$ $NaSO_4{\cdot}H_2O$	Micro-cracking enrichment of Ca Biopiting Hyphae fingerprint crystals Formation elements depletion or redeposition	[18]
				Corroded	$CaCO_3$ $NaAlSi_3O_8$ $CaSiO_3$ n.i.		
			Mn-purple K-rich	Pristine	$CaCO_3$		
				Corroded	$CaCO_3$		
			Fe-brown K-rich	Pristine	-		
				Corroded	SiO_2 n.i.		
	Fungi	40 d.	Colourless K-rich glass		$CaCO_3$ kalicinite, whewellite	Glass higher stability present superficial biofilm and glass with low chemical stability 3D growth of the biofilm	[22]
	Fungi	5 m.	Colourless Na-rich	Pristine	n.i.	Iridescent stains crystal formation slight decrease of the surface smoothness, hyphae fingerprints	[36]
			Colourless Na-rich	Pristine	Ca-rich crystals		
			Colourless Na-rich	Pristine	Ca-rich crystals		

Table 3. *Cont.*

Material	Organisms	Time	Type of Silicate Glass		Alt. Products	Biodeterioration	Ref.
Glazed tiles	Phototrophs	12 m.	Pb-rich	Pristine Corroded	$CaCO_3$ $CaCO_3$	Penetration into fissures, in-prints and surface deposits	[16]
	Fungus	12 m.	Pb-rich	Pristine	$CaCO_3$ $CaSO_4$ Ca-Oxalate	Mobilization of elements Formation of Ca-oxalate crystals	[17]
				Corroded	$CaCO_3$ $CaSO_4$ Ca-Oxalate		

Regarding glazes, only lead–alkali glaze opacified with tin was used in the laboratory experiments; no study was performed with transparent lead glazes. White opacified glazes are very common in glazed tiles production, as for many centuries they were used exclusively in Dutch, Portuguese, Spanish, and Italian glazed tile manufacturing. The laboratory-based experiments used the same glaze composition with two different conservation states: pristine and aged. The effect ageing has also been tested for stained glass by using pristine and corroded glass samples [18]. In the field of conservation studies for understanding susceptibility to biodeterioration depending on the degree of ageing, this is very relevant for preserving cultural heritage.

In general, the selection of microorganisms to be used in the biodeterioration laboratory experiments seems to have been based on the biodiversity found on historical materials. Studies have often used microorganisms directly isolated from colonised materials. Fungi identified from stained glass were the most commonly used microorganisms in laboratory studies [18,36,47], although bacteria have also been tested [51]. This may relate to the fact that fungi were the most identified organisms on stained glass with biological colonisation (Table S1). No phototrophic microorganisms were tested, since these have seldom been identified on stained glass windows. As previously mentioned, only two studies have focused on glazed tiles. One used a mixture of phototrophic microorganisms [16], and the other a fungal species isolated directly from historical glazed tiles [17]. Pure culture or multi-species inoculation have advantages and disadvantages. The use of pure cultures allows understanding of whether a specific microorganism is a biodeteriogen. However, multi-species colonies better mimic the real conditions under which biodeterioration take place, since colonisation usually occurs in a community of organisms.

The majority of the studies were performed under high relative humidity conditions, which are optimal for microbial growth and better simulate the conditions under which biofilms develop over historical glasses [16–18,51]. The environmental conditions were usually kept stable during the experiment. Therefore, no daily temperature or relative humidity variations were tested. Stable conditions favour microbial growth, however the simulation of temperature cycles could be relevant for reproducing physical decay. The variation of these parameters could induce volume oscillations of the microorganisms that can induce physical tensions on the substrate. The time span of the experiments varied from a few weeks to one year, making comparisons of the results of different experiments difficult.

After incubation, morphological and compositional changes were investigated through several analytical techniques. On stained glass, the most common biodeterioration forms reported in the laboratory studies were as follows: micro-cracking, enrichment of some elements, pitting, hyphae fingerprints, depletion or enrichment of elements, and deposition of crystalline compounds (Table 3). These were analogous to the deterioration patterns reported in the analysis of colonised historical samples (Table 2). On glazed tiles, the laboratory-based biodeterioration showed that phototrophic microorganisms were able to grow chasmoendolithically on fissures of the glazes, and imprints were visible on the surface [16]. In contrast, fungi did not cause physical or chemical damage to the substrate [17].

In general, the selection of microorganism to be used in this biodeterioration laboratory assays seems to be based on the biodiversity found on historical materials. Studies often use microorganisms directly isolated from colonized materials. Fungi identified from stained glass are the most commonly used microorganisms in laboratory studies [18,36,47], although bacteria have also been tested [51]. This may relate to the fact that fungi were the most identified organisms on stained glass with biological colonization (Table S1). No phototrophic microorganisms were tested, since these have seldomly been identified on stained glass windows. Regarding glazed tiles there are only two studies: one using a mixture of phototrophic microorganisms [16] and another a fungal specie isolated from historical glazed tiles [17]. The inoculation of pure culture or multi-species has advantages and disadvantages. The use of pure cultures allows understanding if a specific microorganism is a biodeteriogen. However, multi-species mimic better the real conditions in which biodeterioration occurs, since colonization usually occurs in a community of organisms.

The majority of the studies were performed under high relative humidity conditions, which simulate better the conditions in which biofilms develop over historical glasses [16–18,51]. The environmental conditions are usually kept stable during the experiment. Therefore, no daily temperature or relative humidity variations were tested. Although stable conditions favour the microbial growth, the simulation of temperature cycles could be relevant for simulating physical decay, since the variation of these parameters can induce volume variations of the microorganisms. The time span of the experiments varied from a few weeks to one year so, the comparison between the results of different experiments is difficult.

After the incubation, morphological and compositional changes are investigated through several analytical techniques. On stained glass the most common biodeterioration forms reported in the laboratory studies were: micro-cracking, enrichment of some elements, pitting, hyphae fingerprint, depletion or enrichment of elements and deposition of crystalline compounds (Table 3). Similar to the deterioration patterns reported in the analysis of historical samples (Table 2). On glazed tiles, the laboratory-based biodeterioration showed that phototrophic microorganisms were able to grow chasmoendolithically on fissures of the glazes and imprints were visible on the surface [16]. In contrast, fungi could not cause physical damage on the substrate, but seem to cause deposition of crystals on the surface [17].

In most studies, crystalline compounds were formed on the surface of the glass, implying surface corrosion. The identified compounds were $MnCO_3$, $CaCO_3$, $CaSO_4$, $NaSO_4{\cdot}H_2O$, $NaAlSi_3O_8$, $CaSiO_3$, SiO_2, K_2CO_3, and $CaC_2O_4{\cdot}H_2O$. In some studies, the mineralogical composition of the crystals was not determined; only the chemical composition was considered (Table 3). All these compounds were reported in field studies (Table 3). Ca-rich compounds were the most commonly identified, consistent with the fact that crystalline compounds are often detected on real samples with biological colonisation (Table 3). Only Ca-oxalates could be directly associated with biogenic action [60] due to the reaction of oxalic acid with calcium. Two laboratory-based experiments using fungi detected this compound on stained glass [22] and glazed tiles [17]. Although most of these compounds cannot be directly attributed to microbial activity, some authors have claimed that the precipitation was higher on the colonised samples than the controls. The precipitation of salts over glass is considered harmful, since their presence influences the corrosion rate [57], but these studies seemed to demonstrate the ability of microorganisms to also increase the deterioration rate [17,18].

Regarding colorants, Mn-rich compounds were identified in two studies on the biodeterioration of glass by bacteria [51,61] (Table 2). Rodrigues et al. [18] evaluated the biodeterioration of Fe-brown and Mn-purple K-rich glass, and concluded that the purple glass presented with a high degree of pitting with K, Ca, and Na depletions compared to the other studied glass compositions (colourless mixed alkali and Fe-brown K-rich glass). This could indicate a higher susceptibility of these glasses to fungal biodeterioration. Several microorganisms, including bacteria and fungi, have the ability to mobilise Mn [55,62].

However, not only Mn can be mobilised by microorganisms; other metals, many used as colouring agents of glass (e.g., Fe, Co, Cu, Cr), can accumulate in microbial biofilms [63,64]. Furthermore, it is also known that certain colorants can influence the corrosion rate of glass. In Vilarigues et al. [65], glasses doped with Cu, Mn, or Fe and exposed to weathering conditions developed a surface layer richer in the colouring transition-metal ions. The addition of these transition-metal ions, namely Cu, Fe and Mn, influenced the beginning of the corrosion process. Such an effect was previously described on medieval stained glasses coloured with metal oxides that had been naturally weathered [66].

Some studies have also addressed bioreceptivity, which is the material's ability to be colonised by living organisms [13,67]. On glazed tiles, bioreceptivity was tested and it was found that the colonisation rate of phototrophs was mostly affected by the water permeability. Another study that used the same glaze composition as the previously mentioned study detected no difference in the fungal colonisation rated between pristine and aged tiles. Regarding the bioreceptivity, no study has yet assessed the influence of the properties of the glass, such as composition, ageing, or roughness, on the rate of microbial growth. This is probably due to the difficulty in quantifying the growth of fungi and bacteria, which are the main stained glass colonisers. The only study that has investigated bioreceptivity of glass used a modern glass composition to test the effect of superficial porosity in the growth of phototrophic microorganisms [68]. In comparison with stone, the bioreceptivity of glass and glazed ceramic materials has been little investigated [13]. Nevertheless, several real case studies have mentioned differences in the degree of colonisation depending on the glass composition and colour [24,25]. This is also supported by the fact that some colorants seem to have antimicrobial properties, namely cobalt [69,70] and copper [25]. Therefore, the analysis of the effect of the colorants of glass and glazes on the biodeterioration rate and the inhibition of colonisation is an understudied area that could provide relevant information regarding the susceptibility of glasses to microorganisms depending on their colouring elements.

4. Discussion

We were surprised that, until now, the majority of biodiversity and biodeterioration studies have been performed on stained glass works of art coming from Germany. On the other hand, studies on glazed tiles from Portugal are more prevalent in the literature.

Biodiversity data reported from the 21st century literature, indicates that fungi and bacteria are the main colonizer of stained glass windows (Figures 2 and 4). While glazed tiles are also colonized by macroalgae and cyanobacteria. Nevertheless, the authors found that there are several works that identify microorganisms on stained glass and there are less studies published concerning glazed tiles. Still, it is important to study bacteria on glazed tiles since only two studies (Table S2) have been conducted in the 21st century literature. A further research should be performed on green biofilms on stained glass windows, in order to find out if there are phototrophic microorganisms on this glass cultural heritage.

Microbial glass deterioration causes serious damage on glass, not only from the chemical and physical point of view, but also from an aesthetic and iconographic one, since glass loses its transparency. The research of microorganism-glaze-ceramic interface is a crucial step for understanding if microorganisms are indeed capable of causing the detachment of the glaze. Accordingly, biodeterioration should be carefully addressed whenever conservation of historic glass is the issue. Data reported from the 21st century literature regarding bacterial diversity of glazed tiles was determined from only two studies. Both studies applied molecular biology for the characterization of the microbial community [26,29]. The fact that many uncultured species appear in these results and the presence of many unknown species highlights one of the drawbacks of molecular biology (Table S2).

The complexity of biodeterioration studies results from many synergetic and antagonistic factors makes drawing conclusions based solely on the investigation of colonized samples is very difficult. In addition, the analysis of materials with high biodiversity makes it impossible to determine which microorganism was responsible for the biodeteriogenic action.

When laboratory based studies are made the use of pure cultures allows to understand if a specific microorganism is a biodeteriogen. However, multi-species mimic better the real conditions in which biodeterioration occurs, since colonization usually occurs in a community of organisms. Laboratory experiments have the advantage of testing the damaging potential of a specific organism or group of organisms under controlled conditions. However, distinct authors used a different microorganism, time, temperature, light, etc., which make it impossible to compare between distinct studies.

Therefore, a working group should be organized in order to develop guidelines or standards for laboratory-based experiments: for instance, if you have a biofilm with five species you would test the biofim as a whole and each specie separately. Is also important to reproduce the glass or the glaze. It is important to define abiotic factors such as HR, time, temperature, light, the incubation system etc. Microbial glass deterioration causes serious damage to glass, not only from the chemical and physical point of view, but also from an aesthetic and iconographic perspective, since the glass loses its transparency. The research into the microorganism–glaze–ceramic interface is a crucial step for understanding whether microorganisms are indeed capable of causing the detachment of the glaze. Accordingly, biodeterioration should be carefully addressed whenever the conservation of historic glass and glazes tiles is of importance. Conservators must act to achieve the inactivation and removal of the microorganisms in order to halt their progression. The selection of cleaning products and biocides is a complex subject with few optimal solutions, due to the susceptibility of glass to corrosion and the complex biodiversity [28,71]. Additionally, the mechanical removal of biofilms, whether with a simple manual scalpel or through complex laser cleaning, remains in question since microorganisms usually grow close to the glass corrosion layer. In the case of cultural heritage, they are considered part of the original material and also a corrosion-passivating layer; therefore, their removal needs to be carefully evaluated. Protective layers have been investigated, such as antimicrobial [72,73] or anticorrosion [74,75] coatings. However, few or no real case applications have been carried out due to the ethical questions regarding their reversibility, ageing, and interference with aesthetic properties. Further research to understand glass biodeterioration is essential to be able to develop proper conservation guidelines for stained glass and glazed tiles.

The complexity of biodeterioration studies is due to the many synergetic and antagonistic factors that make drawing conclusions based solely on the investigation of colonised samples very difficult. In addition, the analysis of materials with high biodiversity makes it impossible to determine which microorganism was responsible for the biodeteriogenic action. When laboratory-based studies are carried out, the use of pure cultures allows us to understand if a specific microorganism is a biodeteriogen. However, multi-species inoculation better mimics the real conditions under which biodeterioration occurs, since colonisation usually develops in a community of organisms. Laboratory experiments have the advantage of testing the damaging potential of a specific organism or group of organisms under controlled conditions. However, some authors have used different microorganisms, time periods, temperatures, light conditions, etc., which makes it impossible to compare different studies.

Therefore, a working group should be organised in order to develop guidelines or standards for laboratory-based experiments: for instance, outlining procedures for testing biofilms as a whole and each species separately. Is also important to reproduce the glass or the glaze under study. Finally, it is important to define abiotic factors, such as the HR, time period, temperature, light, incubation system, and so forth.

Supplementary Materials: The following are available online at https://www.mdpi.com/article/10.3390/app11209552/s1, Table S1: Fungi reported on historical stained glass windows and glazed tiles in the literature. Table S2: Bacteria reported on historical stained glass windows and glazed tiles in the literature. Table S3: Prototrophs (cyanobacteria and microalgae), Protista and Rotifera reported on historical stained glass windows and glazed tiles in the literature.

Author Contributions: Conceptualization, M.F.M., M.G.V., M.L.C.; methodology, M.F.M.; M.G.V. and M.L.C.; investigation M.F.M., M.G.V. and M.L.C.; writing—review and editing, M.F.M., M.G.V., and M.L.C.; funding acquisition, M.G.V. and M.L.C. All authors have read and agreed to the published version of the manuscript.

Funding: This work was supported by National Funds through FCT-Portuguese Foundation for Science and Technology under the projects UID/Multi/04449/2019 HERCULES/UE; and contract CEECIND/00349/2017 (ML. Coutinho) and CityUMacau Chair in Sustainable Heritage. This work was also founded by VICARTE RESEARCH UNIT (UIDB/00729/2020).

Institutional Review Board Statement: Not applicable.

Informed Consent Statement: Not applicable.

Data Availability Statement: Data is contained within the references or in supplementary material.

Conflicts of Interest: The authors declare no conflict of interest.

References

1. Gnesin, G.G. Glass, Glaze, and Enamel over the Millennia. II. Glazes and Enamels. *Powder Met. Met. Ceram.* **2016**, *54*, 750–756. [CrossRef]
2. Verità, M. Technology and deterioration of vitreous mosaic tesserae. *Stud. Conserv.* **2000**, *45*, 65–76. [CrossRef]
3. Velo-Gala, A.; Mata, J.A.G. Roman window glass: An approach to its study through iconography. *Lvcentvm* **2017**, *36*, 159–176. [CrossRef]
4. Adlington, L.; Freestone, I.; Kunicki-Goldfinger, J.; Ayers, T.; Scott, H.G.; Eavis, A. Regional patterns in medieval European glass composition as a provenancing tool. *J. Archaeol. Sci.* **2019**, *110*, 104991. [CrossRef]
5. Carvalho, R.S. To be part of ... architecture, decoration or iconography. Documenting azulejo as integrated heritage. *ISPRS Ann. Photogramm. Remote Sens. Spat. Inf. Sci.* **2019**, *4*, 39–46. [CrossRef]
6. Shelby, J.E. *Introduction to Glass Science and Technology*; The Royal Society of Chemistry: Cambridge, UK, 1997.
7. Rehren, T.; Freestone, I.C. Ancient glass: From kaleidoscope to crystal ball. *J. Archaeol. Sci.* **2015**, *56*, 233–241. [CrossRef]
8. Frankel, G.S.; Vienna, J.D.; Lian, J.; Scully, J.R.; Gin, S.; Ryan, J.; Wang, J.; Kim, S.H.; Windl, W.; Du, J. A comparative review of the aqueous corrosion of glasses, crystalline ceramics, and metals. *NPJ Mater. Degrad.* **2018**, *2*, 15. [CrossRef]
9. Majérus, O.; Lehuédé, P.; Biron, I.; Alloteau, F.; Narayanasamy, S.; Caurant, D. Glass alteration in atmospheric conditions: Crossing perspectives from cultural heritage, glass industry, and nuclear waste management. *NPJ Mater. Degrad.* **2020**, *4*, 27. [CrossRef]
10. Weaver, J.L.; DePriest, P.T.; Plymale, A.E.; Pearce, C.I.; Arey, B.; Koestler, R.J. Microbial interactions with silicate glasses. *NPJ Mater. Degrad.* **2021**, *5*, 11. [CrossRef]
11. Coutinho, M.L.D.; Miller, A.Z.; Macedo, M.F. Biological colonization and biodeterioration of architectural ceramic materials: An overview. *J. Cult. Herit.* **2015**, *16*, 759–777. [CrossRef]
12. Sanmartín, P.; DeAraujo, A.; VasanthaKumar, A. Melding the Old with the New: Trends in Methods Used to Identify, Monitor, and Control Microorganisms on Cultural Heritage Materials. *Microb. Ecol.* **2018**, *76*, 64–80. [CrossRef]
13. Sanmartín, P.; Miller, A.; Prieto, B.; Viles, H. Revisiting and reanalysing the concept of bioreceptivity 25 years on. *Sci. Total Environ.* **2021**, *770*, 145314. [CrossRef]
14. Messiga, B.; Riccardi, M. Alteration behaviour of glass panes from the medieval Pavia Charterhouse (Italy). *J. Cult. Herit.* **2006**, *7*, 334–338. [CrossRef]
15. Szczepanowska, H.; Cavaliere, A. Fungal deterioration of 18th and 19th century documents: A case study of the Tilghman Family Collection, Wye House, Easton. Maryland. *Int. Biodeterior. Biodegrad.* **2000**, *46*, 245–249. [CrossRef]
16. Coutinho, M.L.; Miller, A.Z.; Rogerio-Candelera, M.A.; Mirão, J.; Alves, L.C.; Veiga, J.P.; Águas, H.; Pereira, S.; Lyubchyk, A.; Macedo, M.F. An integrated approach for assessing the bioreceptivity of glazed tiles to phototrophic microorganisms. *Biofouling* **2016**, *32*, 243–259. [CrossRef] [PubMed]
17. Coutinho, M.; Miller, A.Z.; Phillip, A.; Mirao, J.; Dias, L.; Rogerio-Candelera, M.; Saiz-Jimenez, C.; Martin-Sanchez, P.M.; Cerqueira-Alves, L.; Macedo, M.F. Biodeterioration of majolica glazed tiles by the fungus Devriesia imbrexigena. *Constr. Build. Mater.* **2019**, *212*, 49–56. [CrossRef]
18. Rodrigues, A.; Gutierrez-Patricio, S.; Miller, A.Z.; Saiz-Jimenez, C.; Wiley, R.; Nunes, D.; Vilarigues, M.; Macedo, M.F. Fungal biodeterioration of stained-glass windows. *Int. Biodeterior. Biodegrad.* **2014**, *90*, 152–160. [CrossRef]
19. Moropoulou, A.; Zacharias, N.; Delegou, E.; Maróti, B.; Kasztovszky, Z. Analytical and technological examination of glass tesserae from Hagia Sophia. *Microchem. J.* **2016**, *125*, 170–184. [CrossRef]
20. Tite, M.; Freestone, I.; Mason, R.; Molera, J.; Vendrell-Saz, M.; Wood, N. Lead glazes in antiquity—Methods of Production and reasons for use. *Archeometry* **1998**, *40*, 241–260. [CrossRef]
21. Carmona, N.; Laiz, L.; Gonzalez, J.M.; Garcia-Heras, M.; Villegas, M.-A.; Saiz-Jimenez, C. Biodeterioration of historic stained glasses from the Cartuja de Miraflores (Spain). *Int. Biodeterior. Biodegrad.* **2006**, *58*, 155–161. [CrossRef]

22. Müller, E.; Drewello, U.; Drewello, R.; Weißmann, R.; Wuertz, S. In Situ analysis of biofilms on historic window glass using confocal laser scanning microscopy. *J. Cult. Herit.* **2001**, *2*, 31–42. [CrossRef]
23. Piñar, G.; Garcia-Valles, M.; Gimeno, D.; Fernandez-Turiel, J.-L.; Ettenauer, J.; Sterflinger, K. Microscopic, chemical, and molecular-biological investigation of the decayed medieval stained window glasses of two Catalonian churches. *Int. Biodeterior. Biodegrad.* **2013**, *84*, 388–400. [CrossRef]
24. Pinto, A.M.C.; Sanjad, T.A.; Angélica, R.S.; Da Costa, M.L.; Paiva, R.S.; Palomar, T. 19th century stained-glass windows from Belém do Pará (Brazil): Analytical characterisation and pathology. *Boletín Soc. Española Cerámica Vidr.* **2018**, *57*, 133–141. [CrossRef]
25. Marvasi, M.; Vedovato, E.; Balsamo, C.; Macherelli, A.; Dei, L.; Mastromei, G.; Perito, B. Bacterial community analysis on the Mediaeval stained glass window "Natività" in the Florence Cathedral. *J. Cult. Herit.* **2009**, *10*, 124–133. [CrossRef]
26. Giacomucci, L.; Bertoncello, R.; Salvadori, O.; Martini, I.; Favaro, M.; Villa, F.; Sorlini, C.; Cappitelli, F. Microbial Deterioration of Artistic Tiles from the Façade of the Grande Albergo Ausonia & Hungaria (Venice, Italy). *Microb. Ecol.* **2011**, *62*, 287–298. [CrossRef] [PubMed]
27. Oliveira, M.M.; Sanjad, T.B.C.; Bastos, C.J.P. Biological degradation of glazed ceramic tiles. In *Proceedings of the Historical Constructions*; Lourenço, P.B., Roca, P., Eds.; Universidade do Minho: Guimarães, Portugal, 2001; pp. 337–342. Available online: http://www.hms.civil.uminho.pt/events/historica2001/page%20337-342%20_17_.pdf (accessed on 8 October 2021).
28. Coutinho, M.L.D.; Miller, A.Z.; Martin-Sanchez, P.M.; Mirao, J.; Gomez-Bolea, A.; Machado-Moreira, B.; Cerqueira-Alves, L.; Jurado, V.; Saiz-Jimenez, C.; Lima, A.; et al. A multiproxy approach to evaluate biocidal treatments on biodeteriorated majolica glazed tiles. *Environ. Microbiol.* **2016**, *18*, 4794–4816. [CrossRef] [PubMed]
29. Coutinho, M.L.D.; Miller, A.Z.; Gutierrez-Patricio, S.; Hernandez-Marine, M.; Gómez-Bolea, A.; Rogerio-Candelera, M.A.; Philips, A.; Jurado, V.; Saiz-Jimenez, C.; Macedo, M.F. Microbial communities on deteriorated artistic tiles from Pena National Palace (Sintra, Portugal). *Int. Biodeterior. Biodegrad.* **2013**, *84*, 322–332. [CrossRef]
30. Verde, S.C.; Silva, T.; Corregidor, V.; Esteves, L.; Dias, M.I.; Souza-Egipsy, V.; Ascaso, C.; Wierzchos, J.; Santos, L.; Prudêncio, M.I. Microbiological and compositional features of green stains in the glaze of the Portuguese "Great View of Lisbon" tile panel. *J. Mater. Sci.* **2015**, *50*, 6656–6667. [CrossRef]
31. Pedi, N.; Conceição, E.; Fernandes, M.J.; Massa, D.; Nogeira, E.; Ribeiro, P.; Arcoverde, J.H.; Lemos, S.; Marsden, A.; Neves, R. Fungos isolados em azulejos do convento de Santo António, Recife, Pernambuco. IX Jornadas de Ensino, Pesquisa e Extensão—#JEPEX. 2009. Available online: http://www.eventosufrpe.com.br/jepex2009/cd/resumos/r0550-1.pdf (accessed on 8 October 2021).
32. Cuzman, A.O.; Tiano, P.; Ventura, S.; Frediani, P. Biodiversity on Stone Artifacts. In *The Importance of Biological Interactions in the Study of Biodiversity*; InTech: London, UK, 2011; pp. 367–390.
33. Cecchi, G.; Pantani, L.; Raimondi, V.; Tomaselli, L.; Lamenti, G.; Tiano, P.; Chiari, R. Fluorescence lidar technique for the remote sensing of stone monuments. *J. Cult. Herit.* **2000**, *1*, 29–36. [CrossRef]
34. Tourney, J.; Ngwenya, B.T. The role of bacterial extracellular polymeric substances in geomicrobiology. *Chem. Geol.* **2014**, *386*, 115–132. [CrossRef]
35. Wimpenny, J.; Manz, W.; Szewzyk, U. Heterogeneity in biofilms. *FEMS Microbiol. Rev.* **2000**, *24*, 661–671. [CrossRef]
36. Corrêa Pinto, A.M.; Palomar, T.; Alves, L.C.; da Silva, S.H.M.; Monteiro, R.C.; Macedo, M.F.; Vilarigues, M.G. Fungal biodeterioration of stained-glass windows in monuments from Belém do Pará (Brazil). *Int. Biodeterior. Biodegr.* **2019**, *138*, 106–113. [CrossRef]
37. Schabereiter-Gurtner, C.; Pinar, G.; Lubitz, W.; Rölleke, S. Analysis of fungal communities on historical church window glass by denaturing gradient gel electrophoresis and phylogenetic 18S rDNA sequence analysis. *J. Microbiol. Methods* **2001**, *47*, 345–354. [CrossRef]
38. Crous, P.; Shivas, R.; Wingfield, M.; Summerell, B.; Rossman, A.; Alves, J.; Adams, G.; Barreto, R.; Bell, A.; Coutinho, M.; et al. Fungal Planet description sheets: 128–153. *Pers.-Mol. Phylogeny Evol. Fungi* **2012**, *29*, 146–201. [CrossRef] [PubMed]
39. Corrêa Pinto, A.M. Study and Conservation on Stained-Glass Windows of Historical Buildings from Belém do Pará, Brazil. Ph.D. Thesis, Universidade NOVA de Lisboa, Lisboa, Portugal, 2018.
40. Grinn-Gofroń, A. Airborne Aspergillus and Penicillium in the atmosphere of Szczecin, (Poland) (2004–2009). *Aerobiologia (Bologna)* **2011**, *27*, 67–76. [CrossRef] [PubMed]
41. Gorbushina, A.A.; Palinska, K.A. Biodeteriorative processes on glass: Experimental proof of the role of fungi and cyano-bacteria. *Aerobiologia* **1999**, *15*, 183–191. [CrossRef]
42. Macedo, M.F.; Miller, A.Z.; Dionísio, A.; Saiz-Jimenez, C. Biodiversity of cyanobacteria and green algae on monuments in the Mediterranean Basin: An overview. *Microbiology* **2009**, *155*, 3476–3490. [CrossRef]
43. Keshari, N.; Adhikary, S.P. Diversity of cyanobacteria on stone monuments and building facades of India and their phylogenetic analysis. *Int. Biodeterior. Biodegrad.* **2014**, *90*, 45–51. [CrossRef]
44. Otlewska, A.; Adamiak, J.; Gutarowska, B. Application of molecular techniques for the assessment of microorganism di-versity on cultural heritage objects. *Acta Biochim. Pol.* **2014**, *61*, 217–225. [CrossRef]
45. Diaz-Herraiz, M.; Jurado, V.; Cuezva, S.; Laiz, L.; Pallecchi, P.; Tiano, P.; Sanchez-Moral, S.; Saiz-Jimenez, C. Deterioration of an Etruscan tomb by bacteria from the order Rhizobiales. *Sci. Rep.* **2015**, *4*, 3610. [CrossRef]
46. Rosado, T.; Mirão, J.; Candeias, A.; Caldeira, A.T. Characterizing Microbial Diversity and Damage in Mural Paintings. *Microsc. Microanal.* **2015**, *21*, 78–83. [CrossRef]

47. Drewello, U.; Weißmann, R.; Rölleke, S.; Müller, E.; Wuertz, S.; Fekrsanati, F.; Troll, C.; Drewello, R. Biogenic surface layers on historical window glass and the effect of excimer laser cleaning. *J. Cult. Herit.* **2000**, *1*, S161–S171. [CrossRef]
48. Heaton, N. Lichens and their Action on the Glass and Leadings of Church Windows. *Nat. Cell Biol.* **1923**, *112*, 505–506. [CrossRef]
49. Watanabe, K.; Ohfuji, H.; Ando, J.; Kitagawa, R. Elemental behaviour during the process of corrosion of sekishu glazed roof-tiles affected by *Lecidea* s.lat. sp. (crustose lichen). *Clay Miner.* **2006**, *41*, 819–826. [CrossRef]
50. Watanabe, K.; Ohfuji, H.; Kitagawa, R.; Matsui, Y. Nanoscale pseudobrookite layer in the surface glaze of a Japanese sekishu roof tile. *Clay Miner.* **2009**, *44*, 177–180. [CrossRef]
51. Orial, G.; Warscheid, T.; Bousta, F.; Loisel, C. Incidence bactérienne dans les phénomènes de brunissement des vitraux anciens. *Actual. Chim.* **2007**, *312*, 34–39.
52. Silva, T.P.; Figueiredo, M.O.; Barreiros, M.A.; Prudêncio, M.I. Diagnosis of pathologies in ancient (seventeenth-eighteenth centuries) decorative blue-and-white ceramic tiles: Green stains in the glazes of a panel depicting Lisbon prior to the 1755 earthquake. *Stud. Conserv.* **2014**, *59*, 63–68. [CrossRef]
53. Jongmans, A.G.; Van Breemen, N.; Lundström, U.; Van Hees, P.A.W.; Finlay, R.; Srinivasan, M.; Unestam, T.; Giesler, R.; Melkerud, P.-A.; Olsson, M. Rock-eating fungi. *Nat. Cell Biol.* **1997**, *389*, 682–683. [CrossRef]
54. Rana, G.; Mandal, T.; Mandal, N.K.; Sakha, D.; Meikap, B.C. Calcite Solubilization by Bacteria: A Novel Method of Environment Pollution Control. *Geomicrobiol. J.* **2015**, *32*, 846–852. [CrossRef]
55. López-Arce, P.; Garcia-Guinea, J.; Fierro, J. Manganese micro-nodules on ancient brick walls. *Sci. Total Environ.* **2002**, *302*, 267–274. [CrossRef]
56. Wang, Q.; Ma, G.-Y.; He, L.-Y.; Sheng, X.-F. Characterization of bacterial community inhabiting the surfaces of weathered bricks of Nanjing Ming city walls. *Sci. Total Environ.* **2011**, *409*, 756–762. [CrossRef]
57. Gentaz, L.; Lombardo, T.; Chabas, A.; Loisel, C.; Verney-Carron, A. Impact of neocrystallisations on the SiO_2–K_2O–CaO glass degradation due to atmospheric dry depositions. *Atmos. Environ.* **2012**, *55*, 459–466. [CrossRef]
58. Munier, I.; Lefèvre, R.; Losno, R. Atmospheric factors influencing the formation of neocrystallisations on low durability glass exposed to urban atmosphere. *Int. Congr. Glas.* **2002**, *43C*, 114–124.
59. Garcia-Vallès, M.; Vendrell-Saz, M. The glasses of the transept's rosette of the cathedral of Tarragona: Characterisation, classification and decay. *Bol. la Soc. Esp. Ceram. y Vidr.* **2002**, *41*, 217–224. [CrossRef]
60. Çalişkan, M. The Metabolism of Oxalic Acid. *Turk. J. Zool.* **2000**, *24*, 103–106.
61. Gallien, J.P.; Gouget, B.; Carrot, F.; Orial, G.; Brunet, A. Alteration of glasses by micro-organisms. *Nucl. Instrum. Methods Phys. Res. Sect. B Beam Interact. Mater. Atoms* **2001**, *181*, 610–615. [CrossRef]
62. Miller, A.; Dionísio, A.; Braga, M.S.; Hernández-Mariné, M.; Afonso, M.; Muralha, V.; Herrera, L.; Raabe, J.; Fernandez-Cortes, A.; Cuezva, S.; et al. Biogenic Mn oxide minerals coating in a subsurface granite environment. *Chem. Geol.* **2012**, *322–323*, 181–191. [CrossRef]
63. Gadd, G.M. Metals, minerals and microbes: Geomicrobiology and bioremediation. *Microbiology* **2010**, *156*, 609–643. [CrossRef]
64. Gadd, G.M.; Rhee, Y.J.; Stephenson, K.; Wei, Z. Geomycology: Metals, actinides and biominerals. *Environ. Microbiol. Rep.* **2012**, *4*, 270–296. [CrossRef]
65. Vilarigues, M.; da Silva, R. The effect of Mn, Fe and Cu ions on potash-glass corrosion. *J. Non-Cryst. Solids* **2009**, *355*, 1630–1637. [CrossRef]
66. Sterpenich, J.; Libourel, G. Using stained glass windows to understand the durability of toxic waste matrices. *Chem. Geol.* **2001**, *174*, 181–193. [CrossRef]
67. Guillitte, O. Bioreceptivity: A new concept for building ecology studies. *Sci. Total Environ.* **1995**, *167*, 215–220. [CrossRef]
68. Ferrándiz-Mas, V.; Bond, T.; Zhang, Z.; Melchiorri, J.; Cheeseman, C. Optimising the bioreceptivity of porous glass tiles based on colonization by the alga Chlorella vulgaris. *Sci. Total Environ.* **2016**, *563–564*, 71–80. [CrossRef]
69. Littmann, E.; Autefage, H.; Solanki, A.; Kallepitis, C.; Jones, J.; Alini, M.; Peroglio, M.; Stevens, M. Cobalt-containing bioactive glasses reduce human mesenchymal stem cell chondrogenic differentiation despite HIF-1α stabilisation. *J. Eur. Ceram. Soc.* **2018**, *38*, 877–886. [CrossRef]
70. Raja, F.N.S.; Worthington, T.; Isaacs, M.A.; Chungong, L.F.; Burke, B.; Addison, O.; Martin, R.A.; Burke, B. The Antimicrobial Efficacy of Hypoxia Mimicking Cobalt Oxide Doped Phosphate-Based Glasses against Clinically Relevant Gram Positive, Gram Negative Bacteria and a Fungal Strain. *ACS Biomater. Sci. Eng.* **2018**, *5*, 283–293. [CrossRef]
71. Römich, H.; Jägers, E.; Torge, M.; Müller, W.; Adam, K. Cleaning: A Balancing Act. Available online: https://www.cvma.ac.uk/conserv/cleaning.html (accessed on 8 October 2021).
72. Coutinho, M.L.; Veiga, J.P.; Macedo, M.F.; Miller, A.Z. Testing the Feasibility of Titanium Dioxide Sol-Gel Coatings on Portuguese Glazed Tiles to Prevent Biological Colonization. *Coatings* **2020**, *10*, 1169. [CrossRef]
73. Aversa, R.; Perrotta, V.; Petrescu, R.V.; Carlo, M.; Petrescu, F.I.; Apicella, A. From Structural Colors to Su-per-Hydrophobicity and Achromatic Transparent Protective Coatings: Ion Plating Plasma Assisted TiO_2 and SiO_2 Nano-Film Deposition. *Am. J. Eng. Appl. Sci.* **2016**, *9*, 1037–1045. [CrossRef]
74. Bertoncello, R.; Milanese, L.; Dran, J.C.; Bouquillon, A.; Sada, C. Sol–gel deposition of silica films on silicate glasses: Influence of the presence of lead in the glass or in precursor solutions. *J. Non-Cryst. Solids* **2006**, *352*, 315–321. [CrossRef]
75. Bianco, B.D.; Bertoncello, R. Sol–gel silica coatings for the protection of cultural heritage glass. *Nucl. Instrum. Methods Phys. Res. Sect. B Beam Interact. Mater. Atoms* **2008**, *266*, 2358–2362. [CrossRef]

applied sciences

MDPI

Review

Bioconservation of Historic Stone Buildings—An Updated Review

Benjamín Otto Ortega-Morales [1] and Christine Claire Gaylarde [2,*]

[1] Department of Environmental Microbiology and Biotechnology, Universidad Autónoma de Campeche, Av. Agustín Melgar s/n. C.P., 24039 Campeche, Mexico; beortega@uacam.mx

[2] Department of Microbiology and Plant Biology, Oklahoma University, 770 Van Vleet Oval, Norman, OK 73019, USA

* Correspondence: cgaylarde@gmail.com

Abstract: Cultural heritage buildings of stone construction require careful restorative actions to maintain them as close to the original condition as possible. This includes consolidation and cleaning of the structure. Traditional consolidants may have poor performance due to structural drawbacks such as low adhesion, poor penetration and flexibility. The requirement for organic consolidants to be dissolved in volatile organic compounds may pose environmental and human health risks. Traditional conservation treatments can be replaced by more environmentally acceptable, biologically-based, measures, including bioconsolidation using whole bacterial cells or cell biomolecules; the latter include plant or microbial biopolymers and bacterial cell walls. Biocleaning can employ microorganisms or their extracted enzymes to remove inorganic and organic surface deposits such as sulfate crusts, animal glues, biofilms and felt tip marker graffiti. This review seeks to provide updated information on the innovative bioconservation treatments that have been or are being developed.

Keywords: biocleaning; bioconsolidation; biocalcification; biopolymers; carbonatogenic bacteria; cultural heritage; MICP; microorganisms; stone restoration

Citation: Ortega-Morales, B.O.; Gaylarde, C.C. Bioconservation of Historic Stone Buildings—An Updated Review. *Appl. Sci.* **2021**, *11*, 5695. https://doi.org/10.3390/app11125695

Academic Editors: Maria Filomena Macedo, António Manuel Santos Carriço Portugal, Ana Miller and Ana Catarina Pinheiro

Received: 22 February 2021
Accepted: 16 June 2021
Published: 19 June 2021

1. Introduction

Our stone cultural heritage is subject to weathering over the years. This deterioration is due to physical, chemical and biological factors acting synergistically that cause disfiguration and dissolution of the stone, an increase in porosity and overall weakening of the structure [1]. When the building is of historic and cultural importance, it is necessary that careful restorative action be taken, with the aim of returning the monument, as much as possible, to its original condition. This may involve removal of unwanted surface deposits (cleaning) and strengthening of degraded parts of the structure (consolidation), and ultimately replacement of blocks or whole sections of the structure. Several processes, such as microbial transformation of stone minerals [1], discoloring by microbial pigments including algal, bacterial and fungal biomolecules with staining properties [2–5], and atmospheric pollutant deposition [6], which may all operate simultaneously at surfaces, can make cleaning procedures challenging. Furthermore, cleaning can induce irreversible damage; thus a cost-benefit analysis should be considered in a prior assessment phase, in order to reduce potential risks to artwork [7].

The consolidation and conservation of these buildings requires the use of materials that are compatible with the original structure. Consolidants that have been used in the past, or, indeed, are still in use, include those based on lime water, which is not very efficient [8], and silicic acid, which can lack the necessary flexibility for resisting stresses within the stone [9]. Nanoconsolidants, used as recent improvements in consolidation strategies [10] include the use of nanolime [11] and nanosilica and may be more effective in reducing porosity [12,13]. Nanolime doped with calcium hydroxide/zinc oxide quantum dots has been shown to form a more durable conservation treatment for limestone than silica-based consolidants [14], but it is more costly. The disadvantages of traditional consolidants, for both calcareous and siliceous stone, are poor performance, structural drawbacks such as low adhesion, poor penetration and flexibility and, in the case of organic substances such as alkoxysilanes, the necessity to be dissolved in volatile organic compounds (VOCs), which carry environmental and human health risks [15].

Synthetic resins, such as silane, epoxy, acrylic and polysiloxane, plug the pores when they polymerize, thus causing water retention and internal degradation, while external coatings can peel off. In order to increase the efficiency, several chemical modifications of old consolidants have been produced [16]. However, perhaps more promising are the new, biologically-based consolidants (bioconsolidants), which do not rely on synthetic components, but are naturally-occurring structural plant biopolymers [17,18], or utilize the enzyme activities of microorganisms to produce new materials.

Not only consolidants, but other traditional conservation treatments, can be replaced by more environmentally acceptable, biologically-based, measures to restore stone. This review seeks to provide updated information on these innovative conservation treatments that have been or are being developed.

2. Bioconsolidation Techniques

According to Wheeler [19], a consolidant is a material, or system of more than one material, that penetrates a substratum, improving the inner structure and thereby enhancing the mechanical properties, thus improving the adhesion of surfaces. A good consolidant needs to meet three criteria: (i) compatibility (not causing chemical or physical damage) with the original substratum; (ii) effectiveness by penetrating evenly within the inner structure, and (iii) durability by not yielding noxious by-products after application. Recently, Negri et al. [20] also stressed the need to meet the retractability criterion by not interfering with previous interventions.

Limestone and lime-based mortars have been consolidated using plant biopolymers. Reported biopolymers have been sourced from *Aloe vera*, *Cylindropuntia californica*, *Opuntia engelmannii*, *Opuntia ficus-indica*, *Salvia hispanica* and *Sida angustifolia* [17,21]. Biopolymers are obtained in a range of extraction procedures, soaking leaves and other leafy tissues in batches of water and subsequently leaving them to be released into the bulk water. This water containing the biopolymers is then used for mixing with lime and other materials and used for consolidation or used as a suspension. Consolidants are then applied to the stone surface using techniques such as spraying, brushing or partial immersion, or mixed with other materials to yield composites [22]. Biopolymers alter mechanical properties of building materials and their water absorption and diffusion behavior, thus increasing cohesion and enhancing mechanical properties [23]. Bioconsolidation is also achieved by direct or indirect application of microorganisms and/or their metabolites [24]. Microbially induced carbonate precipitation (MICP), also known as biocalcification, biocarbonatogenesis, or biomineralization, has been suggested as an environmentally friendly method for the consolidation of cultural heritage buildings. It is a natural phenomenon, induced by a wide variety of microorganisms, that can be responsible for chalking (the production of a white, powdery surface) [25], as well as playing an essential role in the formation of stromatolites [26]. This microbially mediated mineralization treatment must produce a coherent calcium carbonate layer that can protect deteriorated stone against water uptake and consolidate its inner structure. Such a natural process should result in a mineral product similar to the initial calcareous stone substrate, a highly desirable result in the eyes of conservators.

There have been several strategies: the application of selected carbonatogenic bacteria to the stone, enrichment of naturally occurring carbonatogenic bacteria, application of cell-free bacterial products, and stimulation of the relevant microorganisms among those already present (autochthonous microorganisms). There have been a number of reviews covering these options in recent years [26–31]. In this bibliographic search, we have attempted to avoid the sometimes lengthy discussions of the history behind the application of the technology and its variants, along with the minutiae of the treatments, preferring to reference such studies and using Table 1 to indicate relevant aspects of the developmental research. We have also considered the less frequently discussed environmental impacts of the new methods; only very recently have the effects on the normal microbial communities of the stone surfaces, and the resistance of the restoration to colonization by potential deteriogens been considered. A perspective on the development of biotechnological sustainable products for ultimate long-term use is also included.

2.1. Microbial Cells in Bioconsolidation Treatments

When considering the application of selected bacteria, a good option is to use organisms isolated from the autochthonous population, which are obviously already adapted to the stone environment and are less likely to interact negatively with the endogenous community [24,32,33]. The first such bacterial strain to be so isolated was a *Bacillus cereus*, which produced carbonate crystals in the alkaline conditions caused by breakdown of amino acids to ammonia [34]. It was successfully tested in the field in 1999 (see Table 1 [35]).

Table 1. Examples of carbonatogenic microorganisms that have been tested for MICP potential.

Organisms	Isolation Source/ Habitat	Testing Method	Reference
Bacillus, Pseudomonas, Brevibacterium, Streptomyces, Stenotrophomonas genera	Degraded limestone monument (Romania) Isolated during the course of the study	Crystal formation by isolated strains on solid medium; crystal identification by FTIR, XRD and SEM	[36]
Acinetobacter sp.	Cave (Yixing Shanjuan, China) Carbonate stone (Spain) Isolated during the course of the study	Consolidation of artificial cracks in masonry cement mortars; compressive strength, water absorption, SEM, XRD, thermogravimetry In vitro consolidation with *Mixococcus xanthus;* chemical analysis of calcarenite stone	[37] [32]
Agrococcus jejuensis sMM51 (Soil, Middle Muschelkalk)	Limestone-associated groundwater, rock and soil (Germany)	Crystal formation in liquid culture; crystal morphology by XRD, EDS and SEM	[38]
Bacillus muralis rLMd (Rock, Lower Muschelkalk) *Bacillus* sp. rMM9 (Rock, Middle Muschelkalk)	The strains are deposited with the Jena Microbial Resource Collection (Jena, Germany)		
Bacillus subtilis 168 (®27370)	ATCC: American Type Culture Collection, Manassas (VA) USA	Crystal formation on solid medium; detailed crystal analysis	[39]
Indigenous bacterial community	Salt-damaged carbonate stone in San Jeronimo Monastery, Spain. Isolated during the course of the study	Consolidation of originating monument; drilling resistance, SEM	[24]
Synechococcus pevalleikii (live and dead)	National marine laboratories, Bharathidasan University, Tiruchirappalli, Tamil Nadu, India	Concrete cubes in vitro. U-V treated cells gave better compressive strength and lower water uptake	[40]
Synechocystis (6803)	PCC: Pasteur Culture Collection (Paris), France	Morphological and spectroscopic changes of mortar surfaces	[41]
Gloeocapsa (73106).	PCC: Pasteur Culture Collection (Paris), France	Measurement of compressive strength, water absorption and porosity of treated mortar	[42]
Pseudomonas sp. (N9), *Bacillus cereus* (T6), *Lysinibacillus sphaericus* (T5), *Bacillus* sp.	Historic white marble (China) Isolated during the course of the study	Crystal formation in solid and liquid culture by XRD analysis	[43]

Table 1. *Cont.*

Organisms	Isolation Source/ Habitat	Testing Method	Reference
Bacillus sp., *Micrococcus* sp.	Black crusts on limestone buildings Isolated during the course of the study	Bio-cementation on limestone slabs; Bio-consolidation of fragmented stones (Church of Santa Maria dei Miracoli, Venice); SEM, EDX	[44]
Bacillus cereus	Natural carbonate rock Isolated during the course of the study	First in situ consolidation attempt (Saint Médard Church, Thouars, France); SEM, water absorption, surface roughness, colorimetry	[45]
Bacillus subtilis LMG 3589	Belgian Co-ordinated Collections of Microorganisms	Consolidation of deteriorated *Globigerina* limestone in Malta; drilling resistance, water absorption, salt deterioration, porosity	[46]
Lysinibacillus sphaericus, Bacillus subtilis, Pseudomonas putida	INQCSS: Instituto Nacional de Controle de Qualidade em Saúde. Brazilian Culture Collection (Rio de Janeiro, Brazil)	Production of $CaCO_3$ in growth medium	[47]
Pseudomonas, Pantoea, Cupriavidus	Ancient marble quarry in Athens, Greece Isolated during the course of the study	SEM, ERD and FTIR analyses of treated marble from same quarry	[48]
Psychrobacillus psycrodurans	Mortar and concrete samples (National University of Colombia—Bogota, buildings) (IBUN: Institute of Biotechnology of the National University of Colombia. Collection of microorganisms (Bogotá, Colombia))	Biocementation tests on mortar cubes; SEM, XRD, compressive strength	[49]
B. licheniformis DSMZ 8782, *B. cereus* 4b, *S. epidermidis* 4a, *M. luteus* BS52, *M. luteus* 6	DSMZ: German Collection of Microorganisms and Cell Cultures GmbH (Brunswick, Germany) Collection of the Enzymology laboratory of B.P. Konstantinov Petersburg, Nuclear Physics Institute NRC "Kurchatov Institute" (Moscow, Russia)	Repairing microcracks in cement, calcium carbonate precipitation	[50]
Sporosarcina pasteurii	DSMZ: German Collection of Microorganisms and Cell Cultures GmbH (Brunswick, Germany)	3-D printing using sand; hardening tests	[51]
Pseudomonas (isolates D2 and F2) and *Acinetobacter* (isolate B14)	Freshwater sessile bacteria Isolated during the course of the study	Histological and fluorescence staining determination of cell viability inside carbonate crystals, and pore size reduction in limestone by image analysis.	[52]

Table 1. *Cont.*

Organisms	Isolation Source/ Habitat	Testing Method	Reference
Penicillium chrysogenum CS1 (Cement Sludge)	Isolated from Cement sludge Isolated during the course of the study	Cementation in sand column to form sandstone; compressive strength	[53]
Colletotrichum acutatum	Diseased fruit crops Isolated during the course of the study	Inoculation on limestone; SEM, EDX, XRD	[54]
Paecilomyces inflatus, *Plectosphaerella cucumerina*.	Stalactite growing from a concrete ceiling Isolated during the course of the study	SEM and XRD of carbonate crystals associated with hyphal growth in broth	[55]

Abbreviations: FTIR: Fourier transform infra-red spectroscopy; XRD: X-ray diffraction analysis; SEM: scanning electron microscopy; EDS: energy dispersive X-ray spectroscopy.

One published example is that of Micallef et al. [45], who effectively treated deteriorated limestone with biocalcifying *B. subtilis* isolated from Maltese hypogea. The treatment conferred uniform bioconsolidation to a depth of 30 mm and the treated stone had high resistance to salt deterioration and low water absorption, with a preserved pore network. Further work on this technology is still necessary. Research by one of the authors is currently underway on a limestone wall in Campeche, southern México (Figure 1). Bioconsolidation of the damaged limestone blocks is underway using a suspension of a locally isolated carbonatogenic bacterial strain applied by spraying. Frames have been set to study the efficiency of consolidation in several quadrants, to determine variability of the results. Efficiency is being measured by surface hardness and cohesion, using peeling tests. Preliminary measurements made two months after application revealed a slight increase of surface hardness and less flake detachment. The experimental study will be continued for up to one year and both short- and medium-term effects of the treatment will be determined under the natural climatic conditions of this region, classified as Köppen climate As (Tropical Savannah).

Many other microorganisms have been tested for their calcifying and consolidation activities, including fungi (see Table 1). An important proportion of microorganisms reported in Table 1 are well characterized isolates that are deposited in established culture collections, while a significant number of bacteria were isolated during the course of the respective studies. The latter implies they are not accessible to other research groups for reproducibility or to undertake new studies. This is important for further research and also in terms of intellectual property rights. Access to microbial strains possessing desired properties such as biocalcifying activity can be exploited by patented biotechnological processes.

Chuo [55] suggested that the most studied bacterium for MICP over the years was *Sporosarcina pasteurii*. Omoregie et al. [56] found that this genus was the most abundant biocalcifying organism isolated from limestone cave samples, but many other species have been isolated, both by these authors and others (for example, [57]); since microorganisms with this ability are common, it is clear that there is a wide range of species available from which to choose the most appropriate for the particular stone and environment. The stone environment is often considered oligotrophic and thus isolation strategies of lithic-dwelling microorganisms should consider the design and use of low nutrient media [58,59].

Figure 1. Limestone wall in Campeche, southern México, exhibiting a range of deterioration pathologies. Formers are being used to demarcate panels prior to treatment by spraying with carbonatogenic bacterial suspension or control solution. Replicate panels will allow statistical analysis. The aim is to increase surface hardness and reduce the detachment rate of flaking material.

Not only is the type of culture medium an important factor in the design of strategies aiming to retrieve a large number of relevant bacteria from environmental samples, but it is also necessary to use a range of isolation methods, including plating, enrichment and micromanipulation or extinction-culturing, either alone or in combination [60]. This combined approach is imperative to obtain more pure cultures of potentially dominant microorganisms derived from the stone environment, allowing a significant advance in the study of the physiology of carbonatogenic microorganisms and the environmental determinants of their activity and abundance. Pure physiologically characterized novel bacterial isolates are a prerequisite for biotechnological development across economic sectors and markets, ranging from those traditional ones already used in the food or pharmaceutical industries to emerging sectors such as cultural heritage protection.

The relatively new techniques of metagenomics and metabolomics could also play a role in selecting suitable organisms. Chimienti et al. [61], using next generation sequencing (NGS), identified the presence of known carbonatogenic bacteria like *Arthrobacter* in the populations colonizing the medieval church of San Leonardo di Siponto in Italy. 16S rRNA and functional gene analyses could show the carbonatogenic potential of metabolic pathways linked to the appropriate biogeochemical cycles. For metabolomics to be successfully applied to the discovery of new potentially useful organisms, however, it is necessary to better understand the biochemical processes involved in carbonatogenesis. Earlier genetic studies have, for example, used mutants impaired in $CaCO_3$ precipitation to indicate a link between biomineralization, fatty acid metabolism, altered phospholipid membrane composition and surface properties [8,31,62–64]. Fatty acid metabolism has also been implicated by work that showed how $CaCO_3$ precipitation in *Lysinibacillus* could modify membrane rigidity by upregulating branched chain fatty acid synthesis [65,66].

The prospects for the future are promising, with the combined use of improved culturing methods and metabolomics-based selection of novel carbonatogenic species. The study of their physiology and ecology should enhance the success of bioconsolidation strategies, in particular for long-term application.

Of course, other factors affect biocalcite precipitation, apart from the particular microbial species. These are cell and nutrient concentrations, source of calcium, presence of other substances, such as surfactants used as dispersants [55] and salt [24], and methods used to inoculate the bacteria onto the stone surfaces. There are several publications around this topic [31,67–69].

The alternative to application of selected carbonatogenic bacteria is to try to stimulate those already present on the stone surface. Jimenez-Lopez et al. [70] tested this approach in vitro, immersing porous limestone slabs in a nutrient medium. The positive result obtained was shown to be due to ammonification of the medium. Later, a similar approach was adopted by Jroudi et al. [71] for the consolidation of badly degraded tuff stone and lime plaster at the Mayan site of Copan (Honduras). The stone was treated with patented (sterile) M-3P nutrient solution containing amino acids and calcium. This resulted in significant changes in the indigenous bacterial community, which were detected using NGS techniques. There were increased levels of *Arthrobacter*, *Micrococcaceae*, *Nocardioides*, *Fictibacillus*, *Streptomyces* and *Rubrobacter*. In the lime plaster, *Bacillus*, *Agrococcus*, and *Microbacterium* were the major genera after treatment. It is notable that most of the increased components of the community after treatment were capable of calcium carbonate biomineralization. Although this may indicate that the treatment could replace application of living microorganisms, the detection of biocalcifying microorganisms does not necessarily imply that these will produce a good consolidating performance, and further testing, such as that suggested in Table 2 below, would be necessary.

Table 2. Properties of representative consolidants and methods of evaluation for stone conservation [72,73].

Consolidating Properties	Consolidation Action	Substrate	Consolidation Treatments	Evaluation of Treatment
Effectiveness	Penetration depth	Marble, limestone, and lime-based mortars	Phosphate treatment based on Hydroxyapatite (HAP) Ethyl silicate (ES)	Scanning electron microscopy (SEM)
Compatibility	Morphology and microstructure of samples resulting from applied products	Marble	Nano-solution of calcium tetrahydrofurfuryloxide (Ca (OTHF)2)	Optical Microscopy (OM) and Scanning Electron Microscopy coupled with Energy Dispersive X-ray spectroscopy (SEM-EDX)
Effectiveness	Mechanical properties: compressive strength, tensile strength, bending strength, modulus of elasticity, ultrasonic pulse velocity, abrasion loss, surface hardness.	Marble, limestone, and lime-based mortars	HAP ES	For example, tensile strength (σ_t) by the Brazilian splitting test using an Amsler-Wolpert loading machine

Table 2. *Cont.*

Consolidating Properties	Consolidation Action	Substrate	Consolidation Treatments	Evaluation of Treatment
Effectiveness	Hardness of the substrate; Increase in mechanical strength; penetration depth of treatment	Limestone/lime based mortar	Nanolime	Drilling Resistance Measurement System (DRMS)
Compatibility	Color changes induced by the treatment	Limestone/ Marble	HAP/nano-solution of calcium tetrahy-drofurfuryloxide (Ca (OTHF)2)	Spectrophotometer Konica Minolta CM-700d to measure CIEL*a*b* coordinates
Compatibility	Newly formed phases and secondary by-products	Marble, limestone, and lime-based mortars	HAP ES	Fourier transform infrared spectroscopy (FT-IR)
Compatibility	Microstructure: variations in open porosity and pore size distribution after the treatment	Marble, limestone, and lime-based mortars	HAP ES	Mercury intrusion porosimetry (MIP)
Compatibility	Thermal behavior: samples subjected to thermal cycles (heating–cooling cycle)	Marble, limestone, and lime-based mortars	HAP ES	Use of dilatometer L75/30/C/W Ceramic Instruments
Durability	Accelerated weathering cycles: Wetting–drying, freezing–thawing and salt crystallization cycles	Marble, limestone, and lime-based mortars	HAP ES	European EN 12371 and Italian UNI 11186 for freeze–thaw test, European EN 12370, RILEM MS-A.1 and RILEM MS-A.2 for salt weathering

2.2. Microbial Products in Bioconsolidation Treatments

Perito et al. [74] showed that dead, as well as live, *B. cereus* cells could produce carbonate crystals in liquid medium, the cell bodies acting as crystallization nuclei. It is not necessary to have whole cells for this; bacterial surfaces such as cell walls or extracellular polymeric substances (EPS) have metal binding properties that can also serve as nucleation sites [75–77]. The ability of cell walls to uptake cations such as Ca^{2+} was demonstrated directly for *B. subtilis* in much earlier work [78].

The advantages of non-cell treatments are that nutrients and appropriate growth conditions are not required and that cell components are smaller than whole cells and therefore able to penetrate further into cracks and pores (but, of course, without clogging the latter). This type of product is, however, more difficult and more expensive to prepare. A mixed inoculant, containing both EPS and living carbonatogenic cells, could give an improved performance [79].

EPS have fundamental cellular functions, including attachment to surfaces and provision of a framework for architectural growth in biofilms. They possess a varied chemistry; the composing biomolecules may include proteins, polysaccharides, lipids plus nucleic acids, lipids alone, and uronic acid. EPS may inhibit or enhance precipitation of calcium carbonate, depending on the functional groups which enhance precipitation by serving as initial nucleation sites, such as anionic groups like sulfate and acidic sugars [80]. Microbially influenced precipitation occurs due to the interactions of extracellular biopolymers and the geochemical environment. This is seen, for example, in the production of moonmilk speleothems, natural calcium carbonate deposits in caves, which has been associated with filamentous actinomycete activity through metabolic profiling of biomineralization pathways [81]. This, indeed, could be an interesting option for further research into bioconsolidation.

There is no consensus on the optimal conditions for biocalcification treatments, regardless of the actual method used, whether with or without cells [82]. This may be due to the fact that optimum conditions will vary depending on the type of stone, the geometry of the structure to be treated and the climate in the particular geographic location.

Conditions will certainly vary depending on the type of stone to be consolidated. Pore structure affects penetration depth and treatment performance [81]. The polymorphs produced by biocalcification are mainly calcite (rhombohedral), aragonite (needles) and vaterite (hexagonal), the final form depending on environmental conditions and bacterial strains. The most common precipitated forms are calcite and vaterite [83–85], calcite being the dominant and most thermodynamically stable [86,87], although it has been suggested that microbially-induced vaterite achieves similar stability to calcite through the incorporation of organic molecules [68,88]. Hydrated phases of $CaCO_3$ have also been reported [89,90]. The nanomechanical properties of the $CaCO_3$ polymorphs could be improved by increased understanding of the biogeochemical processes involved [91].

3. Effectiveness of the Consolidated Stone in the Environment

It is important to know how the new material will perform in its particular environment. Various test methods have been used to evaluate consolidation (Table 2).

In comparison to synthetic consolidants, such as those shown in Table 2, few studies have characterized the performance of bioconsolidation methods; this is an avenue of research that could allow the development of novel approaches and strategies integrating microbiological agents or their metabolites. Novel consolidants need not only to be assessed in terms of the properties shown in Table 2, but also in their response to any microbial insults from the immediate environment.

It is imperative that the effects of the environment on the new consolidant are considered, in order to assess consolidating integrity and durability. This deserves particular attention in the future, necessitating lengthy commitment to such long-term studies.

Biodeterioration Testing and Colonization of Biocalcite

There have been few studies on the colonization of the newly produced material by environmental microorganisms, although a recent article reviews the methodologies employed to assess the durability of cultural heritage stone surfaces in response to microbial colonization [92]. Recolonization after the use of non-biological consolidants on marble, sandstone and plaster in the archeological site of Fiesole, Italy, was shown to depend mainly on the bioreceptivity of the substrate and the climatic conditions [93]. A relatively recent paper suggests that the bioconsolidation treatment based on stimulation of indigenous calcifying microorganisms does not significantly alter the stone microbiota in the long term [94]. The authors showed, using DGGE monitoring of the stone-autochthonous microbiota before and at 5, 12 and 30 months after bioconsolidation with autochthonous bacteria, that the *Actinobacteria* that were initially completely dominant decreased to 44.2% after 5 months. After 12 months, *Cyanobacteria* (22.1%) appeared and remained dominant until thirty months. Thereafter, the population consisted of *Actinobacteria* (42.2%) and *Cyanobacteria* (57.8%). For the fungi, the *Ascomycota* phylum was dominant before treatment (100%), *Basidiomycota* (6.38%) appeared after five months, but disappeared after 12 months. After 30 months the fungal population started to stabilize and *Ascomycota* again dominated (83.33%). Green algae (*Chlorophyta*, Viridiplantae) were rare colonizers of the new material.

The results of this study indicated that final changes to the initial microbial populations were relatively small.

For many years, the only tests carried out on bioconsolidated stone were physical and chemical examinations, which gave information about the strength and chemical structure of the new material. There was no definitive knowledge about the ability of biocalcite to withstand the biological attacks to which it would be subjected in the environment. Indeed, Jroundi et al. [72] state that "It is unknown whether such a treatment is effective and does not produce any deleterious side effects under extreme hot and humid environmental conditions typical for the Maya area where . . . the potential for microbial biodeterioration is very high". In 2015, however, Shirakawa et al. [95] subjected fiber cement panels, treated in various ways to produce a surface layer of biocalcite, to biodeterioration testing in the hot and humid climate of Sao Paulo, Brazil. They found that the calcite formed by treatment with living *Bacillus sphaericus* LMG 222 57 plus urea plus B4 medium was the most resistant to the aesthetic biodeterioration that occurred within the 22 months of exposure. This resistance was correlated with lower water absorption and porosity, together with surface hydrophilicity, all linked to the smaller size of the biocalcite crystals. There was no apparent degradation of the crystals within the timescale of the test. Though promising, longer term experiments, often spanning several years, are necessary to truly know the extent of resistance of biocalcite to microbial attack under various environmental conditions.

4. Other Microbial Applications in Stone Conservation

Novel bioactive molecules based on microbial cells and products have also been used in biocleaning and protection of surfaces from microbial colonization, as discussed in the following sections. These biological techniques can be less invasive, and thus more acceptable, procedures than the traditional ones that use chemical treatments.

5. Biocleaning

Cleaning is one of the most important steps in the restoration process and normally comes before any other. When the building has special features, such as mural paintings, careful successive steps are necessary to remove the deposited polluting layers without affecting the original surface. The chemical and physical procedures traditionally used may have adverse effects on both the materials and the health of the restorer [96]. The recently introduced biological cleaning techniques that have been used on cultural heritage buildings in Italy, Spain and Greece, for example [97], avoid these problems. They may involve living bacterial cells or their hydrolytic enzymes. Both inorganic and organic materials can be removed gently from surfaces using these microbial products at temperatures that do not affect the underlying material.

5.1. Removal of Inorganic Materials—Black Crusts, Nitrate Crusts

Although some black crusts may be layers of dark pigmented cyanobacteria [3], those most commonly recognized on historic stone buildings are composed of gypsum (calcium sulphate dihydrate) and dark particulates from air pollution [98]. Traditionally, these are removed chemically with an ammonium carbonate-EDTA mixture, which may damage the underlying stone if not carefully applied. In 1992, Gauri et al. [99] reported a method for removing these layers from marble using a culture of *Desulfovibrio desulfuricans*, an anaerobic sulfate-reducing species better known for its ability to produce microbially-influenced corrosion of steel structures [100], and references therein). After various efforts to improve the treatment (e.g., [101,102]), it was shown to be effective in removal of the crusts and preservation of the noble patina beneath them and has now been applied successfully to various monuments (e.g., [103,104]). Care must be taken, however, with the duration of the treatment. Ranalli et al. [105] used sulfate-reducing bacteria to remove sulfate crusts and found that prolonged contact between the bacteria and the stone resulted in damaging sulfide precipitation.

Nitrate crusts are one of the types of stone efflorescences which are often considered to be of particular importance in deterioration of indoor artworks, such as mural paintings. Ranalli et al. [106] were among the first to suggest the use of nitrate reducing bacteria (a strain of *Pseudomonas stutzeri* in this case) to remove nitrate salts from stone. There have now been several successful cases of biocleaning of nitrates from stone; the wall paintings in the central vault of Santo Juanes church in Valentia, Spain, and the external wall of Matera cathedral are examples [107]. The latter building was treated with a novel carrier system (Carbogel) containing *Pseudomonas pseudoalcaligenes* that was covered with a PET film. Another novel application system was used on the mural paintings in the Santo Juanes church in Valentia [108]; *Pseudomonas stutzeri* DSMZ 5190 in a Japanese paper and agar layer was heated with infrared to a suitable temperature for optimum bacterial activity. Recently, Romano et al. [109] suggested that extremophilic bacteria might be good candidates for biological removal of efflorescences. Extremophiles live under extreme conditions, very high or very low temperatures, pH values and pressures, or in the presence of normally toxic agents such as irradiation and heavy metals. Often, they belong to the Archaea domain, but they can also be members of the Bacteria or the Eukarya (fungi or algae). The unusual conditions required for their growth means that their application as biocleaning agents should be safe, with no undesirable future effects on the environment or adverse effects on current lifeforms (including humans). This could make them more acceptable for use by conservators and restorers. Romano et al. [109] screened various extremophiles from the culture collection held by CNR, Italy, and selected an aerobic bacterium, *Halomonas campaniensis* 5AGT, as a potential remediation treatment for nitrate crusts. The bacterium lives at high pH values and is non-pathogenic.

Mixed salt crusts also occur and these have been removed by combined biological and chemical treatments ([110], and references therein). These crusts may contain not only sulfates and nitrates, but also carbonates, apatite and protein, hence posing a challenge to traditional techniques that could be overcome by these innovative methods.

5.2. Removal of Organic Materials

Heterotrophic microorganisms (those that utilize organic substrates) are very versatile, producing a wide range of enzymes that can break down many organic materials. Treatment with enzymes such as proteases and collagenases is already used by restorers to remove organic residues from paintings, for example, without affecting the base material [111]. Although microorganisms can be the source of such specific enzymes, it can be seen that the whole cells, containing a mixture of degrading enzymes, might be a more effective treatment in some cases, as well as being more economical. They can also be effective for otherwise difficult materials, such as casein, egg yolk, oil, and animal fat [110]. For example, Ranalli et al. [112] used cells of the bacterium *Pseudomonas stutzeri*, plus a final enzyme treatment, to remove organic residues, including animal glue residues, from wall paintings. The cost of the biological cleaning was much lower than that of conventional methods. More recently, the same group produced a new, improved, system, with an innovative carrier for *P. stutzeri* cells, for cleaning historic wall paintings [113]. Surface contamination of culturally important buildings by traffic fumes may also be susceptible to removal by microbial cells and their enzymes [114].

Rather than using whole bacterial cells, specific degrading enzymes can be isolated from them and used where the polluting material(s) is known. The addition of these enzymes to traditional cleaning methods can also enhance the results. Jeszeová et al. [115] successfully used a mixed enzyme preparation from the bacterium *Exiguobacterium undae*, containing proteases, peptidases, nucleases, peptide ABC transporter substrate-binding proteins and a phosphatase, to remove animal glue from 3 different types of substrate, glass, stone and wood from a range of trees species (*Quercus* sp., *Fagus sylvatica*, and *Picea abies*).

Microorganisms may be used to remove graffiti paint from heritage buildings. The modified alkyd binders and other organic polymers in the paints [116] are susceptible to some polymerases and lipases in microbial cells [116]. Sanmartin et al. [117] isolated a number of paint degrading strains from a wide variety of environmental sources. The most promising bacteria belonged to the genera *Arthrobacter*, *Bacillus*, *Gordonia*, *Microbacterium*, *Pantoea* and *Pseudomonas*, while fungi of the genus *Alternaria* also proved interesting. In 2019, the same group tested the system on stone surfaces, using spray-painted granite and concrete coupons as the test surfaces [118]. The bacteria were grown prior to use in a culture medium enriched with powdered graffiti, to encourage the adaptation of the cells to the target paint. Macroscopic and microscopic examination, together with color and infra-red evaluations, indicated the success of the treatment. Germinario et al. [119,120] examined the effectiveness of lipase enzymes extracted from various bacteria and fungi for removing blue, green, red and black felt tip marker acrylic ink from unglazed ceramic substrates. An oil-in-water emulsion alone could remove the ink, but the addition of lipase increased the efficiency. The more gentle treatment with cells or enzymes, rather than more abrasive chemical removal, is an obvious advantage for use on culturally important monuments.

Microorganisms have also been suggested for removing biofilms, which contain high levels of protein, from historic buildings [121]. Biofilms are produced by microorganisms attaching to and thriving on the stone surface, and contain, besides the cells themselves, their metabolic products, especially EPS, and compounds that may be released by their lysis, such as proteins, lipids and nucleic acids [122]. In their mildest form, biofilms simply disfigure the surface, presenting a disçolo red or dirty appearance; nevertheless, this requires treatment and, if possible prevention.

Valentini et al. [123] used the fungal enzyme glucose oxidase to clean the disfiguring patina from the surface of travertine and peperino stone from the Villa Torlonia in Rome. The hydrogen peroxide formed by the enzyme action was able to remove the thin biofilms, containing algae, cyanobacteria and heterotrophic bacteria, without etching the underlying substratum. The biocleaning method was especially recommended for travertine, whose somewhat lower porosity made it more resistant to peroxide etching. This biocleaning method was shown to be more efficient than the more traditional treatments using saturated $(NH_4)_2CO_3$ solution and EDTA in buffer and use of lipase enzyme.

Microbial cells have also been suggested as a control measure against biofilm formation. This relies on the fact that certain species can inhibit the growth of others, either by simple growth competition or by the production of antibiotic-like substances or other antagonistic metabolites [112,124,125]. This "biological control" is already used successfully in agriculture against plant pathogens (for example, [126,127]). However, the organisms that cause plant diseases are specific pathogens that can be inhibited by other (specific) organisms; cells growing in a biofilm on a stone surface will, unless the circumstances are unusual, consist of several species and genera. It has been suggested that antimicrobial substances produced by members of the genus *Bacillus* could be of interest in protecting stone surfaces from colonization ([110], and references therein), since this bacterial genus can produce a range of antibiotic substances affecting many types of bacteria; however, much work will be necessary before this becomes a truly viable option.

6. Conclusions

The great advantage of biorestoration based on microbial cells or their products, compared with traditional chemical, physical and mechanical methods, is that biological treatments are not destructive of the underlying substrate, simply removing unwanted overlying materials, in the case of biocleaning, or producing new stone, in the case of biocalcification. Biologically-based restoration also has the advantage of being recognized and appreciated by the community of restorers which are often reluctant to use synthetic, man-made materials. This is partially explained by the fact that ancient conservation practices are based on the use of biological materials as consolidants. Analogous behavior has been observed among other communities such as agriculture producers who are often keen to test biological-based methods in control of pests, but reluctant to use synthetic pesticides. The current COVID-19 pandemic has spurred interest in several sectors towards an increased use of green technologies.

Cultural heritage microbiology has relied heavily on fundamental studies at laboratory scale that do not reflect conditions that occur in historic monuments, where processes operate at the monument or archaeological site level, which encompasses a heterogeneous and complex set of conditions that hamper the assessment of microbially-based approaches for conservation. There needs to be convergence of cultural heritage with civil engineering, microbiology and biotechnology to encompass a longer timeframe and greater spatial scale that includes the building/archaeological site levels. Indeed, the best approach to stone conservation is an interdisciplinary one combining engineering, art conservation and biological sciences, including microbiology, biotechnology and environmental sciences [128]. Each discipline brings its own scientific and methodological culture to develop a holistic approach.

Biotechnology has emerged as a field perceived to be a crucial component in the knowledge economy, with potential in many fields, including conservation. Using novel microbial strategies need not only consider technical aspects of performance but also safety for human use, economics and skills. Transdisciplinary approaches are key to advance science and technology in cultural heritage conservation, but they can be challenging given the gaps in knowledge. Few restorers are trained in microbiology and biotechnology skills, but also few biotechnologists are trained in social disciplines such as anthropology and conservation science. For restorers, microbiology and biotechnology skills are key for isolation and assessment of performance of novel biocalcifying organisms and the optimization of culture conditions to cost-effectively produce microbial metabolites such as EPS. EPS with the correct chemistry, which is heavily controlled by cultivation conditions, is fundamental to control and further optimize the bioprecipitation processes, offering EPS and potentially other biopolymers as tools for biotechnological applications. If correctly chosen, biological methods are completely safe, not relying on toxic or potentially poisonous solvents. In addition, it has been shown that these methods can be more economical than chemical treatments. However, as with other biotechnologies, biotechnology for cultural heritage conservation needs to further develop its methods, confirm their reproducibility and show its economic benefits. Process innovation which relies exclusively on improving environmental performance (for environmentally-conscious conservators) does not give enough incentives, in particular for the private sector. Thus, a promising cost-benefit analysis at the economic, environmental and human health levels are necessary before a phase of consolidation and progressive acceptance by the concerned communities, including conservators, governmental agencies and the private sector, are recognized.

Author Contributions: B.O.O.-M. and C.C.G. contributed equally to this paper. Both authors have read and agreed to the published version of the manuscript.

Funding: Benjamín Otto Ortega Morales is grateful for grant CONACYT CB-2015-01 257449 "Influencia de tratamientos con nano y biomateriales en la colonización microbiana de roca monumental".

Institutional Review Board Statement: Not applicable.

Informed Consent Statement: Not applicable.

Data Availability Statement: Not applicable.

Acknowledgments: The authors are grateful to Juan Enrique Pereanez-Sacarías for photo file for Figure 1.

Conflicts of Interest: The authors declare no conflict of interest associated with this article.

References

1. Scheerer, S.; Ortega-Morales, O.; Gaylarde, C. Microbial deterioration of stone monuments—An updated overview. *Adv. Appl. Microbiol.* **2009**, *66*, 97–139.
2. Schabereiter-Gurtner, C.; Piñar, G.; Vybiral, D.; Lubitz, W.; Rölleke, S. Rubrobacter-related bacteria associated with rosy discolouration of masonry and lime wall paintings. *Arch. Microbiol.* **2001**, *176*, 347–354. [CrossRef] [PubMed]
3. Gaylarde, C.C.; Ortega-Morales, B.O.; Bartolo-Pérez, P. Biogenic black crusts on buildings in unpolluted environments. *Curr. Microbiol.* **2007**, *54*, 162–166. [CrossRef]

4. Gómez-Cornelio, S.; Mendoza-Vega, J.; Gaylarde, C.C.; Reyes-Estebanez, M.; Morón-Ríos, A.; De la Rosa, S.D.C.; Ortega-Morales, B.O. Succession of fungi colonizing porous and compact limestone exposed to subtropical environments. *Fungal Biol.* **2012**, *116*, 1064–1072. [CrossRef]
5. Ortega-Morales, B.O.; Gaylarde, C.; Anaya-Hernández, A.; Chan-Bacab, M.J.; De la Rosa-García, S.C.; Arano-Recio, D.; Montero, M.J. Orientation affects Trentepohlia-dominated biofilms on Mayan monuments of the Rio Bec style. *Int. Biodeterior. Biodegrad.* **2013**, *84*, 351–356. [CrossRef]
6. Basu, S.; Orr, S.A.; Aktas, Y.D. A geological perspective on climate change and building stone deterioration in London: Implications for urban stone-built heritage research and management. *Atmosphere* **2020**, *11*, 788. [CrossRef]
7. Palla, F. Blue-Biotechnology and Biocleaning of Historic-Artistic Artifacts. *Conserv. Sci. Cult. Herit.* **2016**, *16*, 185–196.
8. Ziegenbalg, G.; Brummer, K.; Pianski, J. Nano-Lime—A new material for the consolidation and conservation of historic mortars. In Proceedings of the 2nd Historic Mortars Conference HMC10 and RILEM TC 203-RHM Final Workshop, Prague, Czech Republic, 22–24 September 2010.
9. Zárraga, R.; Cervantes, J.; Salazar-Hernandez, C.; Wheeler, G. Effect of the addition of hydroxyl-terminated polydimethylsiloxane to TEOS-based stone consolidants. *J. Cult. Herit.* **2010**, *11*, 138–144. [CrossRef]
10. Favaro, M.; Ossola, F.; Toamsin, P.; Vigato, P.A.; Rossetto, G.; El Habra, N.; Casarin, M. A novel approach to compatible and durable consolidation of limestone. In Proceedings of the 11th International Congress on Deterioration and Conservation of Stone, Torun, Poland, 15–20 September 2008; pp. 865–872.
11. Jang, J.J.; Matero, F.G. Performance evaluation of commercial nanolime as a consolidant for friable lime-based plaster. *J. Am. Inst. Conserv.* **2018**, *57*, 95–111. [CrossRef]
12. Pozo-Antonio, J.S.; Otero, J.; Alonso, P.; Mas i Baeberà, X. Nanolime- and nanosilica-based consolidants applied on heated granite and limestone: Effectiveness and durability. *Constr. Build. Mater.* **2019**, *201*, 852–870. [CrossRef]
13. Tortora, L.; Di Carlo, G.; Mosquera, M.J.; Ingo, G.M. Nanoscience and nanomaterials for the knowledge and conservation of cultural heritage. *Front. Mater.* **2020**, *7*, 372. [CrossRef]
14. Becerra, J.; Zaderenko, A.P.; Ortiz, R.; Karapanagiotis, I.; Ortiz, P. Comparison of the performance of a novel nanolime doped with ZnO quantum dots with common consolidants for historical carbonate stone buildings. *Appl. Clay Sci.* **2020**, *195*, 105732. [CrossRef]
15. Price, C.A.; Doehne, E. *Stone Conservation: An Overview of Current Research*, 2nd ed.; Getty Conservation Institute: Los Angeles, CA, USA, 2010; pp. 1–175.
16. Delgado-Rodrigues, J. Stone consolidation: Research and practice. In Proceedings of the International Symposium on Works of Art and Conservation Science Today, Thessaloniki, Greece, 26–28 November 2010; pp. 1–8.
17. Caneva, G.; Nugari, M. Evaluation of Escobilla's mucilago treatments in the archaeological sites of Joya de Ceren (El Salvador). *Biodeterior. Biodegrad. Lat. Am.* **2005**, *5*, 59–64.
18. Kita, Y. The functions of vegetable mucilage in lime and earth mortars—A review. In Proceedings of the HMC2013-3rd Historic Mortars Conference, Glasgow, Scotland, UK, 11–14 September 2013; pp. 1–6.
19. Wheeler, G. *Alkoxysilanes and the Consolidation of Stone*; The Getty Conservation Institute: Los Angeles, CA, USA, 2005; pp. 55–64.
20. Negri, A.; Nervo, M.; Di Marcello, S.; Castelli, D. Consolidation and Adhesion of Pictorial Layers on a Stone Substrate. The Study Case of the Virgin with the Child from Palazzo Madama, in Turin. *Coatings* **2021**, *11*, 624. [CrossRef]
21. Alisi, C.; Bacchetta, L.; Bojorquez, E.; Falconieri, M.; Gagliardi, S.; Insaurralde, M.; Tatì, A. Mucilages from Different Plant Species Affect the Characteristics of Bio-Mortars for Restoration. *Coatings* **2021**, *11*, 75. [CrossRef]
22. Fierascu, R.C.; Doni, M.; Fierascu, I. Selected aspects regarding the restoration/conservation of traditional wood and masonry building materials: A short overview of the last decade findings. *Appl. Sci.* **2020**, *10*, 1164. [CrossRef]
23. Guihéneuf, S.; Rangeard, D.; Perrot, A.; Cusin, T.; Collet, F.; Prétot, S. Effect of bio-stabilizers on capillary absorption and water vapour transfer into raw earth. *Mater. Struct.* **2020**, *53*, 1–18. [CrossRef]
24. Jroundi, F.; Schiro, M.; Ruiz-Agudo, E.; Elert, K.; Martín-Sánchez, I.; González-Muñoz, M.; Rodriguez-Navarro, C. Protection and consolidation of stone heritage by self-inoculation with indigenous carbonatogenic bacterial communities. *Nat. Commun.* **2017**, *8*, 279. [CrossRef]
25. Sondi, I.; Juračić, M. Whiting events and the formation of aragonite in Mediterranean karstic marine lakes: New evidence on its biologically induced inorganic origin. *Sedimentology* **2010**, *57*, 85–95. [CrossRef]
26. Zhu, T.; Dittrich, M. Carbonate precipitation through microbial activities in natural environment, and their potential in biotechnology: A review. *Front. Bioeng. Biotechnol.* **2016**, *4*, 4. [CrossRef]
27. Dhami, N.K.; Reddy, M.S.; Mukherjee, A. Application of calcifying bacteria for remediation of stones and cultural heritages. *Front. Microbiol.* **2014**, *5*, 304. [CrossRef] [PubMed]
28. Anbu, P.; Kang, C.H.; Shin, Y.J.; So, J.S. Formations of calcium carbonate minerals by bacteria and its multiple applications. *Springerplus* **2016**, *5*, 250. [CrossRef] [PubMed]
29. Nazel, T. Bioconsolidation of stone monuments. An overview. *Rest. Build. Monum.* **2016**, *22*, 37–45. [CrossRef]
30. Castro-Alonso, M.J.; Montañez-Hernandez, L.E.; Sanchez-Muñoz, M.A. Microbially induced calcium carbonate precipitation (MICP) and its potential in bioconcrete: Microbiological and molecular concepts. *Front. Mater.* **2019**, *6*, 126. [CrossRef]
31. Marvasi, M.; Mastromei, G.; Perito, B. Bacterial calcium carbonate mineralization in situ strategies for conservation of stone artworks: From cell components to microbial community. *Front. Microbiol.* **2020**, *11*, 1386. [CrossRef] [PubMed]

32. Jroundi, F.; Gómez-Suaga, P.; Jimenez-Lopez, C.; González-Muñoz, M.T.; Fernández-Vivas, M.A. Stone-isolated carbonatogenic bacteria as inoculants in bioconsolidation treatments for historical limestone. *Sci. Total Environ.* **2012**, *425*, 89–98. [CrossRef]
33. López-Moreno, A.; Sepúlveda-Sánchez, J.D.; Alonso-Guzmán, E.M.; Le Borgne, S. Calcium carbonate precipitation by heterotrophic bacteria isolated from biofilms formed on deteriorated ignimbrite stones: Influence of calcium on EPS production and biofilm formation by these isolates. *Biofouling* **2014**, *30*, 547–560. [CrossRef]
34. Castanier, S.; Le Métayer-Levrel, G.; Orial, G.; Loubière, J.F.; Perthuisot, J.P. Bacterial carbonatogenesis and applications to preservation and restoration of historic property. In *Of Microbes and Art*; Springer: Boston, MA, USA, 2000; pp. 203–218. [CrossRef]
35. Le Metayer-Levrel, G.; Castanier, S.; Orial, G.; Loubière, J.F.; Perthuisot, J.P. Applications of bacterial carbonatogenesis to the protection and regeneration of limestones in buildings and historic patrimony. *Sediment. Geol.* **1999**, *126*, 25–34. [CrossRef]
36. Andrei, A.Ş.; Păuşan, M.R.; Tămaş, T.; Har, N.; Barbu-Tudoran, L.; Leopold, N.; Banciu, H. Diversity and biomineralization potential of the epilithic bacterial communities inhabiting the oldest public stone monument of Cluj-Napoca (Transylvania, Romania). *Front. Microbiol.* **2017**, *8*, 372. [CrossRef]
37. Li, M.; Fang, C.; Kawasaki, S.; Huang, M.; Achal, V. Bio-consolidation of cracks in masonry cement mortars by Acinetobacter sp. SC4 isolated from a karst cave. *Int. Biodeterior. Biodegrad.* **2019**, *141*, 94–100. [CrossRef]
38. Meier, A.; Kastner, A.; Harries, D.; Wierzbicka-Wieczorek, M.; Majzlan, J.; Büchel, G.; Kothe, E. Calcium carbonates: Induced biomineralization with controlled macromorphology. *Biogeosciences* **2017**, *14*, 4867–4878. [CrossRef]
39. Ferral-Pérez, H.; Galicia-García, M.; Alvarado-Tenorio, B.; Izaguirre-Pompa, A.; Aguirre-Ramírez, M. Novel method to achieve crystallinity of calcite by Bacillus subtilis in coupled and non-coupled calcium-carbon sources. *AMB Express* **2020**, *10*, 174. [CrossRef]
40. Tanul, S. Biomineralization Using *Synechococcus Pevalleikii* and Its Applications in Building Material. Master's Thesis, Thapar University, Patialia, India, 2017.
41. Zhu, T.; Lin, Y.; Lu, X.; Dittrich, M. Assessment of cyanobacterial species for carbonate precipitation on mortar surface under different conditions. *Ecol. Eng.* **2018**, *120*, 154–163. [CrossRef]
42. Zhu, T.; Lu, X.; Dittrich, M. Calcification on mortar by live and UV-killed biofilm-forming cyanobacterial *Gloeocapsa* PCC. *Constr. Build. Mater.* **2017**, *146*, 43–53. [CrossRef]
43. Li, Q.; Zhang, B.; Ge, Q.; Yang, X. Calcium carbonate precipitation induced by calcifying bacteria in culture experiments: Influence of the medium on morphology and mineralogy. *Int. Biodeterior. Biodegrad.* **2018**, *134*, 83–92. [CrossRef]
44. Andreolli, M.; Lampis, S.; Bernardi, P.; Calò, S.; Vallini, G. Bacteria from black crusts on stone monuments can precipitate $CaCO_3$ allowing the development of a new bio-consolidation protocol for ornamental stone. *Int. Biodeterior. Biodegrad.* **2020**, *153*, 105031. [CrossRef]
45. Micallef, R.; Vella, D.; Sinagra, E.; Zammit, G. Biocalcifying Bacillus subtilis cells effectively consolidate deteriorated Globigerina limestone. *J. Ind. Microbiol. Biotechnol.* **2016**, *43*, 941–952. [CrossRef]
46. Shirakawa, M.A.; Cincotto, M.A.; Atencio, D.; Gaylarde, C.C.; John, V.M. Effect of culture medium on biocalcification by *Pseudomonas putida*, *Lysinibacillus sphaericus* and *Bacillus subtilis*. *Braz. J. Microbiol.* **2011**, *42*, 499–507. [CrossRef] [PubMed]
47. Daskalakis, M.; Magoulas, A.; Kotoulas, G.; Catsikis, I.; Bakolas, A.; Karageorgis, A.P.; Mavridou, A.; Doulia, D.; Rigas, F. *Pseudomonas*, *Pantoea* and *Cupriavidus* isolates induce calcium carbonate precipitation for biorestoration of ornamental stone. *J. Appl. Microbiol.* **2013**, *115*, 409–423. [CrossRef]
48. Montaño-Salazar, S.M.; Lizarazo-Marriaga, J.; Brandão, P.F.B. Isolation and potential biocementation of calcite precipitation inducing bacteria from Colombian buildings. *Curr. Microbiol.* **2018**, *75*, 256–265. [CrossRef]
49. Golovkina, D.A.; Zhurishkina, E.V.; Ivanova, L.A.; Baranchikov, A.E.; Sokolov, A.Y.; Bobrov, K.S.; Masharsky, A.E.; Tsvigun, N.V.; Kopitsa, G.P.; Kulminskaya, A.A. Calcifying bacteria flexibility in induction of CaCO3 mineralization. *Life* **2020**, *10*, 317. [CrossRef]
50. Keskin, T.; Deniz, I.; Aric, A.; Yilmazsoy, B.T.; Andic-Cakir, O.; Erdogan, A.; Altun, D.; Tokuç, A.; Demirci, B.F.; Sendemir-Urkmez, A.; et al. Development of ecological biodesign products by bacterial biocalcification. *Eur. J. Eng. Nat. Sci.* **2019**, *3*, 17–25.
51. Zamarreno, D.V.; Inkpen, R.; May, E. Carbonate crystals precipitated by freshwater bacteria and their use as a limestone consolidant. *Appl. Environ. Microbiol.* **2009**, *75*, 5981–5990. [CrossRef] [PubMed]
52. Fang, C.; Kumari, D.; Zhu, X.; Achal, V. Role of fungal-mediated mineralization in biocementation of sand and its improved compressive strength. *Int. Biodeterior. Biodegrad.* **2018**, *133*, 216–220. [CrossRef]
53. Li, T.; Hu, Y.; Zhang, B. Biomineralization induced by Colletotrichum acutatum: A potential strategy for cultural relic bioprotection. *Front. Microbiol.* **2018**, *9*, 1884. [CrossRef]
54. Pasquale, V.; Fiore, S.; Hlayem, D.; Lettino, A.; Huertas, F.J.; Chianese, E.; Dumontet, S. Biomineralization of carbonates induced by the fungi Paecilomyces inflatus and Plectosphaerella cucumerina. *Int. Biodeterior. Biodegrad.* **2019**, *140*, 57–66. [CrossRef]
55. Chuo, S.C.; Mohamed, S.F.; Mohd Setapar, S.H.; Ahmad, A.; Jawaid, M.; Wani, W.A.; Yaqoob, A.A.; Ibrahim, M.N.M. Insights into the current trends in the utilization of bacteria for microbially induced calcium carbonate precipitation. *Materials* **2020**, *13*, 4993. [CrossRef]
56. Omoregie, A.I.; Ong, D.E.L.; Nissom, P.M. Assessing ureolytic bacteria with calcifying abilities isolated from limestone caves for biocalcification. *Lett. Appl. Microbiol.* **2018**, *68*, 173–181. [CrossRef]
57. Vincent, J.; Sabot, R.; Lanneluc, I.; Refait, P.; Turcry, P.; Mahieux, P.Y.; Jeannin, M.; Sablé, S. Biomineralization of calcium carbonate by marine bacterial strains isolated from calcareous deposits. *Matér. Tech.* **2020**, *108*, 302. [CrossRef]

58. Wainwright, M. Oligotrophic growth of fungi: Stress or natural state? In *Stress Tolerance of Fungi*; Jennings, D.H., Ed.; Marcel Dekker: New York, NY, USA, 1993; pp. 127–144.
59. Ortega-Morales, B.O.; Narváez-Zapata, J.; Reyes-Estebanez, M.; Quintana, P.; De la Rosa-García, S.C.; Bullen, H.; Gómez-Cornelio, S.; Chan-Bacab, M.J. Bioweathering potential of cultivable fungi associated with semi-arid surface microhabitats of Mayan buildings. *Front. Microbiol.* **2016**, *7*, 201. [CrossRef] [PubMed]
60. Fry, J. Culture-Dependent Microbiology. In *Microbial Diversity and Bioprospecting*; Bull, A., Ed.; ASM Press: Washington, DC, USA, 2004; pp. 80–87.
61. Chimienti, G.; Piredda, R.; Pepe, G.; van der Werf, I.D.; Sabbatini, L.; Crecchio, C.; Ricciuti, P.; D'Erchia, A.M.; Manzari, C.; Pesole, G. Profile of microbial communities on carbonate stones of the medieval church of San Leonardo di Siponto (Italy) by Illumina-based deep sequencing. *Appl. Microbiol. Biotechnol.* **2016**, *100*, 8537–8548. [CrossRef]
62. Barabesi, C.; Galizzi, A.; Mastromei, G.; Rossi, M.; Tamburini, E.; Perito, B. *Bacillus subtilis* gene cluster involved in calcium carbonate biomineralization. *J. Bacteriol.* **2007**, *189*, 228–235. [CrossRef] [PubMed]
63. Marvasi, M.; Casillas-Santiago, L.M.; Henríquez, T.; Casillas-Martinez, L. Involvement of etfA gene during $CaCO_3$ precipitation in Bacillus subtilis biofilm. *Geomicrobiol. J.* **2016**, *34*, 722–728. [CrossRef]
64. Frandi, A.; Zucca, P.; Marvasi, M.; Mastromei, G.; Sanjust, E.; Perito, B. *Bacillus subtilis* fadB (ysiB) gene encodes an enoyl-CoA hydratase. *Ann. Microbiol.* **2011**, *61*, 371–374. [CrossRef]
65. Perito, B.; Casillas, L.; Marvasi, M. Factors affecting formation of large calcite crystals (=1 mm) in Bacillus subtilis 168 biofilm. *Geomicrobiol. J.* **2018**, *35*, 385–391. [CrossRef]
66. Lee, Y.S.; Park, W. Enhanced calcium carbonate-biofilm complex formation by alkali-generating Lysinibacillus boronitolerans YS11 and alkaliphilic *Bacillus* sp. AK13. *AKAMB Express* **2019**, *9*, 49. [CrossRef]
67. Shraddha, G.; Darshan, M. Microbially induced calcite precipitation through urolytic organisms—A review. *Int. J. Life Sci.* **2019**, *7*, 133–139.
68. Yu, X.; Zhan, Q.; Qian, C.; Ma, J.; Liang, Y. The optimal formulation of bio-carbonate and bio-magnesium phosphate cement to reduce ammonia emission. *J. Clean. Prod.* **2019**, *240*, 118156. [CrossRef]
69. Dhami, N.K.; Mukherjee, A.; Reddy, M.S. Micrographical, minerological and nano-mechanical characterisation of microbial carbonates from urease and carbonic anhydrase producing bacteria. *Ecol. Eng.* **2016**, *94*, 443–454. [CrossRef]
70. Jimenez-Lopez, C.; Jroundi, F.; Pascolini, C.; Rodriguez-Navarro, C.; Piñar-Lurrubia, G.; Rodriguez-Gallego, M.; González-Muñoz, M.T. Consolidation of quarry calcarenite by calcium carbonate precipitation induced by bacteria activated among the microbiota inhabiting the stone. *Int. Biodeterior. Biodegrad.* **2008**, *62*, 352–363. [CrossRef]
71. Jroundi, F.; Elert, K.; Ruiz-Agudo, E.; González-Muñoz, M.T.; Rodriguez-Navarro, C. Bacterial diversity evolution in Maya plaster and stone following a bio-conservation treatment. *Front. Microbiol.* **2020**, *11*, 599144. [CrossRef]
72. Sassoni, E.; Franzoni, E. An innovative phosphate-based consolidant for limestone. Part I: Effectiveness and compatibility in comparison with ethyl silicate. *Constr. Build. Mater.* **2014**, *102*, 918–930. [CrossRef]
73. Bonazza, A.; Vidorni, G.; Natali, I.; Ciantelli, C.; Giosuè, C.; Tittarelli, F. Durability assessment to environmental impact of nano-structured consolidants on Carrara marble by field exposure tests. *Sci. Total Environ.* **2017**, *575*, 23–32. [CrossRef] [PubMed]
74. Perito, B.; Marvasi, M.; Barabesi, C.; Mastromei, G.; Bracci, S.; Vendrell, M.; Tiano, P. A Bacillus subtilis cell fraction (BCF) inducing calcium carbonate precipitation: Biotechnological perspectives for monumental stone reinforcement. *J. Cult. Herit.* **2014**, *15*, 345–351. [CrossRef]
75. Douglas, S.; Beveridge, T.J. Mineral formation by bacteria in natural microbial communities. *FEMS Microb. Ecol.* **1998**, *26*, 79–88. [CrossRef]
76. Ercole, C.; Bozzelli, P.; Altieri, F.; Cacchio, P.; Del Gallo, M. Calcium carbonate mineralization: Involvement of extracellular polymeric materials isolated from calcifying bacteria. *Microsc. Microanal.* **2012**, *18*, 829–839. [CrossRef] [PubMed]
77. Oppenheimer-Shaanan, Y.; Sibony-Nevo, O.; Bloom-Ackermann, Z.; Suissa, R.; Steinberg, N.; Kartvelishvily, E.; Brumfeld, V.; Kolodkin-Gal, I. Spatio-temporal assembly of functional mineral scaffolds within microbial biofilms. *NPJ Biofilms Microb.* **2016**, *2*, 15031. [CrossRef] [PubMed]
78. Beveridge, T.J.; Murray, R.G. Sites of metal deposition in the cell wall of Bacillus subtilis. *J. Bacteriol.* **1980**, *141*, 876–887. [CrossRef]
79. Decho, A.W. Overview of biopolymer-induced mineralization: What goes on in biofilms? *Ecol. Eng.* **2010**, *36*, 137–144. [CrossRef]
80. Maciejewska, M.; Adam, D.; Naômé, A.; Martinet, L.; Tenconi, E.; Całusińska, M.; Delfosse, P.; Hanikenne, M.; Baurain, D.; Compère, P.; et al. Assessment of the potential role of Streptomyces in cave moonmilk formation. *Front. Microbiol.* **2017**, *8*, 1181. [CrossRef]
81. Ortega-Villamagua, E.; Gudiño-Gomezjurado, M.; Palma-Cando, A. Microbiologically induced carbonate precipitation in the restoration and conservation of cultural heritage materials. *Molecules* **2020**, *25*, 5499. [CrossRef]
82. De Muynck, W.; Leuridan, S.; Van Loo, D.; Verbeken, K.; Cnudde, V.; De Beile, N.; Verstraete, W. Influence of pore structure on the effectiveness of a biogenic carbonate surface treatment for limestone conservation. *Appl. Environ. Microbiol.* **2011**, *77*, 6808–6820. [CrossRef]
83. Rodriguez-Navarro, C.; Jimenez-Lopez, C.; Rodriguez-Navarro, A.; González-Muñoz, M.T.; Rodriguez-Gallego, M. Bacterially mediated mineralization of vaterite. *Geochim. Cosmochim. Acta* **2007**, *71*, 1197–1213. [CrossRef]

84. Rusznyák, A.; Akob, D.M.; Nietzsche, S.; Eusterhues, K.; Totsche, K.U.; Neu, T.R.; Frosch, T.; Popp, J.; Keiner, R.; Geletneky, J.; et al. Calcite biomineralization by bacterial isolates from the recently discovered pristine karstic herrenberg cave. *Appl. Environ. Microbiol.* **2012**, *78*, 1157–1167. [CrossRef] [PubMed]
85. Dhami, N.K.; Reddy, M.S.; Mukherjee, A. Synergistic role of bacterial urease and carbonic anhydrase in carbonate mineralization. *Appl. Biochem. Biotechnol.* **2014**, *172*, 2552–2561. [CrossRef]
86. Stocks-Fischer, S.; Galinat, J.K.; Bang, S.S. Microbiological precipitation of CaCO. *Soil Biol. Biochem.* **1999**, *31*, 1563–1571. [CrossRef]
87. Okwadha, G.D.O.; Li, J. Optimum conditions for microbial carbonate precipitation. *Chemosphere* **2010**, *81*, 1143–1148. [CrossRef] [PubMed]
88. Jroundi, F.; Gonzalez-Muñoz, M.T.; Garcia-Bueno, A.; Rodriguez-Navarro, C. Consolidation of archaeological gypsum plaster by bacterial biomineralization of calcium carbonate. *Acta Biomater.* **2014**, *10*, 3844–3854. [CrossRef] [PubMed]
89. Gebauer, D.; Gunawidjaja, P.N.; Ko, J.Y.P.; Bacsik, Z.; Aziz, B.; Liu, L.; Hu, Y.; Berström, L.; Tai, C.W.; Sham, T.K.; et al. Proto-calcite and proto-vaterite in amorphous calcium carbonates. *Angew. Chem. Int. Ed.* **2010**, *49*, 8889–8891. [CrossRef]
90. Dhami, N.K.; Reddy, M.S.; Mukherjee, A. Biomineralization of calcium carbonate polymorphs by the bacterial strains isolated from calcareous sites. *J. Microbiol. Biotechnol.* **2013**, *23*, 707–714. [CrossRef]
91. Dhami, N.K.; Mukherjee, A.; Watkin, E.L.J. Microbial diversity and mineralogical-mechanical properties of calcitic cave speleothems in natural and in vitro biomineralization conditions. *Front. Microbiol.* **2018**, *9*, 40. [CrossRef]
92. Favero-Longo, S.E.; Viles, H.A. A review of the nature, role and control of lithobionts on stone cultural heritage: Weighing-up and managing biodeterioration and bioprotection. *World J. Microbiol. Biotechnol.* **2020**, *36*, 100. [CrossRef]
93. Pinna, D.; Galeotti, M.; Perito, B.; Daly, G.; Salvadori, B. In situ long-term monitoring of recolonization by fungi and lichens after innovative and traditional conservative treatments of archaeological stones in Fiesole (Italy). *Int. Biodeterior. Biodegrad.* **2018**, *132*, 49–58. [CrossRef]
94. Jroundi, F.; Gonzalez-Muñoz, M.T.; Sterflinger, K.; Piñar, G. Molecular tools for monitoring the ecological sustainability of a stone bio-consolidation treatment at the Royal Chapel, Granada. *PLoS ONE* **2015**, *10*, e0132465. [CrossRef]
95. Shirakawa, M.A.; John, V.M.; De Belie, N.; Alves, J.V.; Pinto, J.B.; Gaylarde, C.C. Susceptibility of biocalcite-modified fiber cement to biodeterioration. *Int. Biodeterior. Biodegrad.* **2015**, *103*, 215–220. [CrossRef]
96. Balloi, A.; Palla, F. Biocleaning. In *Biotechnology and Conservation of Cultural Heritage*; Palla, F., Barresi, G., Eds.; Springer: Cham, Switzerland, 2017.
97. Bosch-Roig, P.; Ranalli, G. The safety of biocleaning technologies for cultural heritage. *Front. Microbiol.* **2014**, *5*, 155. [CrossRef]
98. Warscheid, T.; Braams, J. Biodeterioration of stone: A review. *Int. Biodeterior. Biodegrad.* **2000**, *46*, 343–368. [CrossRef]
99. Gauri, K.L.; Parks, L.; Jaynes, J.; Atlas, R. Removal of sulfated-crust from marble using sulfate-reducing bacteria. In Proceedings of the International Conference Stone Cleaning and the Nature, Soiling and Decay Mechanisms of Stone, Edinburgh, UK, 14–16 April 1992; pp. 160–165.
100. Scarascia, G.; Lehmann, R.; Machuca, L.L.; Morris, C.; Cheng, K.Y.; Kaksonen, A.; Hong, P.Y. Effect of quorum sensing on the ability of Desulfovibrio vulgaris to form biofilms and to biocorrode carbon steel in saline conditions. *Appl. Environ. Microbiol.* **2020**, *86*, e01664-19. [CrossRef]
101. Bosch-Roig, P.; Lustrato, G.; Zanardini, E.; Ranalli, G. Biocleaning of cultural heritage stone surfaces and frescoes: Which delivery system can be the most appropriate? *Ann. Microbiol.* **2015**, *65*, 1227–1241. [CrossRef]
102. Cappitelli, F.; Zanardini, E.; Ranalli, G.; Mello, E.; Daffonchio, D.; Sorlini, C. Improved methodology for bioremoval of black crusts on historical stone artworks by use of sulfate-reducing bacteria. *Appl. Environ. Microbiol.* **2006**, *72*, 3733–3737. [CrossRef]
103. Gioventù, E.; Lorenzi, P. Bio-removal of black crust from marble surface: Comparison with traditional methodologies and application on a sculpture from the Florence's English Cemetery. *Procedia Chem.* **2013**, *8*, 123–129. [CrossRef]
104. Elhagrassy, A.F.; Hakeem, A. Comparative study of biological cleaning and laser techniques for conservation of weathered stone in Failaka Island, Kuwait. *Sci. Cult.* **2018**, *4*, 43–50.
105. Ranalli, G.; Matteini, M.; Tosini, I.; Zanardini, E.; Sorlini, C. Bioremediation of cultural heritage: Removal of sulphates, nitrates and organic substances. In *Of Microbes and Art*; Springer: Boston, MA, USA, 2000; pp. 231–245. [CrossRef]
106. Ranalli, G.; Chiavarini, M.; Guidetti, V.; Marsala, F.; Matteini, M.; Zanardini, E.; Sorlini, C. The use of microorganisms for the removal of nitrates and organic substances on artistic stoneworks. *Int. Biodeterior. Biodegrad.* **1996**, *40*, 255–261. [CrossRef]
107. Alfano, G.; Lustrato, G.; Belli, C.; Zanardini, E.; Cappitelli, F.; Mello, E.; Sorlini, C.; Ranalli, G. The bioremoval of nitrate and sulfate alterations on artistic stonework: The case-study of Matera Cathedral after six years from the treatment. *Int. Biodeterior. Biodegrad.* **2011**, *65*, 1004–1011. [CrossRef]
108. Roig, P.B.; Ros, J.L.R.; Montes Estellés, R. Biocleaning of nitrate alterations on wall paintings by Pseudomonas stutzeri. *Int. Biodeterior. Biodegrad.* **2013**, *84*, 266–274. [CrossRef]
109. Romano, I.; Abbate, M.; Poli, A.; D'Orazio, L. Bio-cleaning of nitrate salt efflorescence on stone samples using extremophilic bacteria. *Sci. Rep.* **2019**, *9*, 1668. [CrossRef]
110. Soffritti, I.; D'Accolti, M.; Lanzoni, L.; Volta, A.; Bisi, M.; Mazzacane, S.; Caselli, E. The potential use of microorganisms as restorative agents: An update. *Sustainability* **2019**, *11*, 3853. [CrossRef]
111. Mazzuca, C.; Poggi, G.; Bonelli, N.; Micheli, L.; Baglioni, P.; Palleschi, A. Innovative chemical gels meet enzymes: A smart combination for cleaning paper artworks. *J. Colloid Interface Sci.* **2017**, *502*, 153–164. [CrossRef]

112. Ranalli, G.; Alfano, G.; Belli, C.; Lustrato, G.; Colombini, M.P.; Bonaduce, I.; Zanardini, E.; Abbrucato, P.; Cappitelli, F.; Sorlini, C. Biotechnology applied to cultural heritage: Biorestoration of frescoes using viable bacterial cells and enzymes. *J. Appl. Microbiol.* **2005**, *98*, 73–83. [CrossRef]
113. Ranalli, G.; Zanardini, E.; Rampazzi, L.; Corti, C.; Andreotti, A.; Colombini, P.; Bosch-Roig, P.; Lustrato, G.; Giantomassi, C.; Zari, D.; et al. Onsite advanced biocleaning system on historical wall paintings using new agar-gauze bacteria gel. *J. Appl. Microbiol.* **2019**, *126*, 1785–1796. [CrossRef]
114. Kohli, R. Chapter 15—Application of microbial cleaning technology for removal of surface contamination. *Appl. Clean. Tech.* **2019**, *11*, 591–617.
115. Jeszeová, L.; Bauerová-Hlinková, V.; Baráth, P.; Puškárová, A.; Bučková, M.; Kraková, L.; Pangallo, D. Biochemical and proteomic characterization of the extracellular enzymatic preparate of Exiguobacterium undae, suitable for efficient animal glue removal. *Appl. Microbiol. Biotechnol.* **2018**, *102*, 6525–6536. [CrossRef]
116. Bosch-Roig, P.; Pozo-Antonio, J.S.; Sanmartín, P. Identification of the best-performing novel microbial strains from naturally-aged graffiti for biocleaning research. *Int. Biodeterior. Biodegrad.* **2021**, *159*, 105206. [CrossRef]
117. Sanmartín, P.; DeAraujo, A.; Vasanthakumar, A.; Mitchell, R. Feasibility study involving the search for natural strains of microorganisms capable of degrading graffiti from heritage materials. *Int. Biodeterior. Biodegrad.* **2015**, *103*, 186–190. [CrossRef]
118. Sanmartín, P.; Bosch-Roig, P. Biocleaning to remove graffiti: A real possibility? advances towards a complete protocol of action. *Coatings* **2019**, *9*, 104. [CrossRef]
119. Germinario, G.; van der Werf, I.D.; Palazzo, G.; Regidor Ros, J.L.; Montes-Estelles, R.M.; Sabbatini, L. Bioremoval of marker pen inks by exploiting lipase hydrolysis. *Prog. Org. Coat.* **2017**, *110*, 162–171. [CrossRef]
120. Germinario, G.; Garrappa, S.; D'Ambrosio, V.; van der Werf, I.D.; Sabbatini, L. Chemical composition of felt-tip pen inks. *Anal. Bioanal. Chem.* **2018**, *410*, 1079–1094. [CrossRef]
121. Silva, M.; Rosado, T.; Teixeira, D.; Candeias, A.; Caldeira, A.T. Green mitigation strategy for cultural heritage: Bacterial potential for biocide production. *Environ. Sci. Pollut. Res.* **2017**, *24*, 4871–4881. [CrossRef]
122. Karygianni, L.; Ren, Z.; Koo, H.; Thurnheer, T. Biofilm matrixome: Extracellular components in structured microbial communities. *Trends Microbiol.* **2020**, *28*, 668–681. [CrossRef]
123. Valentini, F.; Diamanti, A.; Palleschi, G. New bio-cleaning strategies on porous building materials affected by biodeterioration event. *Appl. Surf. Sci.* **2010**, *256*, 6550–6563. [CrossRef]
124. Masi, M.; Petraretti, M.; De Natale, A.; Pollio, A.; Evidente, A. Fungal metabolites with antagonistic activity against fungi of lithic substrata. *Biomolecules* **2021**, *11*, 295. [CrossRef] [PubMed]
125. Marin, E.; Vaccaro, C.; Leis, M. Biotechnology applied to historic stoneworks conservation: Testing the potential harmfulness of two biological biocides. *Int. J. Conserv. Sci.* **2016**, *7*, 227–238.
126. Huang, R.; Li, G.; Zhang, J.; Yang, L.; Che, H.J.; Jiang, D.H.; Huang, H.C. Control of postharvest Botrytis fruit rot of strawberry by volatile organic compounds of Candida intermedia. *Phytopathology* **2011**, *101*, 859–869. [CrossRef] [PubMed]
127. Lopes, M.R.; Klein, M.N.; Ferraz, L.P.; da Silva, A.C.; Kupper, K.C. Saccharomyces cerevisiae: A novel and efficient biological control agent for Colletotrichum acutatum during pre-harvest. *Microbiol. Res.* **2015**, *175*, 93–99. [CrossRef] [PubMed]
128. Pope, G.; Meierding, T.; Paradise, T. Geomorphology's role in the study of weathering of cultural stone. *Geomorphology* **2002**, *47*, 211–225. [CrossRef]

applied sciences

MDPI

Review

Current Knowledge on the Fungal Degradation Abilities Profiled through Biodeteriorative Plate Essays

João Trovão [1,*] and António Portugal [1,2]

[1] Centre for Functional Ecology, Department of Life Sciences, University of Coimbra, 3004-531 Coimbra, Portugal; aportuga@bot.uc.pt
[2] Fitolab-Laboratory for Phytopathology, Instituto Pedro Nunes, 3030-199 Coimbra, Portugal
* Correspondence: jtrovaosb@gmail.com

Abstract: Fungi are known to contribute to the development of drastic biodeterioration of historical and valuable cultural heritage materials. Understandably, studies in this area are increasingly reliant on modern molecular biology techniques due to the enormous benefits they offer. However, classical culture dependent methodologies still offer the advantage of allowing fungal species biodeteriorative profiles to be studied in great detail. Both the essays available and the results concerning distinct fungal species biodeteriorative profiles obtained by amended plate essays, remain scattered and in need of a deep summarization. As such, the present work attempts to provide an overview of available options for this profiling, while also providing a summary of currently known fungal species putative biodeteriorative abilities solely obtained by the application of these methodologies. Consequently, this work also provides a series of checklists that can be helpful to microbiologists, restorers and conservation workers when attempting to safeguard cultural heritage materials worldwide from biodeterioration.

Keywords: biodeterioration; cultural heritage; deteriorative action; enzymatic activity; fungi

Citation: Trovão, J.; Portugal, A. Current Knowledge on the Fungal Degradation Abilities Profiled through Biodeteriorative Plate Essays. *Appl. Sci.* **2021**, *11*, 4196. https://doi.org/10.3390/app11094196

Academic Editor: Daniela Isola

Received: 26 March 2021
Accepted: 27 April 2021
Published: 5 May 2021

1. Introduction

The Fungal Kingdom comprises a highly diverse eukaryotic group able to inhabit every ecological niche available on the Planet [1]. The growth and biological activity of fungal species in cultural heritage materials is known to develop serious damages by means of biodeterioration (the undesirable modifications of a valuable material occurring by the action of living organisms) [2,3]. Fungi are highly versatile, ubiquitous, chemoheterotrophic microorganisms, being able to grow in a vast number of materials and contributing to the development of various biodeterioration phenomena [2,3]. Such modifications are a result from fungal species settling, development and exploitation of various organic and inorganic compounds present in historic art-pieces and monuments [2–15]. The fungal biodeterioration of books, paper, parchment, textiles, photographs, paintings, sculptures and wooden materials occurs due to the aesthetic modifications, mechanical pressure and exoenzymatic action [2]. Various components of these materials such as cellulose, collagen, linen, glues, inks, waxes and organic binders can be oxidized, hydrolyzed, dissolved, stained or structurally modified as a result of the action of fungal enzymes, pigments and organic acids [2,3,7–11]. A typical and widely known example of these phenomena is known as "foxing", the development of red-brownish localized spots, hypothesized to be a result from fungal proliferation and metabolization of organic acids, oligosaccharides and proteic compounds that can stain and modify the constituent materials of many paper-based and photographic supports [3,8,13]. Another example of microorganism's attack of organic materials is related to the biodeterioration of human remains, mummies and funerary materials, where opportunistic, saprotrophic and highly cellulolytic and proteolytic taxa are able to thrive and trough their actions severely alter them [2,14,15]. Complementarily, historic relics mainly composed of inorganic components such as stone, frescoes, glass and ceramics can also suffer deep aesthetical, physical and chemical modifications resulting from fungal grow and action [2–6,12,16]. In these supports, deterioration is caused by hyphae penetration into the substrate, the production and release of extracellular destructive organic acids, enzymes and metabolites and by the the formation of distinct colored outlines as a result of fungi high pigment contents, contribution to biofilm development and chemical reactions with inorganic compounds [2–6].

Due to the known biodeterioration problems arising from their proliferation, the accurate species identification and a consequent deteriorative profiling of isolates are crucial steps towards the development and the establishment of proper protective measures for the diverse cultural heritage treasures around the world. With the recent development of innovative culture independent methodologies such as -omics technologies, molecular data is becoming increasingly more valuable for the identification of the microbes, the characterization of their metabolic functions and their deteriorative byproducts [17]. Methodologies such as metagenomics, transcriptomics, metabolomics and proteomics revolutionized the field and are increasingly allowing understanding of microbial diversity, but also species specific and holistic contributions to various materials biodeterioration phenomena [17]. These methods are particularly relevant considering that traditional cultivation dependent methodologies hold the disadvantage of being unable to correctly infer microorganism's abundance and only allow the study of active forms, failing to provide information regarding viable non-culturable and non-viable forms [17–27]. Nonetheless, classical culture dependent methodologies still offer an important advantage when compared with modern methodologies, especially when considering that the isolation of microbes allows their natural biodeteriorative profiles to be studied in great detail. Culture media plates modified to specify a positive biodeteriorative ability upon the microorganism development and deteriorative action (see Figure 1 for examples) can provide valuable data that allow the evaluation of the microorganism's putative risks to cultural heritage materials. Moreover, they also offer a highly informative, rapid and low-cost platform [28] that can help in a quick and focused decision-making process aiming to protect valuable artifacts. Currently, plate assays aiming to identify fungal deteriorative characteristics, such as calcium carbonate solubilization, mineralization and various enzymatic activities, have been proposed and somewhat widely used.

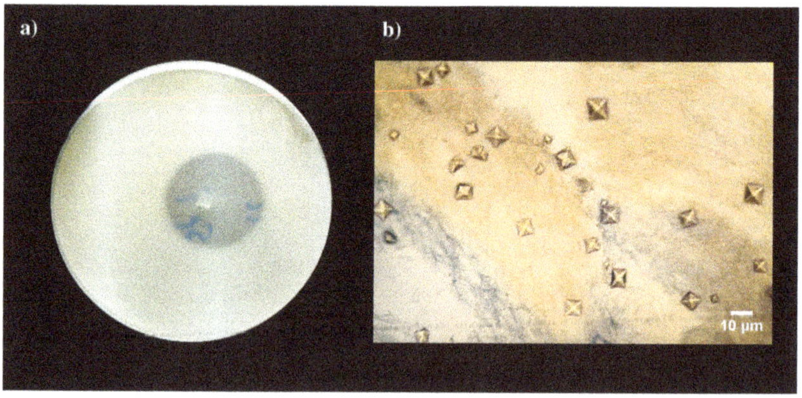

Figure 1. Examples of fungal species biodeteriogenic abilities detected through plate assays: (**a**) Calcium carbonate dissolution visualized by the development of a halo around colonies in $CaCO_3$ glucose agar; and (**b**) calcium oxalates crystals developing around fungal mycelium growing in Malt extract agar containing $CaCO_3$.

Although differences among distinct isolates, assays and incubation conditions are known and expected, the available literature concerning distinctive fungal species deteriorative profiles obtained using such methodologies remains pending a deep summarization. With this in mind, this work aims to provide an overview of available plate assays, as well as the fungal putative biodeteriorative profiles obtained solely through such tests so far. In addition, we also aimed at providing a series of quick and straight forward checklists that can be consulted by microbiologists, restorers and conservation staff, when working to safeguard important cultural heritage materials worldwide. These checklists were also annotated to contain currently accepted fungal names according to Mycobank (www.mycobank.org, last accessed on 26 April 2021) and Index Fungorum (www.indexfungorum.org, last accessed on 26 April 2021) in order to ensure an updated identification for fungi displaying such profiles, and to facilitate information sharing in the future.

2. Calcium Carbonate Solubilization or Dissolution

One of the greatest fungal effects on stone monuments is credited to their secretion of inorganic and organic acids that can alter the material properties [2–4,29–32]. In fact, carbonate weathering has been consistently linked to the excretion and action of these metabolites [33–35]. Evaluation of fungal calcium carbonate solubilization abilities in cultural heritage scenarios has been helpful to study the biodeteriorative contribution of isolates retrieved from air, mural paintings, wooden art objects, frescoes, catacombs, bricks, concrete, buildings and various limestone and plaster monuments and museums [28,33,35–44]. Fungal calcium carbonate solubilization ability screening is usually conducted with $CaCO_3$ glucose agar and adapted formulations [33]. Nonetheless, the application of Malt extract agar and Reasoner's 2A agar amended with $CaCO_3$ (CMEA and CR2A) has also been successfully achieved [35]. Moreover, Kiyuna and colleagues [37] also highlighted the utility of Glucose Yeast extract calcium carbonate agar (GYC) [45] for such evaluation. Positive $CaCO_3$ dissolution is usually evaluated by the visualization of a halo around the growing colony after a period of incubation. In addition, calcium carbonate solubilization screening can also be conducted coupled with the evaluation of the media pH modifications. For this purpose, Creatine Sucrose agar (CREA) [46] followed by the analysis of medium color changes around growing colonies, or liquid media according to the formulations provided by Borrego and colleagues [47] followed by pH analysis, can also be applied. A quick overview of the known fungal species able of $CaCO_3$ dissolution points that isolates from more than fifty species have been found to display this biodeteriorative profile, with the great majority of them being *Aspergillus* and *Penicillium* species (Table 1). Both genera are known important biodeteriogens, producing various acidic molecules and contributing to the deterioration of materials [28,48]. The detection of species from these genera (as well as others for example, from *Pestalotiopsis* and *Talaromyces* among others) might indicate a putative threat to acid susceptible resources, such is the case of stone structures, mural paintings and frescoes.

Table 1. Overview of fungal species for which isolates have been identified as having $CaCO_3$ dissolution abilities in biodeteriorative plate essays.

Current Species Name	Original Study Focus	References
Acremonium charticola (Lindau) W. Gams	Limestone and plaster monuments and museums	[39]
Actinomucor elegans (Eidam) C.R. Benjamin and Hesseltine	Mural paintings	[41]
Alternaria alternata (Fr.) Keissl.	Mural paintings	[41]
Annulohypoxylon stygium (Lév.) Y.M. Ju, J.D. Rogers and H.M. Hsie	Mayan buildings	[35]
Aspergillus amstelodami (L. Mangin) Thom and Church	Wooden art objects	[36]
Aspergillus awamori Nakaz.	Wooden art objects	[36]
Aspergillus europaeus Hubka, A. Nováková, Samson, Houbraken, Frisvad and M. Kolařík	Frescoes and air	[38]
Aspergillus glaucus (L.) Link	Limestone tomb	[43]
Aspergillus nidulans (Eidam) G. Winter	Mural paintings	[41]
Aspergillus niger Tiegh	Wooden art objects, frescoes, air, brick and concrete	[28,36,38,40]
Aspergillus versicolor (Vuill.) Tirab	Wooden art objects, mural paintings, limestone, plaster monuments and museums	[36,39,41]

Table 1. *Cont.*

Current Species Name	Original Study Focus	References
Aspergillus westerdijkiae Frisvad and Samson	Limestone tomb	[43]
Botrytis cinerea Pers.	Limestone tomb	[43]
Cephalotrichum Link [1]	Catacombs	[33]
Cladosporium Link	Etruscan hypogeal tombs	[44]
Cladosporium sphaerospermum Penz.	Mural paintings	[41]
Cyphellophora G.A. de Vries	Etruscan hypogeal tombs	[44]
Cyphellophora olivacea (W. Gams) Réblová & Unter.	Etruscan hypogeal tombs	[44]
Exophiala J.W. Carmich.	Etruscan hypogeal tombs	[44]
Kendrickiella phycomyces (Auersw.) K. Jacobs and M.J. Wingf.	Mural paintings in Tumuli	[37]
Lasiodiplodia theobromae (Pat.) Griffon and Maubl.	Mayan buildings	[35]
Lecanicillium W. Gams and Zare	Limestone and plaster monuments and museums	[39]
Paecilomyces Bainier	Mayan buildings and catacombs	[33,35]
Parengyodontium album (Limber) C.C. Tsang, J.F.W. Chan, W.M. Pong, J.H.K. Chen, A.H.Y. Ngan, M. Cheung, C.K.C. Lai, D.N.C. Tsang, S.K.P. Lau and P.C.Y. Woo	Dolomitic limestone wall	[42]
Penicillium angulare S.W. Peterson, E.M. Bayer and Wicklow	Dolomitic limestone wall	[42]
Penicillium aurantiogriseum Dierckx	Mural paintings	[41]
Penicillium bilaiae Chalab.	Frescoes and air	[38]
Penicillium brevicompactum Dierckx	Air, limestone tomb and dolomitic limestone wall	[28,42,43]
Penicillium chrysogenum Thom	Wooden art objects, air, mural paintings, limestone tomb, dolomitic limestone wall, limestone and plaster monuments and museums	[28,36,39,41–43]
Penicillium commune Thom	Frescoes, air and mural paintings	[38,40,41]
Penicillium crustosum Thom	Dolomitic limestone wall	[42]
Penicillium glabrum (Wehmer) Westling	Wooden art objects, air, dolomitic limestone wall and limestone tomb	[28,36,42]
Penicillium griseofulvum Dierckx	Frescoes and air	[38]
Penicillium lanosum Westling	Frescoes and air	[38]
Penicillium Link	Wooden art objects, air, frescoes, concrete and bricks	[36,40]
Penicillium oxalicum Currie and Thom	Mayan buildings	[35]
Penicillium polonicum K.W. Zaleski	Mural paintings	[41]
Penicillium rubens Biourge	Frescoes and air	[38]
Penicillium scabrosum Frisvad, Samson and Stolk	Dolomitic limestone wall	[42]
Penicillium solitum Westling	Air	[28]
Penicillium viridicatum Westling	Air	[28]
Periconia byssoides Pers.	Dolomitic limestone wall	[42]
Pestalotiopsis maculans (Corda) Nag Raj	Mayan buildings	[35]
Pestalotiopsis microspora (Speg.) G.C. Zhao and Nan Li	Mayan buildings	[35]
Pseudogymnoascus pannorum (Link) Minnis and D.L. Lindner	Limestone and plaster monuments and museums	[39]
Rosellinia De Not.	Mayan buildings	[35]
Sclerotinia Fuckel	Air and frescoes	[36]
Sclerotinia sclerotiorum (Lib.) de Bary	Air and frescoes	[36]
Talaromyces amestolkiae N. Yilmaz, Houbraken, Frisvad and Samson	Dolomitic limestone wall	[42]
Talaromyces sayulitensis Visagie, N. Yilmaz, Seifert and Samson	Air	[28]
Trichocladium canadense S. Hughe	Mayan buildings	[35]
Trichoderma Pers.	Mayan buildings	[35]
Valsaria spartii Maubl.	Dolomitic limestone wall	[42]
Xylaria Hill ex Schrank [2]	Mayan buildings	[35]

Previously identified as: [1] *Doratomyces* sp.; [2] *Hypoxylon* sp.

3. Mineralization or Crystallization Development

Calcium carbonate solubilization by the action of fungal acids can often occur coupled with the recrystallization of minerals in the substrate [2,3,5,6,49–51]. Such mineralization singularities can lead to the development of various biodeterioration phenomena [52,53]. They occur from the reactions of secreted acids (especially oxalic acid) with stone cations [32] and often result in the formation of carbonates and/or calcium magnesium oxalates [5,6]. Characterization of fungal crystallization abilities in cultural heritage scenarios has been helpful to study the biodeteriorative contribution of isolates retrieved from air, limestone monuments, stone stela, wall and mural paintings [35,41–43,53–56]. Fungal mineralization ability screening is usually conducted using B4 (with calcium acetate) or modified B4 (with calcium carbonate) media and adapted formulations [57]. Moreover, $CaCO_3$ modified Malt agar, Nutrient agar (NA) with $CaCl_2$ and the above mentioned CMEA and CR2A media have also been found useful for such purposes [35,42,43,58,59]. Positive mineralization development is usually evaluated by the microscopical visualization of neo-formed minerals around or in fungal hyphae after a period of incubation. Moreover, further characterization of these crystals can also be achieved by applying analytical methodologies such as X-ray powder diffraction (XRD) and/or energy dispersive X-ray spectroscopy (EDS) in conjunction with scanning electron microscopy (SEM) methodologies. So far, circa sixty species have been found to display mineralization abilities in plate essays and, as similarly found for fungal calcium carbonate dissolution, multiple *Aspergillus* and *Penicillium* species have also denoted this biodeteriorative profile (Table 2). Such findings can be correlated with their long-known abilities to secrete oxalic acid, among various other acids [41]. Nonetheless, a relevant number of species from genera *Alternaria*, *Cladosporium*, *Colletotrichum*, *Pestalotiopsis* and *Trichoderma* putatively displaying these biodeteriorative abilities can also be verified. Typical minerals detected include calcium carbonate in the form of calcite and vaterite-calcite, weddellite, whewellite, hydroxyapatite, hydrocerussite, pyromorphite, phosphate and other still unidentified calcium oxalates and minerals. The detection of species from these genera might indicate a putative threat to materials highly susceptible to fungal acidolysis and biomineralization, such is the case of limestone monuments and murals [5,6,60].

Table 2. Overview of fungal species for which isolates have been identified as displaying mineralization abilities in biodeteriorative plate essays.

Current Species Name	Mineral Details	Original Study Focus	References
Actinomucor elegans (Eidam) C.R. Benjamin and Hesseltine	Cc, Wd	Mural paintings	[41]
Aeminium ludgeri J. Trovão, I. Tiago and A. Portugal	Cc	Dolomitic limestone wall	[42]
Alternaria alternata (Fr.) Keissl.	Unk	Air and wall paintings	[56]
Alternaria infectoria E.G. Simmons	Unk	Air and wall paintings	[56]
Annulohypoxylon stygium (Lév.) Y.M. Ju, J.D. Rogers and H.M. Hsie	Cc, Wd	Mayan buildings	[35]
Ascochyta medicaginicola Qian Chen and L. Cai [1]	Unk	Air and wall paintings	[56]
Aspergillus aureolatus Munt.—Cvetk. and Bata	Cc	Air and wall paintings	[56]
Aspergillus europaeus Hubka, A. Nováková, Samson, Houbraken, Frisvad and M. Kolařík	Unk	Air and wall paintings	[56]
Aspergillus flavipes (Bainier and R. Sartory) Thom and Church	Unk	Air and wall paintings	[56]
Aspergillus flavus Link	Cc, Wd	Air and wall paintings	[56]
Aspergillus glaucus (L.) Link	Unk CO	Limestone tomb	[54]
Aspergillus niger Tiegh	Unk, Wh	Limestone monument, air and wall paintings	[53,56]
Aspergillus ostianus Wehmer	Cc, Wd	Air and wall paintings	[56]
Aspergillus pallidofulvus Visagie, Varga, Frisvad and Samson	Cc	Air and wall paintings	[56]
Aspergillus parasiticus Speare	Cc, Wd	Air and wall paintings	[56]

Table 2. *Cont.*

Current Species Name	Mineral Details	Original Study Focus	References
Aspergillus westerdijkiae Frisvad and Samson	Unk CO	Limestone tomb	[54]
Bionectria ochroleuca (Schwein.) Schroers and Samuels	Cc, Wh	Stone stela	[54]
Botryotrichum murorum (Corda) X. Wei Wang and Samson [2]	Cc, Unk	Air, wall paintings and stone stela	[54,56]
Botrytis cinerea Pers.	Unk CO	Limestone tomb	[54]
Chaetomium ancistrocladum Udagawa and Cain	Unk	Air and wall paintings	[56]
Cladosporium cladosporioides (Fresen.) G.A. de Vries	Unk	Air and wall paintings	[56]
Cladosporium oxysporum Berk. and M.A. Curtis	Cc, Wd	Air and wall paintings	[56]
Cladosporium sphaerospermum Penz.	Cc, Wd	Mural paintings	[41]
Cladosporium uredinicola Speg.	Cc	Air and wall paintings	[56]
Colletotrichum acutatum J.H. Simmonds	Cc (Vaterite), Cc, Hap, Phosp	Limestone monument	[53,55]
Colletotrichum gloeosporioides (Penz.) Penz. and Sacc.	Cc (Vaterite)	Limestone monument	[53]
Epicoccum nigrum Link	Cc, Wd	Air and wall paintings	[56]
Fusarium fujikuroi Nirenberg [3]	Cc	Air and wall paintings	[56]
Fusarium proliferatum (Matsush.) Nirenberg ex Gerlach and Nirenberg	Cc	Stone stela	[54]
Lasiodiplodia theobromae (Pat.) Griffon and Maubl.	Unk	Mayan buildings	[35]
Leptosphaeria avenaria G.F. Weber [4]	Cc	Air and wall paintings	[56]
Mucor fragilis Bainier	Cc, Wd	Dolomitic limestone wall	[42]
Paecilomyces Bainier	Unk	Mayan buildings	[35]
Penicillium angulare S.W. Peterson, E.M. Bayer and Wicklow	Cc, Wd, Wh	Dolomitic limestone wall	[42]
Penicillium atrosanguineum B.X. Dong [5]	Cc, Wd	Air and wall paintings	[56]
Penicillium bilaiae Chalab.	Cc, Wd	Air and wall paintings	[56]
Penicillium brevicompactum Dierckx	Cc, Wd	Dolomitic limestone wall	[42]
Penicillium chrysogenum Thom	Cc, Wd, Unk CO	Limestone tomb and mural paintings	[41,43]
Penicillium commune Thom	Cc, Wd	Air, wall and mural paintings	[41,56]
Penicillium crustosum Thom	Cc, Wd	Stone stela	[54]
Penicillium glabrum (Wehmer) Westling	Cc, Wd, Wh	Dolomitic limestone wall	[42]
Penicillium griseofulvum Dierckx	Unk	Air and wall paintings	[56]
Penicillium lanosum Westling	Cc, Wd, Unk	Air and wall paintings	[56]
Penicillium oxalicum Currie and Thom	Cc, Wd, Wh	Mayan buildings and limestone monument	[35,53]
Penicillium polonicum K.W. Zaleski	Cc, Wd	Mural paintings	[41]
Penicillium rubens Biourge	Unk	Air and wall paintings	[56]
Periconia byssoides Pers.	Cc	Dolomitic limestone wall	[42]
Pestalotiopsis maculans (Corda) Nag Raj	Unk	Mayan buildings	[35]
Pestalotiopsis microspora (Speg.) G.C. Zhao and Nan Li	Cc, Wd, Wh	Mayan buildings	[35]
Phialemonium inflatum (Burnside) Dania García, Perdomo, Gené, Cano and Guarro [6]	Cc, Cc (Vaterite), Hyd, Pyr	Stalactite in building	[59]
Plectosphaerella cucumerina (Lindf.) W. Gams	Cc, Hyd, Pyr	Stalactite in building	[59]
Rhizoctonia solani J.G. Kühn [7]	Unk	Air and wall paintings	[56]
Rosellinia De Not.	Cc, Wd	Mayan buildings	[35]
Sclerotinia sclerotiorum (Lib.) de Bary	Cc, Wd	Air and wall paintings	[56]
Stereum hirsutum (Willd.) Pers.	Cc, Wd	Dolomitic limestone wall	[42]
Trichocladium canadense S. Hughes	Unk	Mayan buildings	[35]
Trichoderma atroviride P. Karst.	Cc, Wd	Dolomitic limestone wall	[42]
Trichoderma harzianum Rifai	Cc	Stone stela	[54]
Trichoderma Pers.	Unk	Mayan buildings	[35]
Xylaria Hill ex Schrank [8]	Unk	Mayan buildings	[35]

Previously identified as: [1] *Phoma medicaginis*; [2] *Chaetomium murorum*; [3] *Gibberella moniliformis*; [4] *Phaeosphaeria avenaria*; [5] *Penicillium manginii*; [6] *Paecilomyces inflatus*; [7] *Thanatephorus cucumeris*; [8] *Hypoxylon* sp.; Cc—calcium carbonate; Wd—weddellite; Wh—whewellite; Hap—hydroxyapatite; Hyd—hydrocerussite; Phosp—phosphate; Pyr—pyromorphite; Unk CO—Unidentified calcium oxalates; Unk—Unidentified mineralization's.

4. Enzymatic Action

Fungal ligninolytic action is often considered a threat to wooden structures [61–66]. Cultural heritage materials constructed with these materials can be affected by fungal hyphae penetration but also by the action of various exoenzymes [67]. Moreover, brown and white rot fungi are known to contribute to these substrates' deterioration and degradation in various contexts [68]. So far, fungal ligninolytic ability characterization in cultural heritage scenarios has been helpful to study the biodeteriorative contribution of isolates retrieved from air, wooden materials and art objects [61,69,70]. Ligninolytic ability screening can be conducted using media with Azure B (lignin peroxidase), Phenol Red (Mn peroxidase), Remazol Brilliant Blue R (laccase) [71–74] or, alternatively, by applying Potato Dextrose agar supplemented with guaiacol (PDA-guaiacol) [61,70]. Positive ligninolytic ability is usually evaluated by the clearance of the media specific color (Azure B, Phenol Red and Remazol Brilliant Blue R) or by the development of reddish-brown zones (PDA-guaiacol) after a period of incubation. As pointed by Pangallo and colleagues [36,69], data regarding ligninolytic abilities of filamentous fungi in biodeterioration contexts is still somewhat scarce. Nonetheless, as evidenced by Table 3, almost thirty species have been found to display these biodeteriorative abilities. Moreover, mainly species of genera *Aspergillus*, *Chaetomium*, *Cladosporium* and *Penicillium* represent the bulk of the currently studied lignin deteriorating fungi. As such, the detection of species from these genera might indicate a putative threat to lignin materials, such as the case of some types of paper and wood art pieces and objects.

Table 3. Overview of fungal species for which isolates have been identified as displaying ligninolytic abilities in biodeteriorative plate essays.

Current Species Name	Original Study Focus	References
Alternaria consortialis (Thüm.) J.W. Groves and S. Hughes [1]	Wooden art objects	[36]
Arthrinium phaeospermum (Corda) M.B. Ellis	Wooden art objects	[36]
Aspergillus amstelodami (L. Mangin) Thom and Church [2]	Wooden art objects	[36]
Aspergillus awamori Nakaz.	Wooden art objects	[36]
Aspergillus fischeri Wehmer [3]	Wooden art objects	[36]
Aspergillus flavus Link	Wooden art objects	[36]
Aspergillus fumigatus Fresen.	Wooden art objects	[36]
Aspergillus niger Tiegh	Wooden art objects	[36]
Aspergillus terreus Thom	Wooden art objects and air	[36,69]
Aspergillus ustus (Bainier) Thom and Church	Wooden art objects	[36]
Aspergillus versicolor (Vuill.) Tirab.	Wooden art objects	[36]
Beauveria bassiana (Bals.—Criv.) Vuill.	Wooden art objects	[36]
Chaetomium elatum Kunze	Wooden art objects	[36]
Chaetomium globosum Kunze	Wooden art objects and air	[36,69]
Cladosporium cladosporioides (Fresen.) G.A. de Vries	Wooden objects and air	[69]
Cladosporium herbarum (Pers.) Link	Wooden art objects	[36]
Cladosporium Link	Wooden art objects	[36]
Hypochnicium J. Erikss.	Wooden materials	[61]
"Neocosmospora" solani (Mart.) L. Lombard and Crous [4]	Wooden materials	[70]
Penicillium chrysogenum Thom	Wooden art objects	[36]
Penicillium expansum Link	Wooden art objects	[36]
Penicillium glabrum (Wehmer) Westling	Wooden art objects	[36]
Penicillium herquei Bainier and Sartory	Wooden art objects	[36]
Penicillium Link	Wooden art objects	[36]
Penicillium sacculum E. Dale	Wooden art objects	[36]
Pseudogymnoascus pannorum (Link) Minnis and D.L. Lindner [5]	Wooden art objects	[36]
Trichoderma viride Pers.	Wooden art objects	[36]

Previously identified as: [1] *Alternaria consortiale*; [2] *Eurotium amstelodami*; [3] *Neosartorya fischeri*; [4] *Fusarium solani*; [5] *Chrysosporium pannorum*.

Fungi can also have an important role in the attack of animal-based objects, adhesives and additives. Textile materials such as silk and wool can suffer microbial mediated biodeterioration processes by the action of deteriorating enzymes [75,76]. In particular, the silks fibroin and sericin can both be the target of microbial attack [77]. Moreover, wool keratins can also be the target of attack by microbes [77]. Evaluation of fibroinolytic and keratinolytic action in cultural heritage scenarios has been helpful to study mummies, funeral clothes and accessories biodeterioration [78–80]. Moreover, fungal chitinolytic and pectinolytic action has also been pinpointed as a threat to Ancient Yemeni mummies preserved with diverse organic compounds [81]. Additionally, esterease action profiling has also been helpful to study isolates retrieved from wax seals, air, textiles and human remains [82–84]. Fibroinolytic screening can be conducted using fibroin agar, with the fibroinolytic action being evaluated by the isolates ability to grow in the culture-amended plates [85]. Moreover, keratinolytic action can be evaluated using feather broth and keratin medium and positive ability can be verified by media turbidity changes [79,85]. On the other hand, chitinolytic activity can be evaluated using powdered chitin agar [86] and pectinolytic activities can be evaluated with media containing pectin [87]. Both these deteriorative activities can be estimated and quantified [81]. Additionally, esterease action can be studied using Tributyrin agar and Tween 80 agar [83,84,88]. Their action can be detected by the development of clear zones (Tributyrin agar) or by the precipitation of insoluble salts and compounds (Tween 80) around colonies. As occurring with ligninolytic action, data regarding filamentous fungi fibroinase and keratinolytic action is still infrequent. Twenty-three species were found to be able of fibroinolytic activity, while more than twenty-five were found to have keratinolytic action. Again, *Aspergillus* and *Penicillium* species also dominate these biodeteriorative profiles (Tables 4 and 5). Moreover, various *Alternaria* species also displayed putative keratinolytic abilities. On the other hand, chitinolytic abilities have been identified for *Aspergillus niger* and *Penicillium* sp., while pectinolytic action has been identified in a slightly more diversified range of fungal genera and species (*Aspergillus candidus*, *Mucor circinelloides*, *Penicillium echinulatum*, *Scopulariopsis koningii*, *Stachybotrys chartarum* and *Trichoderma hamatum*) [81]. In parallel, fifty species have been identified as displaying estereolytic action, with a great dominance of *Aspergillus* and *Penicillium* species (Table 6). Understandably, the detection of these fungal species on crypt environments, human remains, buried materials, mummies, wax seals, textiles and clothes denotes a putative threat to these materials [3].

Table 4. Overview of fungal species for which isolates have been identified as displaying fibrinolytic abilities in biodeteriorative plate essays.

Current Species Name	Original Study Focus	References
Alternaria alternata (Fr.) Keissl.	Funeral accessories	[70]
Ascomycota Caval. -Sm.	Funeral clothes	[85]
Aspergillus caninus (Sigler, Deanna A. Sutton, Gibas, Summerb. and Iwen) Houbraken, Tanney, Visagie and Samson [1]	Funeral clothes	[85]
Aspergillus cristatus Raper and Fennell [2]	Funeral clothes	[85]
Aspergillus fumigatus Fresen.	Funeral clothes	[85]
Aspergillus P. Micheli ex Haller	Funeral accessories	[80]
Aspergillus puniceus Kwon-Chung and Fennell	Funeral clothes	[85]
Aspergillus sydowii (Bainier and Sartory) Thom and Church	Funeral clothes	[85]
Aspergillus tubingensis Mosseray	Funeral clothes	[85]
Aspergillus versicolor (Vuill.) Tirab.	Funeral accessories	[80]
Beauveria bassiana (Bals.—Criv.) Vuill.	Funeral clothes	[85]
Myriodontium keratinophilum Samson and Polon.	Funeral clothes	[85]
Penicillium brevicompactum Dierckx	Funeral clothes	[85]
Penicillium commune Thom	Funeral accessories	[80]
Penicillium crocicola T. Yamam.	Funeral clothes	[85]
Penicillium crustosum Thom	Funeral clothes	[85]
Penicillium expansum Link	Funeral clothes	[85]
Penicillium granulatum Bainier	Funeral accessories	[80]
Penicillium Link	Funeral accessories	[80]
Penicillium roseopurpureum Dierckx	Funeral clothes	[85]
Penicillium spinulosum Thom	Funeral clothes	[85]
Sporobolomyces roseus Kluyver and C.B. Niel [3]	Funeral accessories	[80]
Xenochalara juniperi M.J. Wingf. and Crous	Funeral clothes	[85]

Previously identified as: [1] *Phialosimplex caninus*; [2] *Eurotium cristatum*; [3] *Sporidiobolus metaroseus*.

Table 5. Overview of fungal species for which isolates have been identified as displaying keratinolytic abilities in biodeteriorative plate essays.

Current Species Name	Original Study Focus	References
Alternaria consortialis (Thüm.) J.W. Groves and S. Hughes [1]	Wooden art objects	[36]
Arthrinium phaeospermum (Corda) M.B. Ellis	Wooden art objects	[36]
Aspergillus amstelodami (L. Mangin) Thom and Church [2]	Wooden art objects	[36]
Aspergillus awamori Nakaz.	Wooden art objects	[36]
Aspergillus fischeri Wehmer [3]	Wooden art objects	[36]
Aspergillus flavus Link	Wooden art objects	[36]
Aspergillus fumigatus Fresen.	Wooden art objects	[36]
Aspergillus niger Tiegh	Wooden art objects	[36]
Aspergillus terreus Thom	Wooden art objects and air	[36,69]
Aspergillus ustus (Bainier) Thom and Church	Wooden art objects	[36]
Aspergillus versicolor (Vuill.) Tirab.	Wooden art objects	[36]
Beauveria bassiana (Bals.—Criv.) Vuill.	Wooden art objects	[36]
Chaetomium elatum Kunze	Wooden art objects	[36]
Chaetomium globosum Kunze	Wooden art objects and air	[36,69]
Cladosporium cladosporioides (Fresen.) G.A. de Vries	Wooden objects and air	[69]
Cladosporium herbarum (Pers.) Link	Wooden art objects	[36]
Cladosporium Link	Wooden art objects	[36]
Hypochnicium J. Erikss.	Wooden materials	[61]
"Neocosmospora" solani (Mart.) L. Lombard and Crous [4]	Wooden materials	[70]
Penicillium chrysogenum Thom	Wooden art objects	[36]
Penicillium expansum Link	Wooden art objects	[36]
Penicillium glabrum (Wehmer) Westling	Wooden art objects	[36]
Penicillium herquei Bainier and Sartory	Wooden art objects	[36]
Penicillium Link	Wooden art objects	[36]
Penicillium sacculum E. Dale	Wooden art objects	[36]
Pseudogymnoascus pannorum (Link) Minnis and D.L. Lindner [5]	Wooden art objects	[36]
Trichoderma viride Pers.	Wooden art objects	[36]

Previously identified as: [1] *Alternaria consortiale*; [2] *Eurotium amstelodami*; [3] *Neosartorya fischeri*; [4] *Fusarium solani*; [5] *Chrysosporium pannorum*.

Table 6. Overview of fungal species for which isolates have been identified as displaying estereolytic abilities in biodeteriorative plate essays.

Current Species Name	Original Study Focus	References
Acrodontium salmoneum de Hoog	Wax seal	[84]
Agaricaceae Chevall.	Statue and air	[89]
Alternaria Nees	Statue, stone and air	[89,90]
Alternaria tenuissima (Kunze) Wiltshire	Statue and air	[89]
Aspergillus amstelodami (L. Mangin) Thom and Church [1]	Stone	[90]
Aspergillus caespitosus Raper and Thom	Air, textiles and human remains	[83]
Aspergillus calidoustus Varga, Houbraken and Samson	Air, textiles and human remains	[83]
Aspergillus candidus Link	Air, textiles and human remains	[83]
Aspergillus clavatus Desm.	Textiles	[82]
Aspergillus fischeri Wehmer [2]	Air, textiles and human remains	[83]
Aspergillus flavus Link	Stone	[90]
Aspergillus fumigatus Fresen.	Air, statues, textiles and human remains	[83,89]
Aspergillus niger Tiegh	Stone	[90]
Aspergillus repens (Corda) Sacc. [3]	Air, textiles and human remains	[83]
Aspergillus sydowii (Bainier and Sartory) Thom and Church	Air, textiles and human remains	[83]
Aspergillus terreus Thom	Air, textiles and human remains	[83]
Aspergillus ustus (Bainier) Thom and Church	Air, textiles and human remains	[83]
Aspergillus venenatus Jurjević, S.W. Peterson and B.W. Horn	Air, textiles and human remains	[83]
Aspergillus versicolor (Vuill.) Tirab.	Air, statues, textiles and human remains	[83,89]
Aspergillus westerdijkiae Frisvad and Samson	Air, textiles and human remains	[83]
Aureobasidium pullulans (de Bary and Löwenthal) G. Arnaud	Wax seal	[84]
Bulleromyces albus Boekhout and Á. Fonseca [4]	Wax seal	[84]
Cladosporium aggregatocicatricatum Bensch, Crous and U. Braun	Wax seal	[84]
Cladosporium Link	Air, textiles and human remains	[83]
Cladosporium macrocarpum Preuss	Statue and air	[89]
Cladosporium tenuissimum Cooke	Textiles	[82]

Table 6. *Cont.*

Current Species Name	Original Study Focus	References
Coprinellus xanthothrix (Romagn.) Vilgalys, Hopple and Jacq. Johnson	Air, textiles and human remains	[83]
Coprinopsis cinerea (Schaeff.) Redhead, Vilgalys and Moncalvo	Statue and air	[89]
Curvularia Boedijn	Stone	[90]
Fusarium Link	Stone	[90]
Fusarium sporotrichioides Sherb.	Statue and air	[89]
Hypoxylon fragiforme (Pers.) J. Kickx f.	Textiles	[82]
Microascus brevicaulis S.P. Abbott [5]	Air, textiles and human remains	[83]
Nigrospora oryzae (Berk. and Broome) Petch	Air, textiles and human remains	[83]
Penicillium chrysogenum Thom	Air, textiles and human remains	[82,83]
Penicillium commune Thom	Air, textiles and human remains	[83]
Penicillium corylophilum Dierckx	Textiles, statue and air	[82,89]
Penicillium crustosum Thom	Air, statues, textiles and human remains	[83,89]
Penicillium griseofulvum Dierckx	Air, textiles and human remains	[83]
Penicillium hordei Stolk	Air, textiles and human remains	[83]
Penicillium Link	Statue and air	[89]
Penicillium polonicum K.W. Zaleski	Air, textiles and human remains	[83]
Pseudoscopulariopsis hibernica (A. Mangan) Sand. -Den., Gené and Cano [6]	Air, textiles and human remains	[83]
Rhizopus arrhizus A. Fisch. [7]	Air, textiles and human remains	[83]
Rhizopus microsporus Tiegh.	Air, textils, human remains, statue and air	[83,89]
Rhizopus stolonifer (Ehrenb.) Vuill.	Air, textiles and human remains	[83]
Sordaria fimicola (Roberge ex Desm.) Ces. and De Not.	Statue and air	[89]
Talaromyces purpureogenus Samson, N. Yilmaz, Houbraken, Spierenb., Seifert, Peterson, Varga and Frisvad	Stone	[90]
Thyronectria austroamericana (Speg.) Seeler [8]	Statue and air	[89]
Trichoderma paraviridescens Jaklitsch, Samuels and Voglmayr	Statue and air	[89]

Previously identified as: [1] *Eurotium amsteldomi;* [2] *Neosartorya fischeri;* [3] *Eurotium repens;* [4] *Bullera alba;* [5] *Scopulariopsis brevicaulis;* [6] *Scopulariopsis hibernica;* [7] *Rhizopus oryzae;* [8] *Pleonectria austroamericana.*

Fungal lipolytic action can have an important impact on the biodeterioration of parchment and leather related materials [91]. Fungi can attack lipids and take advantage of fatty materials as a mean to obtain carbon (while also contributing to the material deterioration) [92]. Fungal lipolytic ability characterization in cultural heritage scenarios has been helpful to study the biodeteriorative contribution of isolates retrieved from air, textiles, human remains, wax seals, albumen photographical materials, statues, wooden organs and pipes [83,84,89,93,94]. Lipolytic ability screening can be mainly conducted using Spirit Blue agar and Nile blue, and the positive action can be identified by the development of a halo around the colonies, after a period of incubation [89]. Circa sixty species were found to be able of lipolytic action (Table 7). As similarly verified in other deteriorative analyses, *Aspergillus* and *Penicillium* species are still predominant in these profiles. The detection of these fungal species on materials rich in fatty compounds, such as wax seals and photographic materials should be considered putatively hazardous.

Table 7. Overview of fungal species for which isolates have been identified as displaying lipolytic abilities in biodeteriorative plate essays.

Current Species Name	Original Study Focus	References
Acrodontium salmoneum de Hoog	Wax seal	[84]
Acrostalagmus luteoalbus (Link) Zare, W. Gams and Schroers	Albumen photographic materials	[93]
Agaricaceae Chevall.	Statue and air	[89]
Alternaria mali Roberts	Wooden organ and pipes	[94]
Alternaria Nees	Statue and air	[89]
Alternaria tenuissima (Kunze) Wiltshire	Statue and air	[89]
Aspergillus amstelodami (L. Mangin) Thom and Church [1]	Air, textiles, human remains, wooden organ and pipes	[83,94]
Aspergillus caespitosus Raper and Thom	Air, textiles and human remains	[83]
Aspergillus calidoustus Varga, Houbraken and Samson	Air, textiles and human remains	[83]
Aspergillus candidus Link	Air, textiles and human remains	[83]
Aspergillus cristatus Raper and Fennell [2]	Wooden organ and pipes	[94]

Table 7. *Cont.*

Current Species Name	Original Study Focus	References
Aspergillus fischeri Wehmer [3]	Air, textiles, statues and human remains	[89,93]
Aspergillus fumigatus Fresen.	Air, textiles, statues and human remains	[83,89]
Aspergillus repens (Corda) Sacc. [4]	Air, textiles and human remains	[83]
Aspergillus sydowii (Bainier and Sartory) Thom and Church	Air, textiles, human remains, wooden organ and pipes	[83,94]
Aspergillus terreus Thom	Air, textiles and human remains	[83]
Aspergillus ustus (Bainier) Thom and Church	Air, textiles and human remains	[83]
Aspergillus venenatus Jurjević, S.W. Peterson and B.W. Horn	Air, textiles and human remains	[83]
Aspergillus versicolor (Vuill.) Tirab	Air, textiles, human remains, albumen photographic materials, wooden organ and pipes	[83,89,93,94]
Aspergillus westerdijkiae Frisvad and Samson	Air, textiles and human remains	[83]
Aureobasidium pullulans (de Bary and Löwenthal) G. Arnaud	Wax seal	[84]
Bjerkandera adusta (Willd.) P. Karst.	Albumen photographic materials	[83]
Bulleromyces albus Boekhout and Á. Fonseca [5]	Wax seal	[84]
Chaetomium elatum Kunze	Albumen photographic materials	[83]
Cladosporium aggregatocicatricatum Bensch, Crous and U. Braun	Wax seal	[84]
Cladosporium cladosporioides (Fresen.) G.A. de Vries	Wooden organ and pipes	[94]
Cladosporium Link	Air, textiles, human remains and Etruscan hypogeal tombs	[44,83]
Cladosporium macrocarpum Preuss	Statue and air	[89]
Coprinellus xanthothrix (Romagn.) Vilgalys, Hopple and Jacq. Johnson	Air, textiles and human remains	[83]
Coprinopsis cinerea (Schaeff.) Redhead, Vilgalys and Moncalvo	Statue and air	[89]
Cyphellophora G.A. de Vries	Etruscan hypogeal tombs	[44]
Cyphellophora olivacea (W. Gams) Réblová & Unter.	Etruscan hypogeal tombs	[44]
Epicoccum nigrum Link	Wooden organ and pipes	[94]
Exophiala angulospora Iwatsu, Udagawa & T. Takase	Etruscan hypogeal tombs	[44]
Exophiala J.W. Carmich.	Etruscan hypogeal tombs	[44]
Fusarium sporotrichioides Sherb.	Statue and air	[89]
Microascus brevicaulis S.P. Abbott [6]	Air, textiles and human remains	[83]
Mucor plumbeus Bonord.	Albumen photographic materials	[93]
Nigrospora oryzae (Berk. and Broome) Petch	Air, textiles and human remains	[83]
Paecilomyces maximus C. Ram	Wooden organ and pipes	[94]
Penicillium chrysogenum Thom	Air, textiles and human remains	[83]
Penicillium commune Thom	Air, textiles and human remains	[83]
Penicillium corylophilum Dierckx	Statue and air	[89]
Penicillium crustosum Thom	Air, textiles, statues and human remains	[83,89]
Penicillium digitatum (Pers.) Sacc.	Air, textiles and human remains	[83]
Penicillium griseofulvum Dierckx	Air, textiles and human remains	[83]
Penicillium hordei Stolk	Air, textiles and human remains	[83]
Penicillium Link	Albumen photographic materials, statues and air	[89,93]
Penicillium polonicum K.W. Zaleski	Air, textiles and human remains	[83]
Penicillium thomii Maire	Albumen photographic materials	[93]
Phlebia Fr.	Albumen photographic materials	[93]
Pleosporales Luttr. ex M.E. Barr	Albumen photographic materials	[93]
Pseudoscopulariopsis hibernica (A. Mangan) Sand. -Den., Gené and Cano [7]	Air, textiles and human remains	[83]
Rhizopus arrhizus A. Fisch. [8]	Air, textiles and human remains	[83]
Rhizopus microsporus Tiegh.	Air, textiles, human remains, statues and air	[83,89]
Rhizopus stolonifer (Ehrenb.) Vuill.	Air, textiles and human remains	[83]
Sordaria fimicola (Roberge ex Desm.) Ces. and De Not.	Statue and air	[89]
Talaromyces flavus (Klöcker) Stolk and Samson	Air, textiles and human remains	[83]
Thyronectria austroamericana (Speg.) Seeler [9]	Statue and air	[89]
Trichoderma harzianum Rifai	Air, textiles and human remains	[83]
Trichoderma paraviridescens Jaklitsch, Samuels and Voglmayr	Statue and air	[89]
Trichothecium roseum (Pers.) Link	Albumen photographic materials	[93]

Previously identified as: [1] *Eurotium amstelodami*; [2] *Eurotium cristatum*; [3] *Neosartorya fischeri*; [4] *Eurotium repens*; [5] *Bullera alba*; [6] *Scopulariopsis brevicaulis*; [7] *Scopulariopsis hibernica*; [8] *Rhizopus oryzae*; [9] *Pleonectria austroamericana*.

Fungal proteolytic action can contribute to the biodeterioration of proteinaceous materials, such is the case of artistic natural binders. In addition, some conservation approaches also employ similar materials that can be targeted by microbial biodeterioration [38]. Fungal proteolytic ability characterization in cultural heritage scenarios has been helpful to study the biodeteriorative contribution of isolates retrieved from air, funeral clothes and accessories, graphic documents, materials present in libraries and museums, frescoes, textiles, human remains, mummies, mural paintings, cinematographic films, wax seals, paper, parchment, wooden organs and pipes [28,38,41,78–80,83–85,93,95–100]. Proteolytic ability screening can be mainly conducted using Gelatin agar (R2A-Gel), Casein agar (CN), Milk Nutrient agar (MilkNA) and media containing rabbit glue [78,94,96]. After a period of incubation, positive proteolytic ability can be detected by flooding of agar plates with 10% tannin solution and the visualization of the formed hydrolysis zones [101]. Over one hundred and thirty species have been found to be able of promoting protein attack (Table 8). As similarly verified in other enzymatic activities, *Aspergillus* and *Penicillium* species also dominate this biodeteriorative profile. Nonetheless, a significant number of species from genera *Alternaria, Cladosporium* and *Talaromyces* displaying these characteristics can also be confirmed. Detection of these fungal species on proteinaceous materials will putatively result in their accentuated biodeterioration.

Table 8. Overview of fungal species for which isolates have been identified as displaying proteolytic abilities in biodeteriorative plate essays.

Current Species Name	Original Study Focus	References
Acrodontium salmoneum de Hoog	Wax seal	[84]
Acrostalagmus luteoalbus (Link) Zare, W. Gams and Schroers	Albumen photographic materials	[93]
Actinomucor elegans (Eidam) C.R. Benjamin and Hesseltine	Mural paintings	[41]
Alternaria alternata (Fr.) Keissl.	Air, mural paintings, mummies, funeral accessories, cinematographic films	[41,79,80,95,98]
Alternaria mali Roberts	Mummies, wooden organ and pipes	[79,94]
Alternaria Nees	Mummies and funeral clothes	[79,85]
Alternaria solani Sorauer	Mummies	[79]
Alternaria tenuissima (Kunze) Wiltshire	Mummies	[79]
Ascomycota Caval. -Sm.	Funeral clothes	[85]
Aspergillus alliaceus Thom and Church	Air and graphic documents	[98,100]
Aspergillus amstelodami (L. Mangin) Thom and Church	Air, wooden organ and pipes	[28,94]
Aspergillus aureolatus Munt.—Cvetk. and Bata	Frescoes and air	[38]
Aspergillus auricomus (Guég.) Saito	Air and graphic documents	[100]
Aspergillus caespitosus Raper and Thom	Air, textiles and human remains	[83]
Aspergillus candidus Link	Air, graphic documents, photographs, maps, textiles and human remains	[83,84,98–100]
Aspergillus caninus (Sigler, Deanna A. Sutton, Gibas, Summerb. and Iwen) Houbraken, Tanney, Visagie and Samson [1]	Funeral clothes	[85]
Aspergillus chevalieri (L. Mangin) Thom and Church [2]	Air, graphic documents, photographs, maps and materials present in libraries and museums	[96,99,100]
Aspergillus creber Jurjević, S.W. Peterson and B.W. Horn	Air	[28]
Aspergillus cristatus Raper and Fennell [3]	Funeral clothes, wooden organ and pipes	[85,94]
Aspergillus europaeus Hubka, A. Nováková, Samson, Houbraken, Frisvad and M. Kolařík	Frescoes and air	[28,38]
Aspergillus flavipes (Bainier and R. Sartory) Thom and Church	Air	[98]
Aspergillus flavus Link [4]	Air, frescoes, mummies, photographs, maps and materials present in libraries and museums	[38,47,78,79,96,99,100]
Aspergillus fumigatus Fresen.	Air, materials present in libraries and museums, paper and parchment and funeral clothes	[85,96,97]
Aspergillus japonicus Saito	Air and graphic documents	[100]
Aspergillus jensenii Jurjević, S.W. Peterson and B.W. Horn	Air	[28]
Aspergillus nidulans (Eidam) G. Winter	Air, materials present in libraries and museum, mural paintings and cinematographic films	[41,95,96]
Aspergillus niger Tiegh	Frescoes, air, textiles, human remains, photographs, maps, paper and parchment, graphic documents, materials present in libraries and museums	[38,73,96–100]
Aspergillus ostianus Wehmer	Frescoes, air and graphic documents	[38,100]
Aspergillus P. Micheli ex Haller [5]	Air and mural paintings	[41,47]
Aspergillus penicillioides Speg.	Air and graphic documents	[100]
Aspergillus proliferans G. Sm.	Air	[28]
Aspergillus protuberus Munt. -Cvetk.	Air	[28]
Aspergillus pseudodeflectus Samson and Mouch.	Funeral clothes	[85]
Aspergillus puniceus Kwon-Chung and Fennell	Funeral clothes	[85]

Table 8. *Cont.*

Current Species Name	Original Study Focus	References
Aspergillus sydowii (Bainier and Sartory) Thom and Church	Air, mural paintings, funeral clothes, wooden organ and pipes and materials present in libraries and museums	[28,41,85,94,96]
Aspergillus tabacinus Nakaz., Y. Takeda, Simo and A. Watan.	Air and materials present in libraries and museums	[28]
Aspergillus tamarii Kita	Air and materials present in libraries and museums	[96]
Aspergillus terreus Thom	Air and materials present in libraries and museums	[96]
Aspergillus tubingensis Mosseray	Funeral clothes	[85]
Aspergillus unguis (Émile-Weill and L. Gaudin) Thom and Raper	Air and graphic documents	[100]
Aspergillus unilateralis Thrower	Air and graphic documents	[100]
Aspergillus ustus (Bainier) Thom and Church	Cinematographic films	[95]
Aspergillus versicolor (Vuill.) Tirab	Air, mural paintings, frescoes, textiles, human remains, funeral accessories, cinematographic materials, wooden organ and pipes and materials present in libraries and museums	[28,38,41,78,80,83,94–96]
Aspergillus westerdijkiae Frisvad and Samson	Air, textiles and human remains	[83]
Aureobasidium pullulans (de Bary and Löwenthal) G. Arnaud	Wax seal and mummies	[79,84]
Beauveria bassiana (Bals.—Criv.) Vuill.	Funeral clothes	[85]
Bjerkandera adusta (Willd.) P. Karst.	Frescoes, air albumen photographic materials	[38,93]
Botryotrichum murorum (Corda) X. Wei Wang and Samson [6]	Frescoes and air	[38]
Bulleromyces albus Boekhout and Á. Fonseca [7]	Wax seal	[84]
Chaetomium ancistrocladum Udagawa and Cain	Frescoes and air	[38]
Chaetomium elatum Kunze	Albumen photographic materials	[93]
Chaetomium globosum Kunze	Mural paintings	[41]
Cladosporium aggregatocicatricatum Bensch, Crous and U. Braun	Wax seal	[84]
Cladosporium cladosporioides (Fresen.)	Mural paintings, frescoes, air, mummies, graphic documents, maps, photographs, cinematographic films, wooden organ and pipes	[38,41,79,94,95,99,100]
Cladosporium cucumerinum Ellis and Arthur	Mural paintings	[41]
Cladosporium herbarum (Pers.) Link [8]	Mummies and mural painitngs	[41,81]
Cladosporium Link	Etruscan hypogeal tombs, air and mummies	[44,47,79]
Cladosporium macrocarpum Preuss	Air and frescoes	[78]
Cladosporium oxysporum Berk. and M.A. Curtis	Frescoes and air	[38]
Cladosporium pseudocladosporioides Bensch, Crous and U. Braun	Mummies	[79]
Cladosporium sphaerospermum Penz.	Mummies and mural paintings	[41,79]
Cladosporium tenuissimum Cooke	Mural paintings	[41]
Cladosporium uredinicola Speg.	Mummies, frescoes and air	[38,79]
Coprinellus xanthothrix (Romagn.) Vilgalys, Hopple and Jacq. Johnson	Air, textiles and human remains	[83]
Curvularia australiensis (Bugnic. ex M.B. Ellis) Manamgoda, L. Cai and K.D. Hyde [9]	Air and graphic documents	[100]
Curvularia Boedijn	Air	[47]
Curvularia pallescens Boedijn	Air and graphic documents	[98,100]
Cyphellophora G.A. de Vries	Etruscan hypogeal tombs	[44]
Didymella glomerata (Corda) Qian Chen and L. Cai [10]	Cinematographic films	[95]
Dipodascus Lagerh. [11]	Air	[47]
Epicoccum Link	Air and graphic documents	[100]
Epicoccum nigrum Link	Mummies, wooden organ and pipes	[79,94]
Fusarium chlamydosporum Wollenw. and Reinking	Mural paintings	[41]
Fusarium Link	Air and graphic documents	[57,100]
Fusarium oxysporum Schltdl.	Air	[98]
Hormodendrum pyri W.H. English	Wax seal	[84]
Leptosphaeria avenaria G.F. Weber [12]	frescoes and air	[38]
Meyerozyma guilliermondii (Wick.) Kurtzman and M. Suzuki [13]	Air	[98]
Microascus brevicaulis S.P. Abbott [14]	Air, textiles and human remains	[83]
Mucor racemosus Bull.	Cinematographic films, paper and parchment	[95,97]
Mucor spinosus Schrank	Paper and parchment	[97]
Myriodontium keratinophilum Samson and Polon.	Funeral clothes	[85]
Nigrospora oryzae (Berk. and Broome) Petch [15]	Air, textiles and human remains	[83,98]
Paecilomyces maximus C. Ram [16]	Wooden organ and pipes	[94]
Paecilomyces variotii Bainier	Air	[98]
Penicillium aurantiogriseum Dierckx	Mural paintings	[41]
Penicillium bilaiae Chalab.	Frescoes and air	[38]
Penicillium brevicompactum Dierckx	Air and funeral clothes	[28,85]
Penicillium camemberti Thom	Paper and parchment	[97]
Penicillium canescens Sopp	Air and graphic documents	[100]
Penicillium chrysogenum Thom	Air, mural paintings, textiles, human remains, mummies, paper, photographs, maps, parchment, graphic documents and cinematographic films	[28,41,79,83,95,97,99,100]
Penicillium citreonigrum Dierckx	Air	[28]
Penicillium citrinum Thom	Air, photographs, maps and graphic documents	[28,99,100]
Penicillium commune Thom	Air, textiles, human remains, mural paintings, funeral accessories and clothes, paper and parchment	[41,80,83,85,87]
Penicillium crocicola T. Yamam.	Funeral clothes	[85]
Penicillium crustosum Thom	Funeral clothes, wooden organ and pipes	[85,94]

Table 8. *Cont.*

Current Species Name	Original Study Focus	References
Penicillium decumbens Thom	Air, paper and parchment	[28,97]
Penicillium digitatum (Pers.) Sacc.	Air, textiles and human remains	[28,83]
Penicillium expansum Link	Air, mummies and funeral clothes	[28,79,85]
Penicillium glabrum (Wehmer) Westling	Air	[28]
Penicillium granulatum Bainier	Funeral accessories	[80]
Penicillium griseofulvum Dierckx	Air, frescoes, graphic documents, textiles and human remains	[38,83,100]
Penicillium hordei Stolk	Air, textiles and human remains	[83]
Penicillium janczewskii K.W. Zaleski	Air, maps, photographs and graphic documents	[99,100]
Penicillium Link	Air, frescoes, albumen photographic materials and funeral accessories	[57,78,80,93]
Penicillium olsonii Bainier and Sartory	Mural paintings	[41]
Penicillium polonicum K.W. Zaleski	Air, textiles, mural paintings and human remains	[41,83]
Penicillium raistrickii G. Sm.	Mummies	[79]
Penicillium roseopurpureum Dierckx	Mummies and funeral clothes	[79,85]
Penicillium simplicissimum (Oudem.) Thom [17]	Air, maps, photographs and graphic documents	[99,100]
Penicillium solitum Westling	Air	[28]
Penicillium spinulosum Thom	Funeral clothes	[85]
Penicillium viridicatum Westling	Air, paper and parchment	[28,97]
Penicillium waksmanii K.W. Zaleski	Air and graphic documents	[100]
Pestalotia De Not.	Air and graphic documents	[100]
Phoma herbarum Westend.	Paper and parchment	[97]
Pleurotus ostreatus (Jacq.) P. Kumm. [18]	Albumen photographic materials	[93]
Pseudozyma prolifica Bandoni	Mural paintings	[31]
Rhizopus microsporus Tiegh. [19]	Air, textiles and human remains	[83]
Sarocladium W. Gams and D. Hawksw. [20]	Air, graphic documents and albumen photographic materials	[93,98,100]
Sporobolomyces roseus Kluyver and C.B. Niel [21]	Funeral Accessories	[80]
Stemphylium vesicarium (Wallr.) E.G. Simmons	Mummies	[79]
Talaromyces aculeatus (Raper and Fennell) Samson, N. Yilmaz, Frisvad and Seifert	Mural paintings	[41]
Talaromyces rugulosus (Thom) Samson, N. Yilmaz, Frisvad and Seifert	Wooden organ and pipes	[94]
Talaromyces sayulitensis Visagie, N. Yilmaz, Seifert and Samson	Air	[28]
Trichoderma harzianum Rifai	Air, textiles and human remains	[83]
Trichoderma longibrachiatum Rifai	Cinematographic films	[95]
Trichoderma viride Pers.	Air	[98]
Trichothecium roseum (Pers.) Link	Albumen photographic materials	[93]
Xenochalara juniperi M.J. Wingf. and Crous	Funeral clothes	[85]

Previously identified as: [1] *Phialosimplex caninus*; [2] *Eurotium chevalieri*; [3] *Eurotium cristatum*; [4] *Aspergillus oryzae*; [5] *Emericella* sp.; [6] *Chaetomium murorum*; [7] *Bullera alba*; [8] *Cladosporium herbarium*; [9] *Bipolaris australiensis*; [10] *Phoma glomerata*; [11] *Geotrichum* sp.; [12] *Phaeosphaeria avenaria*; [13] *Candida guilliermondii*; [14] *Scopulariopsis brevicaulis*; [15] *Nigrospora sphaerica*; [16] *Paecilomyces formosus*; [17] *Penicillium janthinellum*; [18] *Pleurotus pulmonarius*; [19] *Rhizopus microsporus var. oligosporus*; [20] *Cephalosporium* sp.; [21] *Sporidiobolus metaroseus*.

Fungal cellulolytic action is known to contribute to the biodeterioration of paper, canvas oil paintings, binders and photographic materials [3]. Moreover, cellulolytic action abilities characterization in cultural heritage scenarios has been helpful to study the biodeteriorative contribution of isolates retrieved from air, albumen photographic materials, mummies, funeral accessories, wooden art objects, organs and pipes, wax seals, graphic documents, stone, drawings, lithographs, paintings, textiles, human remains, maps, photographs, paper and other materials present in libraries and museums [28,36,69,79,81–83,96–100,102–105]. Cellulolytic ability screening can be conducted using Czapek-Dox agar supplemented with hydroxyethyl cellulose [69], Congo Red agar [79], Mandels and Reese medium with carboxymethyl cellulose (CMC) [106] or media containing sterilized filter paper [47]. Positive evaluation of cellulolytic ability can be assessed by the visualization of hydrolyzed areas or after congo red application and treatment. Over one hundred and fifty fungal species have been found to have cellulolytic abilities (Table 9). The great majority of species belonged to genera *Alternaria*, *Aspergillus*, *Chaetomium*, *Cladosporium*, *Penicillium* and *Talaromyces*. As such, detection of these fungal species on cellulolytic materials including paper, paintings and photographic materials, should be perceived as putatively threatening from a biodeterioration standpoint.

Table 9. Overview of fungal species for which isolates have been identified as displaying cellulolytic abilities in biodeteriorative plate essays.

Current Species Name	Original Study Focus	References
Acremonium alabamense Morgan-Jones	Air	[98]
Acrostalagmus luteoalbus (Link) Zare, W. Gams and Schroers	Albumen photographic materials	[93]
Alternaria alternata (Fr.) Keissl.	Air, mummies and funeral accessories	[79,80,98]
Alternaria chartarum Preuss [1]	Mummies	[81]
Alternaria consortialis (Thüm.) J.W. Groves and S. Hughes [2]	Wooden art objects	[36]
Alternaria mali Roberts	Mummies, wooden organ and pipes	[79,94]
Alternaria Nees	Air and graphic documents, stone and mummies	[47,79,90,91]
Alternaria solani Sorauer	Mummies	[79]
Alternaria tenuissima (Kunze) Wiltshire	Air and mummies	[79,102]
Arthrinium arundinis (Corda) Dyko and B. Sutton	Drawings and lithographs	[94]
Arthrinium phaeospermum (Corda) M.B. Ellis	Wooden art objects	[36]
Ascomycota Caval. -Sm.	Funeral clothes and oainting	[85,103]
Ascotricha Berk.	Painting	[103]
Ascotricha chartarum Berk.	Painting	[103]
Aspergillus alliaceus Thom and Church	Air and graphic documents	[98,100]
Aspergillus amstelodami (L. Mangin) Thom and Church [3]	Air, stone, wooden organ, pipes and objects	[28,36,69,90,94]
Aspergillus auricomus (Guég.) Saito	Air and graphic documents	[100]
Aspergillus caespitosus Raper and Thom	Air, textiles and human remains	[83]
Aspergillus calidoustus Varga, Houbraken and Samson	Air, textiles and human remains	[28,83]
Aspergillus candidus Link	Air, textiles, mummies, graphic documents, maps, photographs and human remains	[81,83,98–100]
Aspergillus chevalieri (L. Mangin) Thom and Church [4]	Air, photographs, maps and graphic documents	[99,100]
Aspergillus clavatus Desm.	Air and textiles	[82,102]
Aspergillus creber Jurjević, S.W. Peterson and B.W. Horn	Air	[28]
Aspergillus cristatus Raper and Fennell [5]	Funeral clothes, wooden organ and pipes	[85,94]
Aspergillus fischeri Wehmer [6]	Air, textiles and human remains	[83]
Aspergillus flavipes (Bainier and R. Sartory) Thom and Church	Air	[98,102]
Aspergillus flavus Link [7]	Air, stone, paintings, graphic documents, photographs, maps, paintings, library and museums	[28,47,90,96,98–100,102,104]
Aspergillus fumigatus Fresen.	Air, textiles, paper, parchment, funeral clothes, libraries, museums and human remains	[83,85,96,97]
Aspergillus japonicus Saito	Air and graphic documents	[100]
Aspergillus jensenii Jurjević, S.W. Peterson and B.W. Horn	Air	[28]
Aspergillus melleus Yukawa	Air	[28]
Aspergillus niger Tiegh	Air, stone, paintings, libraries, museums, graphic documents, photographs, maps, drawings and lithographs	[28,90,96,98–100,102,104,105]
Aspergillus ostianus Wehmer	Air and graphic documents	[100]
Aspergillus P. Micheli ex Haller	Air	[47]
Aspergillus penicillioides Speg.	Air and graphic documents	[100]
Aspergillus protuberus Munt. -Cvetk.	Air	[28]
Aspergillus pseudodeflectus Samson and Mouch.	Funeral clothes	[85]
Aspergillus sydowii (Bainier and Sartory) Thom and Church	Air, textiles, human remains, funeral clothes, libraries, museums, wooden organ and pipes	[28,83,85,94,96]
Aspergillus tabacinus Nakaz., Y. Takeda, Simo and A. Watan.	Air	[28]
Aspergillus tamarii Kita	Air and materials present in libraries and museums	[96]

Table 9. *Cont.*

Current Species Name	Original Study Focus	References
Aspergillus terreus Thom	Air, textiles, wooden art objects, libraries, museums and human remains	[36,69,83,96,102]
Aspergillus unguis (Émile-Weill and L. Gaudin) Thom and Raper	Air and graphic documents	[100]
Aspergillus unilateralis Thrower	Air and graphic documents	[100]
Aspergillus ustus (Bainier) Thom and Church	Mummies and wooden art objects	[36,81]
Aspergillus venenatus Jurjević, S.W. Peterson and B.W. Horn	Air, textiles and human remains	[83]
Aspergillus versicolor (Vuill.) Tirab	Air, wooden art objects, textiles, human remains, albumen photographic materials, drawings, lithographs, wooden organ and pipes	[28,36,83,93,94,102,105]
Aspergillus westerdijkiae Frisvad and Samson	Air, textiles and human remains	[83]
Aureobasidium pullulans (de Bary and Löwenthal) G. Arnaud	Mummies	[81]
Bjerkandera adusta (Willd.) P. Karst.	Paper, parchment and albumen photographic materials	[93,97]
Chaetomium elatum Kunze	Air, wooden art objects and albumen photographic materials	[36,93,102]
Chaetomium globosum Kunze	Wooden art, air, paintings, libraries, museums, drawings and lithographs	[36,69,96,103,105]
Chaetomium thermophilum La Touche	Mummies	[81]
Cladosporium aggregatocicatricatum Bensch, Crous and U. Braun	Wax seal	[84]
Cladosporium angustisporum Bensch, Summerell, Crous and U. Braun	Drawings and lithographs	[105]
Cladosporium cladosporioides (Fresen.)	Air, wooden art objects, mummies, libraries, museums, graphic documents, photographs, maps, drawings. lithographs, wooden organ and pipes	[36,69,79,94,96,98–100,102,105]
Cladosporium herbarum (Pers.) Link [8]	Air and mummies	[81,102]
Cladosporium Link	Etruscan hypogeal tombs, air, textiles, human remains, wooden art objects, mummies, drawings and lithographs	[36,44,57,79,83,105]
Cladosporium oxysporum Berk. and M.A. Curtis	Air and materials present in libraries and museums	[96,102]
Cladosporium pseudocladosporioides Bensch, Crous and U. Braun	Mummies	[79]
Cladosporium sphaerospermum Penz.	Mummies, air and materials present in libraries and museums	[79,82,92,96,105]
Cladosporium tenuissimum Cooke	Textiles, mummies, drawings and lithographs	[79,82,105]
Cladosporium uredinicola Speg.	Mummies	[79]
Collariella bostrychodes (Zopf) X. Wei Wang and Samson [9]	Air and materials present in libraries and museums	[96]
Colletotrichum kahawae J.M. Waller and Bridge	Drawings and lithographs	[105]
Coniochaeta cipronana Coronado-Ruiz, Avendaño, Escudero-Leyva, Conejo-Barboza, P. Chaverri and Chavarría	Drawings and lithographs	[105]
Coprinellus xanthothrix (Romagn.) Vilgalys, Hopple and Jacq. Johnson	Air, textiles and human remains	[83]
Curvularia australiensis (Bugnic. ex M.B. Ellis) Manamgoda, L. Cai and K.D. Hyde [10]	Air and graphic documents	[100]
Curvularia Boedijn	Air and stone	[47,90]
Curvularia clavata B.L. Jain	Air	[102]
Curvularia eragrostidis (Henn.) J.A. Mey.	Air and graphic documents	[98,100]
Curvularia lunata (Wakker) Boedijn	Air and materials present in libraries and museums	[86,102]
Curvularia pallescens Boedijn	Air and graphic documents	[98,100,102]

Table 9. *Cont.*

Current Species Name	Original Study Focus	References
Cyphellophora G.A. de Vries	Etruscan hypogeal tombs	[44]
Cyphellophora olivacea (W. Gams) Réblová & Unter.	Etruscan hypogeal tombs	[44]
Dichotomopilus indicus (Corda) X. Wei Wang and Samson [11]	Air and materials present in libraries and museums	[96]
Dipodascus Lagerh. [12]	Air	[47]
Epicoccum Link	Air and graphic documents	[100]
Epicoccum nigrum Link	Mummies, wooden organ and pipes	[79,94]
Fomes fomentarius (L.) Fr.	Textiles	[82]
Fulvia fulva (Cooke) Cif. [13]	Air	[102]
Fusarium Link	Air, stone and graphic documents	[47,90,100,102]
Fusarium oxysporum Schltdl.	Air and wooden materials	[70,98]
Fusarium proliferatum (Matsush.) Nirenberg ex Gerlach and Nirenberg	Air	[98]
Hormodendrum pyri W.H. English	Wax seal	[84]
Hypochnicium J. Erikss.	Wooden materials	[61]
Meyerozyma guilliermondii (Wick.) Kurtzman and M. Suzuki [14]	Air	[98]
Microascus brevicaulis S.P. Abbott [15]	Air, textiles and human remains	[83]
Mortierella Coem.	Wooden materials	[61]
Mucor circinelloides Tiegh.	Mummies	[81]
Mucor P. Micheli ex L.	Air	[102]
Mucor plumbeus Bonord.	Albumen photographic materials	[93]
Mucor racemosus Bull.	Air, paper and parchment	[97,98]
Mucor spinosus Schrank	Paper and parchment	[97]
"*Neocosmospora*" *solani* (Mart.) L. Lombard and Crous [16]	Wooden materials	[70]
Neurospora crassa Shear and B.O. Dodge	Air	[98,102]
Neurospora sitophila Shear and B.O. Dodge [17]	Air and materials present in libraries and museums	[96,98]
Nigrospora oryzae (Berk. and Broome) Petch	Air, textiles and human remains	[83]
Nigrospora oryzae (Berk. and Broome) Petch	Air and materials present in libraries and museums	[96,98]
Paecilomyces maximus C. Ram [18]	Wooden organ and pipes	[94]
Paecilomyces variotii Bainier	Air	[98]
Penicillium aurantiogriseum Dierckx	Air	[102]
Penicillium brevicompactum Dierckx	Air	[28]
Penicillium camemberti Thom	Paper and parchment	[97]
Penicillium canescens Sopp	Air and graphic documents	[28,100]
Penicillium carneum (Frisvad) Frisvad	Air	[28]
Penicillium chrysogenum Thom	Wooden objects, air, textiles, mummies, human remains, drawings, maps, photographs, lithographs, graphic documents and materials present in libraries and museums	[28,36,69,79,82,83,96–100,102,105]
Penicillium citreonigrum Dierckx	Air	[28]
Penicillium citrinum Thom	Air, paintings, graphic documents, photographs, maps and materials present in libraries and museums	[28,96,99,100,102,104]
Penicillium commune Thom	Air, textiles, paper, parchment, funeral clothes and human remains	[83,85,97,102]
Penicillium corylophilum Dierckx	Textiles, air and materials present in libraries and museums	[82,96]
Penicillium crocicola T. Yamam.	Funeral clothes	[85]
Penicillium crustosum Thom	Air, textiles, funeral clothes, human remains and wooden organ and pipes	[83,85,94]
Penicillium decumbens Thom [19]	Air, paper and parchment	[28,97,102]
Penicillium digitatum (Pers.) Sacc.	Air, textiles and human remains	[28,83,102]

Table 9. *Cont.*

Current Species Name	Original Study Focus	References
Penicillium expansum Link	Air, wooden art objects and funeral clothes	[28,36,69,85]
Penicillium glabrum (Wehmer) Westling	Wooden objects, air and materials present in libraries and museums	[28,69,96]
Penicillium griseofulvum Dierckx	Air, textiles, graphic documents and human remains	[83,100,102]
Penicillium herquei Bainier and Sartory	Wooden objects and air	[36,69]
Penicillium janczewskii K.W. Zaleski	Air, photographs, maps and graphic documents	[99,100]
Penicillium Link	Air, mummies, wooden art objects, albumen photographic materials, funeral accessories and paintings	[36,47,80,81,93,98–100,103]
Penicillium raistrickii G. Sm.	Mummies	[79]
Penicillium roseopurpureum Dierckx	Mummies and funeral clothes	[79,85]
Penicillium rubens Biourge	Painting	[103]
Penicillium sacculum E. Dale	Wooden objects and air	[36,69]
Penicillium sanguifluum (Sopp) Biourge	Air	[28]
Penicillium simplicissimum (Oudem.) Thom	Air, photographs, maps, graphic materials and materials present in libraries and museums	[96,99,100]
Penicillium solitum Westling	Air	[28]
Penicillium thomii Maire	Albumen photographic materials	[93]
Penicillium ulaiense H.M. Hsieh, H.J. Su and Tzean	Air	[28]
Penicillium viridicatum Westling	Paper and parchment	[97]
Penicillium waksmanii K.W. Zaleski	Air and graphic documents	[100]
Penicillium westlingii K.W. Zaleski	Drawings and lithographs	[105]
Periconia epigraphicola Coronado-Ruiz, R. E. Escudero-Leyva, G. Conejo-Barboza, P. Chaverri and M. Chavarría	Drawings and lithographs	[105]
Pestalotia De Not.	Air and graphic documents	[100]
Phlebia Fr.	Albumen photographic materials	[93]
Phoma herbarum Westend.	Paper and parchment	[97]
Pleosporales Luttr. ex M.E. Barr	Albumen photographic materials	[93]
Pleurotus ostreatus (Jacq.) P. Kumm. [20]	Albumen photographic materials	[93]
Pseudallescheria fimeti (Arx, Mukerji and N. Singh) McGinnis, A.A. Padhye and Ajello	Painting	[103]
Pseudallescheria Negr. and I. Fisch.	Painting	[103]
Pseudogymnoascus pannorum (Link) Minnis and D.L. Lindner [21]	Wooden objects and air	[36,69]
Pseudoscopulariopsis hibernica (A. Mangan) Sand. -Den., Gené and Cano [22]	Air, textiles and human remains	[83]
Rhizopus arrhizus A. Fisch. [23]	Air, textiles and human remains	[83]
Rhizopus microsporus Tiegh.	Air, textiles and human remains	[83]
Sarocladium W. Gams and D. Hawksw. [24]	Air and graphic documents	[98,100]
Sporobolomyces roseus Kluyver and C.B. Niel [25]	Funeral accessories	[80]
Stachybotrys chartarum (Ehrenb.) S. Hughes	Mummies	[81]
Stemphylium vesicarium (Wallr.) E.G. Simmons	Mummies	[79]
Syncephalastrum racemosum Cohn ex J. Schröt.	Air and materials present in libraries and museums	[96]
Talaromyces amestolkiae N. Yilmaz, Houbraken, Frisvad and Samson	Air	[28]
Talaromyces purpureogenus Samson, N. Yilmaz, Houbraken, Spierenb., Seifert, Peterson, Varga and Frisvad	Stone	[90]
Talaromyces rugulosus (Thom) Samson, N. Yilmaz, Frisvad and Seifert	Wooden organ and pipes	[94]
Talaromyces sayulitensis Visagie, N. Yilmaz, Seifert and Samson	Air	[28]

Table 9. *Cont.*

Current Species Name	Original Study Focus	References
Talaromyces verruculosus (Peyronel) Samson, N. Yilmaz, Frisvad and Seifert	Air	[28]
Trichoderma harzianum Rifai	Air, textiles and human remains	[83]
Trichoderma longibrachiatum Rifai	Drawings and lithographs	[105]
Trichoderma viride Pers.	Wooden art objects and air	[36,69,98]
Trichothecium roseum (Pers.) Link	Albumen photographic materials	[93]

Previously identified as: [1] *Ulocladium chartarum;* [2] *Alternaria consortiale;* [3] *Eurotium amsteldomi;* [4] *Eurotium chevalieri;* [5] *Eurotium cristatum;* [6] *Neosartorya fischeri;* [7] *Aspergillus oryzae;* [8] *Cladosporium herbarium;* [9] *Chaetomium bostrychodes;* [10] *Bipolaris australiensis;* [11] *Chaetomium indicum;* [12] *Geotrichum* sp.; [13] *Cladosporium fulvum;* [14] *Candida guilliermondii;* [15] *Scopulariopsis brevicaulis;* [16] *Fusarium solani;* [17] *Chrysonilia sitophila;* [18] *Paecilomyces formosus;* [19] *Penicillium funiculosum;* [20] *Pleurotus pulmonarius;* [21] *Chrysosporium pannorum;* [22] *Scopulariopsis hibernica;* [23] *Rhizopus oryzae;* [24] *Cephalosporium* sp.; [25] *Sporidiobolus metaroseus.*

5. Conclusions

As pointed and reviewed by Pyzik and colleagues [107] the application of high-throughput Next-Generation sequencing technologies has highlighted that cultural heritage materials are inhabited by various unknown microorganisms still pending taxonomic description and their biodeteriorative profiling. The material biodeterioration is known to sometimes be caused by a predominant or specific microbial group, while more often the complex biodeterioration processes are a result of the synergistic action of a group of organisms resulting from various colonization events influenced by the impacts of multiple external factors throughout a time frame [77]. Cultivation methodologies often face limitations in what regards the ability for distinct organisms to be effectively cultivated and their original biodeteriorative characteristics replicated under laboratory conditions [108]. Understandably the application of more modern molecular techniques in cultural heritage biodeterioration studies has been increasingly being used and updated for the last two decades [27,108]. Although with their own set of limitations, culture-dependent methodologies still offer three main advantages: (1) The isolation of microbes for further differential analysis; (2) the possibility to isolate, characterize and describe previously unknown taxa; and (3) the development and improvement of biological and genetic databases. These aspects are especially important when considering that even the biodeteriorative role (but also their taxonomic classification) of long known species might also need to be constantly revised, updated and reevaluated [107,109]. For instance, the inclusion of the *Fusarium solani* Species Complex in the genus *Neocosmospora* was recently reevaluated and continues to be the focus of additional studies [110].

Fungi are constantly regarded as one if not the most important microorganism groups causing cultural heritage materials biodeterioration [2,3,30]. This review highlighted that, so far, isolates from more than two-hundred fungal species have been showed to exhibit biodeteriorative abilities when studied by specific plate essays. Based on the available studies performed so far, it is possible to verify that *Aspergillus* and *Penicillium* species dominate the biodeteriorative abilities usually screened in biodeterioration contexts. With this in mind, it should be reinforced that the detection of these species in various cultural heritage materials can, under specific conditions, result in severe biodeterioration of the substrate. Nonetheless, a careful analysis of these checklists, as well as, the biodeteriorative screening of obtained isolates, wherever possible, is strongly advised. Not all isolates might display deteriorative action or display similar degradative rates and thus a proper and specific analysis in each case and/or the implementation of additional tests (e.g., molecular identification of genes involved in biodeterioration (see for example [111])) is also recommended. In conjunction with molecular approaches not relying in cultivation, they can provide a holistic evaluation of a specific biodeterioration phenomena. As pointed by Sterflinger and colleagues [112], understanding deterioration mechanisms and the main microbial perpetrators is still one of the major challenges in historic and cultural materials biodeterioration research. As such, the information summarized in this work provides a contribution that can help microbiologists, restorers, conservators and curators in their attempt to preserve cultural heritage materials for future generations.

Author Contributions: Conceptualization, J.T. writing—original draft preparation, J.T.; writing—review and editing, J.T. and A.P.; supervision, A.P.; funding acquisition, A.P. All authors have read and agreed to the published version of the manuscript.

Funding: This work was financed by IPN—Financiamento Base FITEC approved under the National Call with reference no. 01/FITEC/2018 to obtain multi-year base financing under the INTERFACE Program and by FEDER- Fundo Europeu de Desenvolvimento Regional funds through the COMPETE 2020-Operational Programme for Competitiveness and Internationalisation (POCI), and by Portuguese funds through FCT- Fundação para a Ciência e a Tecnologia in the framework of the project POCI-01-0145-FEDER-PTDC/EPH-PAT/3345/2014. This work was carried out at the R&D Unit Centre for Functional Ecology—Science for People and the Planet (CFE), with reference UIDB/04004/2020, financed by FCT/MCTES through national funds (PIDDAC). João Trovão was supported by POCH—Programa Operacional Capital Humano (co-funding by the European Social Fund and national funding by MCTES), through a "FCT- Fundação para a Ciência e Tecnologia" PhD research grant (SFRH/BD/132523/2017).

Institutional Review Board Statement: Not applicable.

Informed Consent Statement: Not applicable.

Data Availability Statement: Not applicable.

Conflicts of Interest: The authors declare no conflict of interest.

References

1. Deacon, J.W. *Fungal Biology*, 4th ed.; Blackwell Pub.: Malden, MA, USA, 2006.
2. Sterflinger, K. Fungi: Their Role in Deterioration of Cultural Heritage. *Fungal Biol. Rev.* **2010**, *24*, 47–55. [CrossRef]
3. Sterflinger, K.; Piñar, G. Microbial Deterioration of Cultural Heritage and Works of Art—Tilting at Windmills? *Appl. Microbiol. Biotechnol.* **2013**, *97*, 9637–9646. [CrossRef]
4. Dakal, T.C.; Cameotra, S.S. Microbially Induced Deterioration of Architectural Heritages: Routes and Mechanisms Involved. *Environ. Sci. Eur.* **2012**, *24*, 36. [CrossRef]
5. Gadd, G.M. Geomicrobiology of the Built Environment. *Nat. Microbiol.* **2017**, *2*, 1–9. [CrossRef]
6. Gadd, G.M. Fungi, Rocks, and Minerals. *Elements* **2017**, *13*, 171–176. [CrossRef]
7. Cappitelli, F.; Sorlini, C. From Papyrus to Compact Disc: The Microbial Deterioration of Documentary Heritage. *Crit. Rev. Microbiol.* **2005**, *31*, 1–10. [CrossRef]
8. Sterflinger, K.; Pinzari, F. The Revenge of Time: Fungal Deterioration of Cultural Heritage with Particular Reference to Books, Paper and Parchment. *Environ. Microbiol.* **2012**, *14*, 559–566. [CrossRef]
9. Paiva de Carvalho, H.; Mesquita, N.; Trovão, J.; Fernández Rodríguez, S.; Pinheiro, A.C.; Gomes, V.; Alcoforado, A.; Gil, F.; Portugal, A. Fungal Contamination of Paintings and Wooden Sculptures inside the Storage Room of a Museum: Are Current Norms and Reference Values Adequate? *J. Cult. Herit.* **2018**, *34*, 268–276. [CrossRef]
10. Poyatos, F.; Morales, F.; Nicholson, A.W.; Giordano, A. Physiology of Biodeterioration on Canvas Paintings. *J. Cell Physiol.* **2018**, *233*, 2741–2751. [CrossRef] [PubMed]
11. Kosel, J.; Ropret, P. Overview of Fungal Isolates on Heritage Collections of Photographic Materials and Their Biological Potency. *J. Cult. Herit.* **2021**, *48*, 277–291. [CrossRef]
12. Coutinho, M.L.; Miller, A.Z.; Macedo, M.F. Biological Colonization and Biodeterioration of Architectural Ceramic Materials: An Overview. *J. Cult. Herit.* **2015**, *16*, 759–777. [CrossRef]
13. Sclocchi, M.C.; Kraková, L.; Pinzari, F.; Colaizzi, P.; Bicchieri, M.; Šaková, N.; Pangallo, D. Microbial Life and Death in a Foxing Stain: A Suggested Mechanism of Photographic Prints Defacement. *Microb. Ecol.* **2017**, *73*, 815–826. [CrossRef]
14. Piñar, G.; Piombino-Mascali, D.; Maixner, F.; Zink, A.; Sterflinger, K. Microbial Survey of the Mummies from the Capuchin Catacombs of Palermo, Italy: Biodeterioration Risk and Contamination of the Indoor Air. *FEMS Microbiol. Ecol.* **2013**, *86*, 341–356. [CrossRef] [PubMed]
15. Ruga, L.; Orlandi, F.; Romano, B.; Fornaciari, M. The Assessment of Fungal Bioaerosols in the Crypt of St. Peter in Perugia (Italy). *Int. Biodeterior. Biodegrad.* **2015**, *98*, 121–130. [CrossRef]
16. Rodrigues, A.; Gutierrez-Patricio, S.; Miller, A.Z.; Saiz-Jimenez, C.; Wiley, R.; Nunes, D.; Vilarigues, M.; Macedo, M.F. Fungal Biodeterioration of Stained-Glass Windows. *Int. Biodeterior. Biodegrad.* **2014**, *90*, 152–160. [CrossRef]
17. Beata, G. The Use of -Omics Tools for Assessing Biodeterioration of Cultural Heritage: A Review. *J. Cult. Herit.* **2020**, *45*, 351–361. [CrossRef]

18. Amann, R.I.; Ludwig, W.; Schleifer, K.H. Phylogenetic Identification and in Situ Detection of Individual Microbial Cells without Cultivation. *Microbiol. Rev.* **1995**, *59*, 143–169. [CrossRef]
19. Dakal, T.C.; Arora, P.K. Evaluation of Potential of Molecular and Physical Techniques in Studying Biodeterioration. *Rev. Environ. Sci. Biotechnol.* **2012**, *11*, 71–104. [CrossRef]
20. González Grau, J.M.; Sáiz-Jiménez, C. Unknown Microbial Communities on Rock Art Paintings. Consequences for Conservation and Future Perspectives. *Coalition* **2005**, *10*, 4–7.
21. Gonzalez, J.M.; Saiz-Jimenez, C. Microbial Diversity in Biodeteriorated Monuments as Studied by Denaturing Gradient Gel Electrophoresis. *J. Sep. Sci.* **2004**, *27*, 174–180. [CrossRef]
22. González, J.M.; Sáiz-Jiménez, C. Application of Molecular Nucleic Acid-Based Techniques for the Study of Microbial Communities in Monuments and Artworks. *Int. Microbiol.* **2005**, *8*, 189–194. [PubMed]
23. Laiz, L.; Piñar, G.; Lubitz, W.; Sáiz-Jiménez, C. The Colonisation of Building Materials by Microorganisms as Revealed by Culturing and Molecular Methods. *Mol. Biol. Cult. Herit.* **2003**. [CrossRef]
24. Mihajlovski, A.; Seyer, D.; Benamara, H.; Bousta, F.; Di Martino, P. An Overview of Techniques for the Characterization and Quantification of Microbial Colonization on Stone Monuments. *Ann. Microbiol.* **2015**, *65*, 1243–1255. [CrossRef]
25. Otlewska, A.; Adamiak, J.; Gutarowska, B. Application of Molecular Techniques for the Assessment of Microorganism Diversity on Cultural Heritage Objects. *Acta Biochim. Pol.* **2014**, *61*. [CrossRef]
26. Sanmartín, P.; DeAraujo, A.; Vasanthakumar, A. Melding the Old with the New: Trends in Methods Used to Identify, Monitor, and Control Microorganisms on Cultural Heritage Materials. *Microb. Ecol.* **2018**, *76*, 64–80. [CrossRef] [PubMed]
27. Ward, D.M.; Weller, R.; Bateson, M.M. 16S RRNA Sequences Reveal Numerous Uncultured Microorganisms in a Natural Community. *Nature* **1990**, *345*, 63–65. [CrossRef]
28. Savković, Ž.; Stupar, M.; Unković, N.; Ivanović, Ž.; Blagojević, J.; Vukojević, J.; Ljaljević Grbić, M. In Vitro Biodegradation Potential of Airborne *Aspergilli* and *Penicillia*. *Sci. Nat.* **2019**, *106*, 8. [CrossRef] [PubMed]
29. Warscheid, T.; Braams, J. Biodeterioration of Stone: A Review. *Int. Biodeterior. Biodegrad.* **2000**, *46*, 343–368. [CrossRef]
30. Sterflinger, K. Fungi as Geologic Agents. *Geomicrobiol. J.* **2000**, *17*, 97–124. [CrossRef]
31. Scheerer, S.; Ortega-Morales, O.; Gaylarde, C. Chapter 5 Microbial Deterioration of Stone Monuments—An Updated Overview. *Adv. Appl. Microbiol.* **2009**, *66*, 97–139. [CrossRef]
32. Liu, X.; Koestler, R.J.; Warscheid, T.; Katayama, Y.; Gu, J.-D. Microbial Deterioration and Sustainable Conservation of Stone Monuments and Buildings. *Nat. Sustain.* **2020**, *3*, 991–1004. [CrossRef]
33. Albertano, P.; Urzì, C. Structural Interactions among Epilithic Cyanobacteria and Heterotrophic Microorganisms in Roman Hypogea. *Microb. Ecol.* **1999**, *38*, 244–252. [CrossRef] [PubMed]
34. Burford, E.P.; Kierans, M.; Gadd, G.M. Geomycology: Fungi in Mineral Substrata. *Mycologist* **2003**, *17*, 98–107. [CrossRef]
35. Ortega-Morales, B.O.; Narváez-Zapata, J.; Reyes-Estebanez, M.; Quintana, P.; la Rosa-García, D.; Bullen, H.; Gómez-Cornelio, S.; Chan-Bacab, M.J. Bioweathering Potential of Cultivable Fungi Associated with Semi-Arid Surface Microhabitats of Mayan Buildings. *Front. Microbiol.* **2016**, *7*, 201. [CrossRef] [PubMed]
36. Pangallo, D.P.; Chovanová, K.C.; Šimonovičová, A.Š.; Ferianc, P.F. Investigation of Microbial Community Isolated from Indoor Artworks and Air Environment: Identification, Biodegradative Abilities, and DNA Typing. *Can. J. Microbiol.* **2009**, *55*, 277–287. [CrossRef] [PubMed]
37. Kiyuna, T.; An, K.-D.; Kigawa, R.; Sano, C.; Miura, S.; Sugiyama, J. Bristle-like Fungal Colonizers on the Stone Walls of the Kitora and Takamatsuzuka Tumuli Are Identified as *Kendrickiella Phycomyces*. *Mycoscience* **2012**, *53*, 446–459. [CrossRef]
38. Unković, N.; Dimkić, I.; Stupar, M.; Stanković, S.; Vukojević, J.; Grbić, M.L. Biodegradative Potential of Fungal Isolates from Sacral Ambient: In Vitro Study as Risk Assessment Implication for the Conservation of Wall Paintings. *PLoS ONE* **2018**, *13*, e0190922. [CrossRef]
39. Ponizovskaya, V.B.; Rebrikova, N.L.; Kachalkin, A.V.; Antropova, A.B.; Bilanenko, E.N.; Mokeeva, V.L. Micromycetes as Colonizers of Mineral Building Materials in Historic Monuments and Museums. *Fungal Biol.* **2019**, *123*, 290–306. [CrossRef]
40. Gámez-Espinosa, E.; Bellotti, N.; Deyá, C.; Cabello, M. Mycological Studies as a Tool to Improve the Control of Building Materials Biodeterioration. *J. Build. Eng.* **2020**, *32*, 101738. [CrossRef]
41. Ma, W.; Wu, F.; Tian, T.; He, D.; Zhang, Q.; Gu, J.-D.; Duan, Y.; Ma, D.; Wang, W.; Feng, H. Fungal Diversity and Its Contribution to the Biodeterioration of Mural Paintings in Two 1700-Year-Old Tombs of China. *Int. Biodeterior. Biodegrad.* **2020**, *152*, 104972. [CrossRef]
42. Trovão, J.; Tiago, I.; Catarino, L.; Gil, F.; Portugal, A. In Vitro Analyses of Fungi and Dolomitic Limestone Interactions: Bioreceptivity and Biodeterioration Assessment. *Int. Biodeterior. Biodegrad.* **2020**, *155*, 105107. [CrossRef]
43. Trovão, J.; Gil, F.; Catarino, L.; Soares, F.; Tiago, I.; Portugal, A. Analysis of Fungal Deterioration Phenomena in the First Portuguese King Tomb Using a Multi-Analytical Approach. *Int. Biodeterior. Biodegrad.* **2020**, *149*, 104933. [CrossRef]
44. Isola, D.; Zucconi, L.; Cecchini, A.; Caneva, G. Dark-Pigmented Biodeteriogenic Fungi in Etruscan Hypogeal Tombs: New Data on Their Culture-Dependent Diversity, Favouring Conditions, and Resistance to Biocidal Treatments. *Fungal Biol.* **2021**. [CrossRef]
45. De Ley, J.; Swings, J.; Gosselé, F. Genus I. *Acetobacter* Beijerinck 1898, 215[AL]. In *Bergey's Manual of Systematic Bacteriology*; Williams and Wilkins: Baltimore, MD, USA, 1984; Volume 1, pp. 268–274.
46. Samson, R.A.; Houbraken, J.; Thrane, U.; Frisvad, J.C.; Andersen, B. *Food and Indoor Fungi*, 2nd ed.; CBS-KNAW Fungal Biodiversity Centre: Utrecht, The Netherlands, 2010.

47. Borrego, S.; Guiamet, P.; Gómez de Saravia, S.; Batistini, P.; Garcia, M.; Lavin, P.; Perdomo, I. The Quality of Air at Archives and the Biodeterioration of Photographs. *Int. Biodeterior. Biodegrad.* **2010**, *64*, 139–145. [CrossRef]
48. Gutarowska, B. Metabolic Activity of Moulds as a Factor of Building Materials Biodegradation. *Pol. J. Microbiol.* **2010**, *59*, 119–124. [CrossRef]
49. Ahmad, A.; Rautaray, D.; Sastry, M. Biogenic Calcium Carbonate: Calcite Crystals of Variable Morphology by the Reaction of Aqueous Ca^{2+} Ions with Fungi. *Adv. Funct. Mater.* **2004**, *14*, 1075–1080. [CrossRef]
50. Burford, E.P.; Hillier, S.; Gadd, G.M. Biomineralization of Fungal Hyphae with Calcite (CaCO3) and Calcium Oxalate Mono- and Dihydrate in Carboniferous Limestone Microcosms. *Geomicrobiol. J.* **2006**, *23*, 599–611. [CrossRef]
51. Gadd, G.M.; Bahri-Esfahani, J.; Li, Q.; Rhee, Y.J.; Wei, Z.; Fomina, M.; Liang, X. Oxalate Production by Fungi: Significance in Geomycology, Biodeterioration and Bioremediation. *Fungal Biol. Rev.* **2014**, *28*, 36–55. [CrossRef]
52. Del Monte, M.; Sabbioni, C.; Zappia, G. The Origin of Calcium Oxalates on Historical Buildings, Monuments and Natural Outcrops. *Sci. Total Environ.* **1987**, *67*, 17–39. [CrossRef]
53. Li, T.; Hu, Y.; Zhang, B.; Yang, X. Role of Fungi in the Formation of Patinas on Feilaifeng Limestone, China. *Microb. Ecol.* **2018**, *76*, 352–361. [CrossRef]
54. Savković, Ž.; Unković, N.; Stupar, M.; Franković, M.; Jovanović, M.; Erić, S.; Šarić, K.; Stanković, S.; Dimkić, I.; Vukojević, J.; et al. Diversity and Biodeteriorative Potential of Fungal Dwellers on Ancient Stone Stela. *Int. Biodeterior. Biodegrad.* **2016**, *115*, 212–223. [CrossRef]
55. Li, T.; Hu, Y.; Zhang, B. Biomineralization Induced by *Colletotrichum acutatum*: A Potential Strategy for Cultural Relic Bioprotection. *Front. Microbiol.* **2018**, *9*, 1884. [CrossRef]
56. Unković, N.; Erić, S.; Šarić, K.; Stupar, M.; Savković, Ž.; Stanković, S.; Stanojević, O.; Dimkić, I.; Vukojević, J.; Ljaljević Grbić, M. Biogenesis of Secondary Mycogenic Minerals Related to Wall Paintings Deterioration Process. *Micron* **2017**, *100*, 1–9. [CrossRef]
57. Boquet, E.; Boronat, A.; Ramos-Cormenzana, A. Production of Calcite (Calcium Carbonate) Crystals by Soil Bacteria Is a General Phenomenon. *Nature* **1973**, *246*, 527–529. [CrossRef]
58. Guggiari, M.; Bloque, R.; Aragno, M.; Verrecchia, E.; Job, D.; Junier, P. Experimental Calcium-Oxalate Crystal Production and Dissolution by Selected Wood-Rot Fungi. *Int. Biodeterior. Biodegrad.* **2011**, *65*, 803–809. [CrossRef]
59. Pasquale, V.; Fiore, S.; Hlayem, D.; Lettino, A.; Huertas, F.J.; Chianese, E.; Dumontet, S. Biomineralization of Carbonates Induced by the Fungi *Paecilomyces inflatus* and *Plectosphaerella cucumerina*. *Int. Biodeterior. Biodegrad.* **2019**, *140*, 57–66. [CrossRef]
60. Gadd, G.M. Geomycology: Biogeochemical Transformations of Rocks, Minerals, Metals and Radionuclides by Fungi, Bioweathering and Bioremediation. *Mycol. Res.* **2007**, *111*, 3–49. [CrossRef] [PubMed]
61. Liu, Z.; Wang, Y.; Pan, X.; Ge, Q.; Ma, Q.; Li, Q.; Fu, T.; Hu, C.; Zhu, X.; Pan, J. Identification of Fungal Communities Associated with the Biodeterioration of Waterlogged Archeological Wood in a Han Dynasty Tomb in China. *Front. Microbiol.* **2017**, *8*. [CrossRef]
62. Gutarowska, B.; Celikkol-Aydin, S.; Bonifay, V.; Otlewska, A.; Aydin, E.; Oldham, A.L.; Brauer, J.I.; Duncan, K.E.; Adamiak, J.; Sunner, J.A.; et al. Metabolomic and High-Throughput Sequencing Analysis—Modern Approach for the Assessment of Biodeterioration of Materials from Historic Buildings. *Front. Microbiol.* **2015**, *6*, 979. [CrossRef] [PubMed]
63. Ortiz, R.; Navarrete, H.; Navarrete, J.; Párraga, M.; Carrasco, I.; de la Vega, E.; Ortiz, M.; Herrera, P.; Blanchette, R.A. Deterioration, Decay and Identification of Fungi Isolated from Wooden Structures at the Humberstone and Santa Laura Saltpeter Works: A World Heritage Site in Chile. *Int. Biodeterior. Biodegrad.* **2014**, *86*, 309–316. [CrossRef]
64. Piñar, G.; Dalnodar, D.; Voitl, C.; Reschreiter, H.; Sterflinger, K. Biodeterioration Risk Threatens the 3100 Year Old Staircase of Hallstatt (Austria): Possible Involvement of Halophilic Microorganisms. *PLoS ONE* **2016**, *11*, e0148279. [CrossRef] [PubMed]
65. Pournou, A. *Biodeterioration of Wooden Cultural Heritage: Organisms and Decay Mechanisms in Aquatic and Terrestrial Ecosystems*; Springer International Publishing: Cham, Switzerland, 2020; ISBN 978-3-030-46503-2.
66. Alfieri, P.V.; García, R.A.; Rosato, V.G.; Correa, M.V. Biodeterioration and Biodegradation of Wooden Heritage: Role of Fungal Succession. *Int. J. Conserv. Sci.* **2016**, *7*, 607–614.
67. Blanchette, R.A. A Review of Microbial Deterioration Found in Archaeological Wood from Different Environments. *Int. Biodeterior. Biodegrad.* **2000**, *46*, 189–204. [CrossRef]
68. Goodell, B.; Winandy, J.E.; Morrell, J.J. Fungal Degradation of Wood: Emerging Data, New Insights and Changing Perceptions. *Coatings* **2020**, *10*, 1210. [CrossRef]
69. Pangallo, D.; Šimonovičová, A.; Chovanová, K.; Ferianc, P. Wooden Art Objects and the Museum Environment: Identification and Biodegradative Characteristics of Isolated Microflora. *Lett. Appl. Microbiol.* **2007**, *45*, 87–94. [CrossRef]
70. Liu, Z.; Fu, T.; Hu, C.; Shen, D.; Macchioni, N.; Sozzi, L.; Chen, Y.; Liu, J.; Tian, X.; Ge, Q.; et al. Microbial Community Analysis and Biodeterioration of Waterlogged Archaeological Wood from the Nanhai No. 1 Shipwreck during Storage. *Sci. Rep.* **2018**, *8*, 7170. [CrossRef] [PubMed]
71. Archibald, F.S. A New Assay for Lignin-Type Peroxidases Employing the Dye Azure B. *Appl. Environ. Microbiol.* **1992**, *58*, 3110–3116. [CrossRef]
72. Manji, S.; Ishihara, A. Screening of Tetrachlorodibenzo-p-Dioxin-Degrading Fungi Capable of Producing Extracellular Peroxidases under Various Conditions. *Appl. Microbiol. Biotechnol.* **2004**, *63*, 438–444. [CrossRef]
73. Falade, A.O.; Eyisi, O.A.L.; Mabinya, L.V.; Nwodo, U.U.; Okoh, A.I. Peroxidase Production and Ligninolytic Potentials of Fresh Water Bacteria *Raoultella Ornithinolytica* and *Ensifer Adhaerens*. *Biotechnol. Rep.* **2017**, *16*, 12–17. [CrossRef]

74. Kiiskinen, L.-L.; Rättö, M.; Kruus, K. Screening for Novel Laccase-Producing Microbes. *J. Appl. Microbiol.* **2004**, *97*, 640–646. [CrossRef]
75. Szostak-Kotowa, J. Biodeterioration of Textiles. *Int. Biodeterior. Biodegrad.* **2004**, *53*, 165–170. [CrossRef]
76. Gutarowska, B.; Pietrzak, K.; Machnowski, W.; Milczarek, J.M. Historical Textiles—A Review of Microbial Deterioration Analysis and Disinfection Methods. *Text. Res. J.* **2017**, *87*, 2388–2406. [CrossRef]
77. Mazzoli, R.; Giuffrida, M.G.; Pessione, E. Back to the Past: "Find the Guilty Bug—Microorganisms Involved in the Biodeterioration of Archeological and Historical Artifacts". *Appl. Microbiol. Biotechnol.* **2018**, *102*, 6393–6407. [CrossRef]
78. Pangallo, D.; Kraková, L.; Chovanová, K.; Šimonovičová, A.; De Leo, F.; Urzì, C. Analysis and Comparison of the Microflora Isolated from Fresco Surface and from Surrounding Air Environment through Molecular and Biodegradative Assays. *World J. Microbiol. Biotechnol.* **2012**, *28*, 2015–2027. [CrossRef]
79. Kraková, L.; Šoltys, K.; Puškárová, A.; Bučková, M.; Jeszeová, L.; Kucharík, M.; Budiš, J.; Orovčík, L.; Szemes, T.; Pangallo, D. The Microbiomes of a XVIII Century Mummy from the Castle of Krásna Hôrka (Slovakia) and Its Surrounding Environment. *Environ. Microbiol.* **2018**, *20*, 3294–3308. [CrossRef] [PubMed]
80. Kisová, Z.; Planý, M.; Pavlović, J.; Bučková, M.; Puškárová, A.; Kraková, L.; Kapustová, M.; Pangallo, D.; Šoltys, K. Biodeteriogens Characterization and Molecular Analyses of Diverse Funeral Accessories from XVII Century. *Appl. Sci.* **2020**, *10*, 5451. [CrossRef]
81. Naji, K.M.; Abdullah, Q.Y.M.; AL-Zaqri, A.Q.M.; Alghalibi, S.M. Evaluating the Biodeterioration Enzymatic Activities of Fungal Contamination Isolated from Some Ancient Yemeni Mummies Preserved in the National Museum. *Biochem. Res. Int.* **2014**, *2014*, e481508. [CrossRef]
82. Kavkler, K.; Gunde-Cimerman, N.; Zalar, P.; Demšar, A. Fungal Contamination of Textile Objects Preserved in Slovene Museums and Religious Institutions. *Int. Biodeterior. Biodegrad.* **2015**, *97*, 51–59. [CrossRef]
83. Šimonovičová, A.; Kraková, L.; Pangallo, D.; Majorošová, M.; Piecková, E.; Bodoriková, S.; Dörnhoferová, M. Fungi on Mummified Human Remains and in the Indoor Air in the Kuffner Family Crypt in Sládkovičovo (Slovakia). *Int. Biodeterior. Biodegrad.* **2015**, *99*, 157–164. [CrossRef]
84. Šoltys, K.; Planý, M.; Biocca, P.; Vianello, V.; Bučková, M.; Puškárová, A.; Sclocchi, M.C.; Colaizzi, P.; Bicchieri, M.; Pangallo, D.; et al. Lead Soaps Formation and Biodiversity in a XVIII Century Wax Seal Coloured with Minium. *Environ. Microbiol.* **2020**, *22*, 1517–1534. [CrossRef] [PubMed]
85. Pangallo, D.; Kraková, L.; Chovanová, K.; Bučková, M.; Puškarová, A.; Šimonovičová, A. Disclosing a Crypt: Microbial Diversity and Degradation Activity of the Microflora Isolated from Funeral Clothes of Cardinal Peter Pázmány. *Microbiol. Res.* **2013**, *168*, 289–299. [CrossRef]
86. Hsu, S.C.; Lockwood, J.L. Powdered Chitin Agar as a Selective Medium for Enumeration of Actinomycetes in Water and Soil. *Appl. Microbiol.* **1975**, *29*, 422–426. [CrossRef]
87. Eggins, H.O.W.; Pugh, G.J.F. Isolation of Cellulose-Decomposing Fungi from the Soil. *Nature* **1962**, *193*, 94–95. [CrossRef]
88. Paterson, R.R.M.; Bridge, P.D.; International Mycological Institute. *Biochemical Techniques for Filamentous Fungi*; CAB International: Wallingford, UK, 1994.
89. Pangallo, D.; Bučková, M.; Kraková, L.; Puškárová, A.; Šaková, N.; Grivalský, T.; Chovanová, K.; Zemánková, M. Biodeterioration of Epoxy Resin: A Microbial Survey through Culture-Independent and Culture-Dependent Approaches. *Environ. Microbiol.* **2015**, *17*, 462–479. [CrossRef]
90. Shivani, D.; Kumar, J.S. Extracellular Enzymatic Profile of Fungal Deteriogens of Historical Palace of Ujjain. *Int. J. Curr. Microbiol. Appl. Sci.* **2015**, *4*, 122–132.
91. Strzelczyk, A.B. Observations on Aesthetic and Structural Changes Induced in Polish Historic Objects by Microorganisms. *Int. Biodeterior. Biodegrad.* **2004**, *53*, 151–156. [CrossRef]
92. Caneva, G.; Nugari, M.P.; Salvadori, O.; International Centre for the Study of the Preservation and the Restoration of Cultural Property. *Biology in the Conservation of Works of Art*; ICCROM—International Centre for the Study of the Preservation and Restoration of Cultural Property: Rome, Italy, 1991.
93. Puškárová, A.; Bučková, M.; Habalová, B.; Kraková, L.; Maková, A.; Pangallo, D. Microbial Communities Affecting Albumen Photography Heritage: A Methodological Survey. *Sci. Rep.* **2016**, *6*, 20810. [CrossRef]
94. Štafura, A.; Nagy, Š.; Bučková, M.; Puškárová, A.; Kraková, L.; Čulík, M.; Beronská, N.; Nagy, Š.; Pangallo, D. The Influence of Microfilamentous Fungi on Wooden Organ Pipes: One Year Investigation. *Int. Biodeterior. Biodegrad.* **2017**, *121*, 139–147. [CrossRef]
95. Abrusci, C.; Martín-González, A.; Del Amo, A.; Catalina, F.; Collado, J.; Platas, G. Isolation and Identification of Bacteria and Fungi from Cinematographic Films. *Int. Biodeterior. Biodegrad.* **2005**, *56*, 58–68. [CrossRef]
96. Rojas, T.I.; Aira, M.J.; Batista, A.; Cruz, I.L.; González, S. Fungal Biodeterioration in Historic Buildings of Havana (Cuba). *Grana* **2012**, *51*, 44–51. [CrossRef]
97. Kraková, L.; Chovanová, K.; Selim, S.A.; Šimonovičová, A.; Puškarová, A.; Maková, A.; Pangallo, D. A Multiphasic Approach for Investigation of the Microbial Diversity and Its Biodegradative Abilities in Historical Paper and Parchment Documents. *Int. Biodeterior. Biodegrad.* **2012**, *70*, 117–125. [CrossRef]
98. Anaya, M.; Borrego, S.F.; Gámez, E.; Castro, M.; Molina, A.; Valdés, O. Viable Fungi in the Air of Indoor Environments of the National Archive of the Republic of Cuba. *Aerobiologia* **2016**, *32*, 513–527. [CrossRef]
99. Borrego, S.; Molina, A.; Santana, A. Mold on Stored Photographs and Maps: A Case Study. *Top. Photogr. Preserv.* **2015**, *16*, 109–120.

100. Borrego, S.; Molina, A.; Santana, A. Fungi in Archive Repositories Environments and the Deterioration of the Graphics Documents. *EC Microbiol.* **2017**, *11*, 205–226.
101. Saran, S.; Isar, J.; Saxena, R.K. A Modified Method for the Detection of Microbial Proteases on Agar Plates Using Tannic Acid. *J. Biochem. Biophys. Methods* **2007**, *70*, 697–699. [CrossRef] [PubMed]
102. Borrego, S.; Perdomo, I. Aerobiological Investigations inside Repositories of the National Archive of the Republic of Cuba. *Aerobiologia* **2012**, *28*, 303–316. [CrossRef]
103. Okpalanozie, O.E.; Adebusoye, S.A.; Troiano, F.; Polo, A.; Cappitelli, F.; Ilori, M.O. Evaluating the Microbiological Risk to a Contemporary Nigerian Painting: Molecular and Biodegradative Studies. *Int. Biodeterior. Biodegrad.* **2016**, *114*, 184–192. [CrossRef]
104. Boniek, D.; Bonadio, L.; Santos de Abreu, C.; Dos Santos, A.F.; de Resende Stoianoff, M.A. Fungal Bioprospecting and Antifungal Treatment on a Deteriorated Brazilian Contemporary Painting. *Lett. Appl. Microbiol.* **2018**, *67*, 337–342. [CrossRef]
105. Coronado-Ruiz, C.; Avendaño, R.; Escudero-Leyva, E.; Conejo-Barboza, G.; Chaverri, P.; Chavarría, M. Two New Cellulolytic Fungal Species Isolated from a 19 Th -Century Art Collection. *Sci. Rep.* **2018**, *8*, 7492. [CrossRef] [PubMed]
106. Teather, R.M.; Wood, P.J. Use of Congo Red-Polysaccharide Interactions in Enumeration and Characterization of Cellulolytic Bacteria from the Bovine Rumen. *Appl. Environ. Microbiol.* **1982**, *43*, 777–780. [CrossRef] [PubMed]
107. Pyzik, A.; Ciuchcinski, K.; Dziurzynski, M.; Dziewit, L. The Bad and the Good—Microorganisms in Cultural Heritage Environments—An Update on Biodeterioration and Biotreatment Approaches. *Materials* **2021**, *14*, 177. [CrossRef]
108. Piñar, G.; Sterflinger, K. Natural Sciences at the Service of Art and Cultural Heritage: An Interdisciplinary Area in Development and Important Challenges. *Microb. Biotechnol.* **2021**. [CrossRef] [PubMed]
109. Leplat, J.; François, A.; Bousta, F. *Parengyodontium album*, a Frequently Reported Fungal Species in the Cultural Heritage Environment. *Fungal Biol. Rev.* **2020**, *34*, 126–135. [CrossRef]
110. O'Donnell, K.; Al-Hatmi, A.M.S.; Aoki, T.; Brankovics, B.; Cano-Lira, J.F.; Coleman, J.J.; de Hoog, G.S.; Pietro, A.D.; Frandsen, R.J.N.; Geiser, D.M.; et al. No to *Neocosmospora*: Phylogenomic and Practical Reasons for Continued Inclusion of the *Fusarium Solani* Species Complex in the Genus *Fusarium*. *mSphere* **2020**, *5*. [CrossRef] [PubMed]
111. Kraková, L.; Chovanová, K.; Puškarová, A.; Bučková, M.; Pangallo, D. A Novel PCR-Based Approach for the Detection and Classification of Potential Cellulolytic Fungal Strains Isolated from Museum Items and Surrounding Indoor Environment. *Lett. Appl. Microbiol.* **2012**, *54*, 433–440. [CrossRef] [PubMed]
112. Sterflinger, K.; Little, B.; Pinar, G.; Pinzari, F.; de los Rios, A.; Gu, J.-D. Future Directions and Challenges in Biodeterioration Research on Historic Materials and Cultural Properties. *Int. Biodeterior. Biodegrad.* **2018**, *129*, 10–12. [CrossRef]

applied sciences

MDPI

Review

Biologically Derived Gels for the Cleaning of Historical and Artistic Metal Heritage

Arianna Passaretti [1,2], Luana Cuvillier [1,2], Giorgia Sciutto [3], Elodie Guilminot [4] and Edith Joseph [1,2,*]

1 Haute Ecole Arc Conservation-Restauration, HES-SO, Espace de l'Europe 11, 2000 Neuchatel, Switzerland; arianna.passaretti@he-arc.ch (A.P.); luana.cuvillier@he-arc.ch (L.C.)
2 Laboratory of Technologies for Heritage Materials, Institute of Chemistry, University of Neuchâtel, Av. Bellevaux 51, 2000 Neuchatel, Switzerland
3 Department of Chemistry, University of Bologna, Ravenna Campus, via Guaccimanni, 48120 Ravenna, Italy; giorgia.sciutto@unibo.it
4 Arc'Antique conservation and research laboratory, 26 Rue de la Haute Forêt, 44300 Nantes, France; Elodie.GUILMINOT@loire-atlantique.fr
* Correspondence: edith.joseph@unine.ch or edith.joseph@he-arc.ch

Abstract: In the general global rise of attention and research to seek greener attitudes, the field of cultural heritage (CH) makes no exception. In the last decades, an increasing number of sustainable and biologically based solutions have been proposed for the protection and care of artworks. Additionally, the safety of the target artwork and the operator must be kept as core goals. Within this scenario, new products and treatments should be explored and implemented in the common conservation praxes. Therefore, this review addressing metal heritage is aimed to report biologically derived gel formulations already proposed for this specific area as reliable tools for cleaning. Promising bio-gel-based protocols, still to be implemented in metal conservation, are also presented to promote their investigation by stakeholders in metal conservation. After an opening overview on the common practices for cleaning metallic surfaces in CH, the focus will be moved onto the potentialities of gel-alternatives and in particular of ones with a biological origin. In more detail, we displayed water-gels (i.e., hydrogels) and solvent-gels (i.e., organogels) together with particular attention to bio-solvents. The discussion is closed in light of the state-of-the-art and future perspectives.

Keywords: metal; conservation; cleaning; hydrogels; organogels; bio-solvents; cultural heritage

Citation: Passaretti, A.; Cuvillier, L.; Sciutto, G.; Guilminot, E.; Joseph, E. Biologically Derived Gels for the Cleaning of Historical and Artistic Metal Heritage. *Appl. Sci.* **2021**, *11*, 3405. https://doi.org/10.3390/app11083405

Academic Editor: Maria Filomena Macedo

Received: 26 February 2021
Accepted: 30 March 2021
Published: 10 April 2021

1. Metal Care in Cultural Heritage: An Overview

In the motley world of cultural heritage, metals are among the main actors since the emergence of metalworking of gold, silver, and copper [1]. Due to their inherent chemo-physical properties and aesthetic potential, metals have had a broad spectrum of applications in art through the ages: sculptures, musical instruments, jewelry, utensils, scientific instruments, decorative arts, and so forth [2–7]. Such a massive presence in cultural heritage has inevitably yielded a plethora of studies on metal preservation to safeguard this inheritance through generations.

The core problem for the preservation of metals is the spontaneous process of corrosion towards which they naturally tend [8]. Indeed, corrosion is a thermodynamically favored phenomenon, which alters metals' original features and represents a tremendous economic loss in whichever sector, as well as a conservation issue in the heritage field [9]. Different aspects can influence corrosion processes: along with the basic dual presence of water and oxygen, the characteristics of the surrounding environment rule and meddle the phenomenon. Factors such as burial soil, marine location, microbiological activity, atmospheric pollutants, and, in general, alloys' complexity and exposure conditions play an important role in this detrimental process [10–12].

From a preventive perspective, despite the utopic willingness to artificially control the micro-environment, not all the active agents in corrosion can be avoided or adjusted, especially if we consider outdoor artifacts. Therefore, in metal care, the first phase consists of tackling the corrosion layer present. This usually means either stabilizing or removing the corrosion compounds, depending on the substrate nature and the type of alteration exhibited, in addition to an evaluation according to restoration principles (e.g., noble or unstable patina, corrosion as proof of authenticity or anesthetic degradation) [13]. Both issues can be addressed from different angles. For stabilization, the current techniques go from electrolytic reduction, dichlorination, and immersion in an alkaline sulfide solution for archaeological iron artifacts, to the use of chemical chelating agents or corrosion inhibitors such as benzotriazole (BTA) for copper-based relics [14–17]. On the other side, for corrosion removal, we can cite potentially invasive mechanical methods such as wet or dry abrasion (sanding paper or cloth, sandblasting, scalpel, laser, or cryogenics), as well as chemical approaches, exploiting electrochemical reduction or synthetic complexing agents such as ethylenediaminetetraacetic acid disodium salt dihydrate (Na_2EDTA), tetraethylenepentamine (TEPA) and triammonium citrate (TAC) [18].

In the following step, a solid solution is to prevent the interaction between the metal surface and the external detrimental factors. The use of organic coatings is thus a favored method, being able to achieve a film with the required physical, mechanical, and aesthetical characteristics [19]. The typical products employed are both from natural or synthetic origin—e.g., beeswax, epoxy resin, acrylic-based varnish [20]—however, the latter has acquired more relevance since they are designed with the desired features and fine-tuned to be the most environmental-proof as possible [21]. These compounds are also sensitive to detrimental factors through time, even if to a lesser degree or merely not like metals. Therefore, there might be the necessity of removing—and replacing—them once their preservative purpose is compromised (e.g., film cracking and detachment, harmful reaction with the underlying substrate itself, visual appearance no more acceptable according to conservation standards) [22,23].

To achieve this purpose of removal, the most employed approach is the dissolution of the no more-desired coating present [24–27]. Solubility follows the basic principle of like dissolves like: "chemically similar" substances—i.e., mainly in terms of intermolecular forces — are soluble in each other. Despite its simplicity, the rule allows to navigate when choosing solvents; also, conservators often rely on helpful solubility diagrams, such as the Teas Chart [28], for the selection of appropriate solvents according to the case treated.

However, solvents can turn to be unsafe for users due to their flammability or by prolonged inhalation for instance. This scenario, unfortunately, fits with the role of conservators, who often employ in a confined workplace solvents such as xylene, toluene, isopropanol, ethanol, acetone, white spirit, turpentine [29]. Nevertheless, their dangerous repercussions are also visible on a vast scale in nature pollution and consequently global human health [30], thereby addressing the world of conservation to responsible greener solutions [31,32].

Furthermore, in art conservation, one of the leading priorities is the safety of the artwork itself. This entails the importance not only of suitable products but also of an appropriate application methodology [33,34]. In the case of metal heritage, the presence of a corrosion layer turns the substrate into highly porous and drastically more water/solvent-sensitive. Moreover, solvents can be potentially corrosive if not properly handled and retained [35,36]. Furthermore, it might happen that the coating, once solubilized, is not taken away and gets re-deposited on the surface.

To minimize such potential inherent drawbacks of solvents and comply with the need for a proper application procedure, delivery systems are a valid alternative for art conservation, to assure a controlled release of treating loaded solutions.

2. Gels: A Reliable Delivery System

Nowadays, various methodologies are at the disposition of conservators to obviate the use of neat solvents or unrestrained treating solutions [37–41]; however, the most reliable, tailorable, and performing are certainly gels [42].

Gels can be defined as biphasic systems made of at least two components: a polymer and a fluid phase. The polymer, acting as a thickening agent, forms a three-dimensional network in the liquid medium [43]. This cross-linked system can trap the liquid, limiting simultaneously evaporation and release. The liquid is retained in the network using interactions between the polymer chains. Due to their chemical nature, these interactions can be reversible thanks to weak bonds—i.e., hydrogen bonds, van der Waals—or irreversible in presence of covalent bonding. The different bonding leads to the formation of physical and chemical gels [42], respectively, with different chemical features [42]. In particular, the inherent forces within chemical gels are stronger than the adhesive forces between gel and substrate to be treated. This implies an easier removal of such formulation, thereby avoiding extra steps to get rid of potential residues [44].

Finally, gel systems can be classified according to the fluid contained inside the polymer matrix: in presence of water, they are called "hydrogels", while non-aqueous formulations, hence loaded with organic solvents, are defined as "organogels" or "solvent-gels" [42].

Cleaning Gels in Cultural Heritage

At the beginning of the nineties, Richard Wolbers introduced gels to the world of art conservation [45]. His work focused on the treatment of paintings [46], which are water- and solvent-sensitive targets; thus, gel formulations seemed optimal for higher-controlled and less invasive interventions. Considering the promising evidences, this technique became rapidly captivating in several fields, including wall paintings and stone care [47–49]. Moreover, the employment of gels turned to be essential when dealing with paper or wood substrates, for which the conventional procedure by immersion into water solution was not ideal [50,51].

Nowadays gel-based cleaning protocols are recognized as well-grounded and effective methodologies, thanks to their high retention and interesting rheological properties that allow a precise and selective treatment confined to the upmost layer [52,53]. In addition, such systems can adsorb the undesired substances to be removed (i.e., corrosion products and degraded protective materials) in the polymeric matrix. Finally, they drastically limit solvent fumes released in the air, thereby creating a less hazardous work environment for operators.

Nevertheless, the use of gels is still anecdotic on metal artworks, even though there is the need for localized treatments, considering how heterogenous they can be: composite artifacts, gilded artworks, painted metals, etc.

In a comparative study on gel and swab cleaning on brass, AFM (Atomic force microscopy) pictures proved the existence of scratches after the swab action due to particles originally present on the metal. On the contrary, the gel performance, although slightly less efficient, was smoother and non-invasive, leaving no physical changes on the surface [54]. These observations led to the setting-up of a project named "Gels-Métaux", initiated in France and expanding to a larger-scale, calling forth an international appeal. The project tackles various tricky topics: gel preparation protocols, compatibility between formulations and active agents (i.e., chelators), and introducing gels into new branches of metal conservation like archaeological findings [55].

Notably, some studies regarding the use of synthetic gels have been published for metal conservation. In 2014, a set of indoor bronze sculptures were successfully treated via Carbopol-gel applications to remove layer-by-layer dirt, protective waxes, and corrosion phases [56]. While water-based formulations made of the synthetic polymer poly (vinyl alcohol) (PVA) were designed by Baglioni et al. as part of the NANORESTART project and assessed on copper alloys [16,57]. Despite the captivating potentialities, high attention must be addressed to their waste management and the correlated ecological impact. For instance, PVA is readily water-soluble; however, inadequate disposal can cause pollution of groundwaters. Its biodegradation is performed by very peculiar bacteria, barely represented in nature [58,59]. Another critical point is the sustainability of gelling agents' production. Although bio-derived from cellulose, often the synthesis of cellulose-based polymers cannot be considered green, since it may require reagents hazardous for the aquatic environment (e.g., chloroacetic acid used for the synthesis of carboxymethyl cellulose) [60].

In contrast, scarcer is the literature on green or bio-based alternatives. Therefore, an overview of both hydro- and solvent-gels is proposed here, highlighting their suitability for metal heritage.

3. Bio-Based Hydrogels for Metal Preservation

A wide variety of hydrogels based on natural polymers have been exploited, especially in the biomedical field, including polysaccharides and proteins (collagen and gelatine) [61]. However, in art conservation, mainly polysaccharides have been employed, namely agar, gellan gum, xanthan gum, and chitosan (Figure 1). The resulting formulations all belong to the class of physical gels. Indeed, the water is physically retained within the gel network by hydrogen bonding of the polar groups of the bio-polymers [62]. Although the literature mentioning their application on metals is quite unusual, bio-based systems appear as perfectly fitting it.

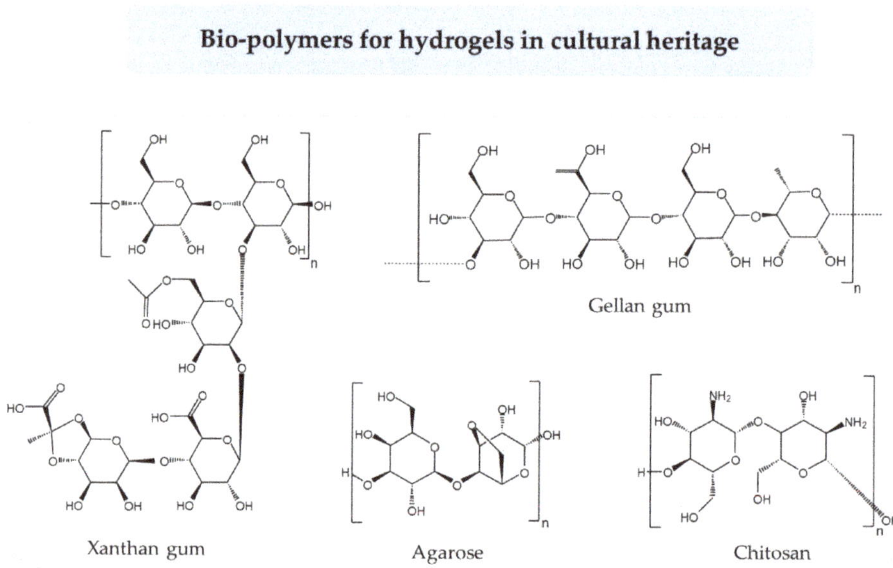

Figure 1. Bio-polymers used for hydrogel formulations in cultural heritage: structural formula of xanthan gum, gellan gum, agarose and chitosan.

3.1. *Physical Hydrogels*

3.1.1. Agar

Rigid gels are valid delivery systems due to their capability to retain solvents that allow to focus the cleaning on the uppermost layer, thereby limiting negative actions on eventual sensitive underlying materials. Moreover, their compact consistency makes them appreciated for the easy removability or even peelability [63]. However, their main advantage is also their main weakness. Being rigid, it is difficult to achieve good performances on rough substrates where the contact gel-surface is difficult to maintain. In art conservation, the leading thickening agent in this category is uncontestably agar. Agar is a gelling powder, extracted from red seaweed membranes, in particular *Gelidium* sp., *Pterocladia* sp. and *Gracilaria* sp. [64], which forms a rigid hydrogel after being heated above 90 °C and cooled down [63]. Agar is usually prepared in concentration from 2 to 6% *w*/*v*.

Agar is broadly available, with various purity and composition according to different assigned purposes in the food industry, biomedicine, or art conservation [65,66]. The primary compound is agarose, which is more often used impurified together with agaropectin, being less expensive [63].

The heating process can be a turn-off for conservators, especially when they are required to work out of their lab, with no heating devices. In addition, an agar gel cannot be formed with acidic or basic solutions, which, however, can be absorbed by immersion once the gel is created. Finally, it gives particularly satisfying results when directly applied hot on the target, as in a liquid state, it still has the chance to conform to the artifact surface [55].

A gel called Nevek® has been recently developed and is mentioned as particularly suitable for the cleaning of metallic surfaces [55]. It is a ready-to-use 5% double gellified agar with the addition of 5% ethanol to avoid microbial growth during storage.

In previous literature, several methodologies based on agar were tested on metals. In the case of archaeological silver horns, an agar gel was prepared and impregnated with the active compound before application [67]. Meanwhile, on Islamic inlays copper-alloys, agar powder was mixed to a complexing solution, buffered at pH 7, and then warmed up to gel [68]. In 2013, an agarose rigid gel was successfully exploited on Japanese lacquer objects, to face the complex combination of metals present, namely tin, lead, and iron. The use of a hydrogel system addressed the electrolytic reduction treatment exclusively on the corroded areas, without affecting and altering the surrounding water-sensitive lacquers [69]. Recently agar formulations resulted to be efficient carriers for the electrochemical cleaning of silver threads and silver leaves [70,71]. However, they were not performing at the best on non-flat objects due to their rigidity and thus their tendency to crack [71].

Still, attention should be drawn to the recent shortage of agar supplies, because of its increasing exploitation in numerous fields and the unpredictability of its production due to the dependence on marine conditions (i.e., warmth, infesters). Although the usage of agar and agarose is moderate in art conservation compared to other sectors, it is important to be aware of these issues and try to act accordingly, by choosing other gels whenever possible [64].

3.1.2. Gellan Gum

Gellan gum is a polysaccharide produced by the bacterium *Sphingomonas elodea* [66]. Similar to agar, it forms a rigid gel, compatible with impregnation of acidic and basic solutions once prepared and best-performing if applied warm [55]. Though some references mention the incompatibility of gellan gum with complexing agents [68].

A commercial derivative-form, the Phytagel®, which can be prepared cold, has been used on copper artifacts [72]. It was selected thanks to its adherence on vertical surfaces, ease of removal, and compatibility with the chosen active agents (i.e., microorganisms). In addition to the transparency considered an asset as it allowed to observe and monitor the treatment, its gelling properties are displayed even with low concentration (0.5 to 5 g/L) and are stable in a wide pH range [72].

3.1.3. Xanthan Gum

Xanthan gum is a thickening agent biosynthesized by the bacterium *Xanthomonas Campestris* through the fermentation of sucrose or glucose [61]. It is compatible with aqueous solutions in a wide range of pH, although appears to be less effective in gelling basic solutions [55]. The inherent high viscosity makes its removal complicated from metallic substrates [55,68], requiring the use of water afterward for a perfect elimination, which would annul the primary goal of the use of gels.

Although never tried out on metallic substrates, the removal of Paraloid® B72, the most common coating for metals in CH [20], was achieved on a painting thanks to Vanzan® N-FC, a commercially available xanthan gum, with only 5% of solvent in the liquid aqueous matrix. This is an encouraging step towards the removal of acrylic coatings on metals with the use of aqueous natural gels that have the advantage of being safer, both for the environment and conservators [73].

3.1.4. Chitosan

Chitosan is the N-deacetylated form of chitin, which is, in turn, the most abundant animal polysaccharide present in the exoskeleton of crustaceans and the cell wall of fungi [61]. Currently, chitosan is exploited mostly for drug delivery [61]. In general, it can gel in presence of both alkaline and acidic solutions [74].

In art conservation, it is rather uncommon, but it was used to develop chitosan-based coatings for corrosion mitigation, of copper alloys [25]. The research aimed to design a non-toxic, sustainable, easily-applicable, and, whenever necessary, removable matrix to incorporate corrosion inhibitors for copper—mercaptobenzothiazole (MBT) and BTA [25]. Remarkably, the chitosan polymer allowed the presence of a reduced quantity of toxic inhibitors and diminished their leaching in the environment. The transparent and colorless appearance of the chitosan-based lacquer meets the aesthetic standards in metal care. In addition, it showed a great adhesion on metals [25]. Furthermore, the development of chitosan-based active protective coatings has been carried out for aluminum alloys as well, by doping the matrix with either Ce^{3+} ions or MBT [75,76]. In CH, this could be of particular interest for aeronautical heritage, considering for instance relics from World War II [77].

Finally, chitosan-hydrogels have been investigated for their capacity to be loaded, and thereafter release BTA on metallic surfaces [78].

3.2. Chemical Hydrogels

Lai et al. achieved chemical cross-linking of a bio-based polymer containing chitosan, L-cysteine, and itaconic anhydride. Interestingly, the end-product presents metallic ions uptake properties. The synthesis is described as simple: after mortar-grounding the three components, the mixture is placed into a mold and left in immersion in an acetic acid solution for twenty-four hours [79].

Chemical hydrogels can be easily realized using the so-called Michael addition reaction [61,79]. The covalent bonding between the different polymer chains is hence obtained thanks to amino groups [79]. As opposed to the chemical hydrogels formulated by Baglioni et al. [39,44,57], these natural chemical hydrogels come from renewable sources and are biodegradable [80]. These qualities combined with the inherently favorable consistency of chemical gels, make them promising actors for the treatment of metal surfaces in cultural heritage.

4. Organogels for Organic Coatings Removal

Nowadays organogels as cleaning systems have been explored mostly on paintings and graphic documents, being multi-component artifacts and strongly sensitive to solvents [81–83]. On the contrary, in the literature rare is evidence of their application in the branch of metal conservation [84,85]. The main critical aspect of such implementation appears to be the rigidity of such systems. Most of the time, indeed, it implicates a non-even contact and adhesion on the surface, which, for metals, can be complex in terms of shape (e.g., curves, vertical position) and morphology (e.g., due to corrosion). However, the high modularity of gels eases the feasibility of satisfactory results on metals alike on other substrates.

Organogels in CH are frequently designed with non-green polymers and/or solvents [82,83,85,86]. Indeed, little is the attention addressed to the development of sustainable solvent-gel formulations, which can be achieved by the selection of specific green solvents (Sections 4.1 and 4.2) and thickening agents (Section 4.3.1). Despite the clear principle of assembling two components to form a gel, these components must be compatible to create well-polymerized gels that are satisfactory in terms of stability and performance [39].

The acceptance of green organogels should be addressed to the sustainability of all the components employed (i.e., polymer matrix, solvents, and additional constituents). Utmost care should be paid in the choice of the materials, considering features ranging from the original derivation of the sources to their recyclability and degradability [87–89].

4.1. Solvent Greenness

Solvents are indispensable chemical tools in art conservation, as much as in a broad spectrum of sectors: pharmaceutics, fuel, food industry, household products, cosmetics, paints, etc. In other words, solvents are part of everyday life. Therefore, it appears obvious to put these substances under review, not only for possible consequences on direct users but also for the potential impact on a wider scale in the environment, and consequently, on global health. The topic is crucial and reached the attention of renowned institutions such as the World Health Organization [90] that pointed out problems derived from certain substances and drew up localized restrictions on their use, to mitigate the derived hazard.

Besides the spheres of human and nature health, in the field of cultural heritage, the evaluation of the potential harm on the pieces of art themselves is crucial, even overriding compared to the abovementioned factors.

From this perspective, it is legitimate that there is growing concern around turning the page from several commonly used solvents towards the so-called "green" solvents. This term gathers a wide range of greener alternatives that can be selected and recognized using twelve factors listed in the journal *Green Chemistry* [91]. They face aspects such as renewable origin, recyclability, relative energy consumption and biodegradability; toxicity and flammability; availability on a large scale, and affordability; but most importantly, they should display properties and performances analogous to traditional solvents.

Hence, water could be ascribed among the greenest solvents, being fully human and environmentally harmless, readily supplied, and not expensive. Nonetheless, several reactions cannot be faced with water alone due to solubility factors. In addition, for metals, it is important to remind that water is a key player in the spontaneous natural process of corrosion, implying the willingness of substituting it with more substrate-friendly organic alternatives.

Besides water, an organic solvent satisfying even a few of the Green Chemistry principles can be rated as "green" weighed against its conventional counterparts. Thus, a proper comparative assessment should be carried out, considering the three macro-subjects of health, safety, and nature, according to one of the numerous scoring-guides proposed in the literature [90,92].

4.2. Bio-Based Solvents

Solvents obtained from biomass are named "bio-solvents". They are produced from readily renewable feedstocks, by fermentation or chemical transformation of biomass derivatives [90]. This implies the valorization of sources otherwise considered as wastes, increasing the greenness score of this class of chemicals.

Due to the strong lack of previous literature regarding bio-solvents in the field of metal heritage, the present review proposes a reasoned selection of bio-solvents.

4.2.1. Bioethanol and Derivatives

Ethanol is a ubiquitous solvent in several fields, in cultural heritage it can be used to work with natural resins and varnishes.

While the petrochemical industry produces ethanol by ethylene hydration, the bio-based version is naturally obtained from yeast's metabolic activity from agricultural wastes rich in sugars and lignocellulosic materials [90]. Its production via fermentation allows benefiting from sources judged as discards conversely, hence accomplishing the green principle of sustainable origin. Therefore, bioethanol is a perfect example of a drop-in eco-friendly replacement of possible large spread in any area of application since it is a well-known solvent with a green origin. Because of its evident identical features to ethanol, bioethanol is one of the most popular bio-based solvents and alcohols.

Furthermore, bioethanol can be employed as raw material for the synthesis of other organic solvents such as acetone and ethyl acetate [90]. The bio-derived acetone is free from some non-renewable contaminants, such as benzene, phthalates, and other phenols, possibly coming from petroleum [93]. On the other side, ethyl acetate has begun to be selected as a favored choice to replace several polar solvents [94], including acetone itself. In art conservation, in general, ethyl acetate is used normally to dissolve synthetic lacquers (e.g., acrylic- and nitrocellulose-based), guaranteeing low vapor pressure, toxicity and cost, in addition to a pleasant odor which is leading it to a massive use in the cosmetic industry [95].

4.2.2. n-Butyl Alcohol and Derivatives

n-Butyl alcohol (or n-butanol) is a colorless solvent, which, apart from a petrochemical origin, can be bio-sourced therefore acquiring the title of "bio-butanol". It can be produced by bacterial fermentation of renewable feedstocks sugar- or lignocellulose-storing and, remarkably nowadays, this green process is economically equal to the non-sustainable counterpart [90].

The inherent low acute toxicity makes n-butyl alcohol accepted in the food and personal care industry, as well as in the field of art conservation [90,96]. Indeed, its efficacy as a lacquer solvent is well-known for example on hard copal and shellac. However, its fumes can have a toxic effect by lengthened inhalation; thus, despite the bio-origin, this solvent is harmful under the irrevocable criterion of human health. Nevertheless, its employment is not only limited as a solvating agent but also in the manner of raw material to produce other widely used substances, including n-butyl acrylate and n-butyl acetate.

n-Butyl acrylate is widely used as a precursor to several synthetic binders and adhesives, used in contemporary art and conservation praxis [97,98]. On the other hand, n-butyl acetate is well-recognized as a strong solvent for cellulose nitrate-based and other synthetic resins, and most interestingly its vapors are considered not toxic [96]. As confirmation, it is an active compound in personal care goods and household cleaners, as well as an artificial flavoring in the food industry [90].

The employment of n-butyl acetate for the formulation of cleaning organogels is noteworthy. A series of methyl methacrylate-based (MMA) gels has been loaded with various organic solvents and performances compared for precise removal of unwanted adhesives and varnishes from easel paintings [82].

4.2.3. Ethyl Lactate

Lactates are progressively becoming protagonists in numerous industrial sectors as environmentally-benign solvents [99]. They derive from lactic acid, which can be manufactured by renewable technologies engaging biomass as a primary source. Depending on the microorganisms involved, it can be obtained either as pure (R)- or (S)- enantiomer or as a racemate through fermentation of hexoses and other carbohydrates [100].

Because of the chemical properties and the natural derivation, the lactates family is gaining attention to substitute more hazardous and toxic substances like toluene, xylene, or methyl ethyl ketone (MEK) [101,102]. Besides the clear green impact on the formulations and syntheses, for which these solvents were normally necessary, the replacement with lactates ensures a dually safer workplace and final goods for the health of operators and end-users [103].

Among the most employed lactates, besides propyl and butyl, ethyl lactate (ethyl 2-hydroxypropanoate) stands out. It accomplishes many green principles [99]: it is produced by esterification of bio-lactic acid, hence its origin is sustainable and renewable; it has low production cost, low toxicity, and readily biodegradability into water and carbon dioxide [103,104].

As a solvent, ethyl lactate appears a reliable alternative due to its low vapor pressure, high boiling point, and ability to dissolve both polar and non-polar species [102,104]. Thus, its features make it apt for the formulation of industrial dyes, paints, and varnishes [90,102].

Appealed by these assets, several research and articles have been published on the potential insertion of ethyl lactate in the broad field of art conservation.

Compared to other solvents for the preparation of acrylic resins—i.e., Paraloid® B72 and B44 –, ethyl lactate is not sufficiently volatile for this purpose [105]. The fact that it remains partially trapped inside the adhesive itself has posed concerns on the potential consequences on the mechanical properties of the film and the treated artifact.

On the other side, quite a few studies have been carried out on its cleaning efficacy. Its value is proved and recognized in the removal of graffiti inks from underlying original supports [106]. Notably, ethyl lactate demonstrated not only a good solvent strength on graffiti but also a satisfactory selectivity towards specific binders without impacting the subjacent original paint. As a confirmation of its well-noted efficacy, it is reported that ethyl lactate is bulkily sold to companies for the cleaning of graffiti, composites, and analogs [90].

Similarly, ethyl lactate led to satisfactory removal of natural terpenic varnishes from oil paintings [107]. Remarkably, in this study, the solvent was retained into a bio-derived gel system to face the low volatility of ethyl lactate and mitigate the quantity of possible residues on the treated surface.

4.2.4. Gamma-Valerolactone

γ-valerolactone (GVL) is one of the most common lactones. It is a natural polar chemical that can be obtained from levulinic acid, eco-friendly derived from lignocellulosic and carbohydrates biomass [90,108]. Besides being bio-sourced, the spectrum of its green features also includes biodegradability and low toxicity. This latter quality brought GVL to become a valid candidate to replace dipolar aprotic solvents nowadays outlawed because of their toxicity [108]. In the industrial world, γ-valerolactone is experimented as a green fossil-fuel additive and it is notable for its use in food and cosmetic goods because of its safety and pleasant scent [99].

In the perspective of metal conservation, GVL's polarity makes it a suitable solvent to treat equally polar media. In parallel, the very low volatility and the non-toxicity shows compelling advantages for restorers and conservators, yielding a safer workplace with scarce volatile organic compounds (VOC) emissions.

Thanks to these characteristics, a few years ago, γ-valerolactone was exploited for the first time in the field of cultural heritage. The research involved the cleaning of paintings through the removal of natural (i.e., dammar) and synthetic (i.e., Paraloid®) varnishes from the surface [81]. Due to its low vapor pressure and potential harmfulness on such sensitive works of art [107], γ-valerolactone was applied in a gelled formulation, to better control its activity and release. The encouraging results have recently driven research to further investigate the potentiality of this solvent in art conservation with the use of a cutting-edge delivery material [109].

4.2.5. Biodiesel and Derivatives

To obviate the traditional reliance on fossil-fuels, biodiesel has been developed as an innovative green option. Nowadays it can be extracted from vegetable oils or algae, labeling it as renewable, bio-sourced, and biodegradable [90,99].

From the chemical point of view, biodiesel is a blend of alkyl esters and long-chain fatty acids [84]: this makes it an appropriate solvent for non-polar substances.

In the field of metal conservation, biodiesel evidenced the capability of acting as a cleaning agent for the removal of wax from a bronze sculpture [84]. Regrettably, besides the solvent efficiency on non-polar coatings, the research pointed out the non-volatility of biodiesel, which appears as the main drawback of such application. Even if englobed in a gel system, the solvent tends towards leaving residues on the artifact. Given the fact that biodiesel could be a potential carbon source for microorganisms, eventual biodegrading bacteria and fungi present could induce a biodeterioration of organic materials such as canvas, textile, or wood. Therefore, despite its satisfactory application in this specific case, this protocol gives the impression of being unadvisable for the more sensitive substrates mentioned above, no less than metals in a negative state of conservation (e.g., highly-porous corroded).

Derived as a by-product of biodiesel, glycerol arises as a promising green solvent: renewable, biodegradable, non-volatile, non-toxic, and affordable [90,99]. It is classified as an intermediately polar solvent [108]; however, its use in industrial activities and reactions have been often limited due to its high viscosity [99]. Even in combination with choline chloride, creating a Deep Eutectic Solvent (DES) [90], the derived chemical is still too viscous for certain applications [99], in addition to the inherent non-volatility of DES systems.

In art conservation, it used to be employed chiefly in paintings as a moistening and plasticizing medium [96]. However, for the same reasons narrowing down its application in industry, glycerol does not seem an appealing solvent in this field. Indeed, the non-volatility, one of its main green advantages, turns into a weakness in conservation practices, being unable to evaporate from the dissolved substance or the targeted surface.

4.3. *Bio-Based Organogels for Cleaning*

4.3.1. Biopolymers

Biopolymers are naturally occurring polymers (e.g., polyhydroxyalkanoates—PHAs) or derived from bio-based monomers (e.g., polylactide—PLA) [110]. Their most evident green quality is that starting resources are not petroleum-based. Biodegradability is not implicit, indeed only those biologically synthesized are inherently biodegradable. That is the reason for the increasing number of publications on life cycle analysis (LCA) of polymers [87].

Biopolymers appear as valid candidates to replace petrochemical counterparts; however, the generally high cost of production draws some boundaries for wider diffusion. They are broadly employed for biomedical applications (e.g., suture threads [111], drug carriers [112]) because of their natural deterioration and human-compatibility, without any side effects. However, they have also spread in fields ranging from food industry to waste management, passing through cosmetics and household products [113–117].

To develop reliable bio-organogels for metal conservation, a deep investigation on these innovative polymers should be carried out to assess their suitability as thickening agents and the fulfillment of conservation ethics criteria. Regrettably, limited research was carried out in this field and to the best of the authors' knowledge, only one biopolymer has been explored so far: polyhydroxybutyrate (PHB).

PHB belongs to the family of polyhydroxyalkanoates (PHAs), and therefore, to the esters class. PHAs can be produced by more than 300 bacteria and archaea species, in both oxic or anoxic environments. The synthesis occurs either during the growth phase or under stress conditions (e.g., depletion of vital nutrients) while in excess of carbon [118]. PHA is hoarded as intracellular granules in the cytoplasm and kept as energy storage.

PHAs generate high appeal in many industrial sectors thanks to their inherent chemo-physical features, like well-known and spread petrochemical polymers such as polypropylene (PP) and polystyrene (PS). Notably, PHAs' natural degradation into carbon dioxide, water, and biomass requires about two months [119]. Nonetheless, as reported in up-to-date research, the current cost of PHAs is nearly six times higher compared to fossil-sourced plastics, clearly restricting these biopolymers to a wider use [118].

PHB is a semi-crystalline polymer synthesized by many microorganisms (e.g., *Alcaligenes* spp., *Bacillus* spp., *Azotobacter* spp., *Pseudomonas* spp.) from renewable feedstocks rich in sugars [120,121]. From a green perspective, its core advantages are bio-origin, biodegradability, and biocompatibility. From a hands-on point of view, it owns thermoplastic processability and solubility in polar solvents but not in water [122]. In the sporadic applications in art conservation, the form poly-3-hydroxybutyrate (P3HB) has been chosen [84,107,123,124] (Figure 2).

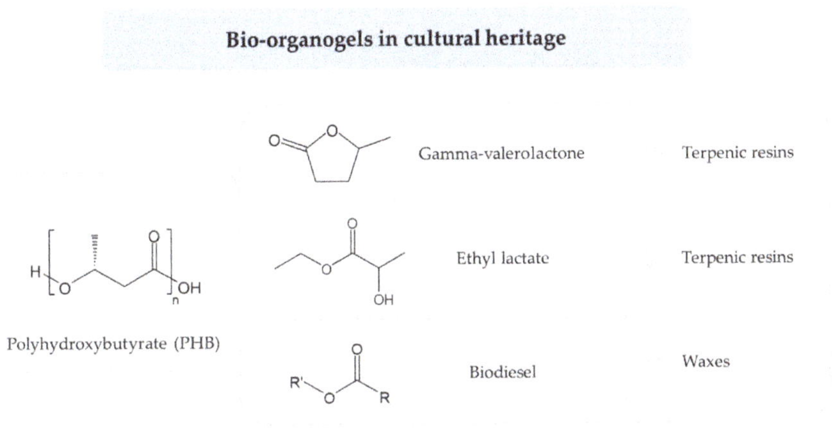

Figure 2. Bio-organogels (i.e., bio-polymer and bio-solvents) employed in cultural heritage and the respective organic coatings targeted. The structural formula of polyhydroxybutyrate, γ-valerolactone, ethyl lactate, and biodiesel.

4.3.2. Bio-Organogels in Art Conservation

In the first case study here reported, PHB was employed together with alginic acid to create a functionalized organogel for the treatment of archaeological wooden findings [123]. The core idea was the addition of iron chelators to the gel, to examine its potentialities to develop innovative in situ applications. Regrettably, to enhance polyhydroxybutyrate solubility, chloroform was employed to obtain lower molar mass polymer chains. This intermediate step would represent a critical drawback for large-scale use in conservation, due to the severe health danger associated with chloroform [125]. Its use and the unclear outcomes highlighted the need of improving the preparation protocol and exploring different chelating agents.

The first fully bio-based organogel in CH was made from polyhydroxybutyrate (PHB) as a thickening agent and γ-valerolactone (GVL) as an organic solvent, with or without the addition of triethyl citrate (TEC) as a plasticizer to test the different mechanical properties of the prepared gels [124]. The resulting PHB-GVL and PHB-GVL-TEC systems led to the removal of terpenic resin from oil paintings, leaving only little residues of GVL that did not appear concerning in terms of retention into underlying layers. Finally, no improvements were observed with the addition of the TEC plasticizer, making feasible and reliable the formulation with only PHB and GVL components [81].

However, the unwanted deposition of solvent on the object is a weakness, especially when referring to rigid gels. It is a crucial point that attracted the attention of supplementary studies [126]. An additional implementation of the PHB-GVL formulation has recently been published and involves the use of cutting-edge electrospun material, interposed between gel and painting, able to further minimize the residues of γ-valerolactone (GVL) on the target [109].

Another PHB-organogel was designed exploiting ethyl lactate (EL) as solvent. Indeed, the bio-based solvent was able to dissolve the polyhydroxybutyrate, thereby creating the so-called PHB-EL gelled system [107]. The obtained formulation performed satisfactorily to remove dammar from the surface of oil paintings [107]. Notably, its action was comparable both to traditionally applied free solvents and to the PHB-gel carrying dimethyl carbonate (DMC) as a green—but not bio-based—alternative solvent. The drop in mechanical stress is remarkable relating to ethyl lactate englobed in the gel or used neat for the same purpose of cleaning. Still impressive is the little amount of EL residues after treatment despite its inherent low vapor pressure, ascertained by solid-phase microextraction (SPME) analysis [107].

Finally, the last fully bio-derived polyhydroxybutyrate-gel involves biodiesel (BD) as a key solvent [84]. As biodiesel itself is not able to create a gel out of PHB, the formulation included also dimethyl carbonate (DMC) to dissolve the PHB polymer. Interestingly, the PHB-DMC/BD system was the only one applied on bronze objects coated with waxes. Previous tests demonstrated that dimethyl carbonate alone does not affect non-polar materials (i.e., wax); therefore, the sole action of biodiesel for the removal of such coating was proved. However, the research points out also the well-known non-volatility of biodiesel, which appears as the main drawback of such a procedure. Indeed, even if retained in a gelled system, biodiesel tends towards leaving residues on the treated surface, implicating an additional cleaning with free DMC to remove these undesired remains.

5. Future Perspectives

In 2020, the authors of this review conducted a survey, spread through numerous worldwide networks to ask conservators about their normal practices in metal heritage preservation. Remarkably, despite the global trend into more sustainable products, no feedback reported the use of green or bio-solvents, as well as the exploitation of gels for the treatment of neither the metallic substrates themselves nor the protective organic coatings present. Numerous were the factors deducible to explain the rare use for the bio-conservation of metals, including the relatively high price, the unawareness of either risk related to traditional methods or existence of reliable alternatives, and the effort required to change long-established habits in favor of unprecedented methodologies.

Therefore, for the extensive practice of bio-based gels in metal conservation, the first crucial point appears the close cooperation between scientists and stakeholders, to formulate and present ground-breaking safe alternatives that conservators will be prone to rely upon. Furthermore, the proposed bio-gels should have a non-prohibitive cost and be easily set up by conservators in their workplaces or supplied by research labs.

In light of the current state of research on biologically derived solutions for art conservation, it is not far-fetched to say that completely eco-friendly, user-harmless, and metal-safe gel formulations might be available soon for cultural heritage professionals.

Author Contributions: Conceptualization, E.J., E.G., and G.S.; methodology, A.P., L.C., E.G., G.S., and E.J.; resources, E.G. and G.S.; data curation, A.P., and L.C.; writing—original draft preparation, A.P., L.C., and E.J.; supervision, E.J., E.G., and G.S.; project administration, E.J.; funding acquisition, E.J. All authors have read and agreed to the published version of the manuscript.

Funding: This research was funded by the Swiss National Science Foundation (SNSF); grant number 205121_188755, HELIX (Investigating metal bioremediation for the preservation of historical metal artworks) project 2020–2024.

Institutional Review Board Statement: Not applicable.

Informed Consent Statement: Not applicable.

Data Availability Statement: Not applicable.

Acknowledgments: The authors acknowledge Stephan von Reuss for his co-supervision of the Ph.D. projects of Luana Cuvillier and Arianna Passaretti.

Conflicts of Interest: The authors declare no conflict of interest.

References

1. Kristiansen, K.; Larsson, T.B. L'âge du Bronze, une période historique. *Ann. Hist. Sci. Soc.* **2005**, *60*, 975–1007. [CrossRef]
2. Bertolotti, G.; Bersani, D.; Lottici, P.P.; Alesiani, M.; Malcherek, T.; Schlüter, J. Micro-Raman study of copper hydroxychlorides and other corrosion products of bronze samples mimicking archaeological coins. *Anal. Bioanal. Chem.* **2012**, *402*, 1451–1457. [CrossRef]
3. Felix, V.S.; Pereira, M.O.; Freitas, R.P.; Aranha, P.J.M.; Heringer, P.C.S.; Anjos, M.J.; Lopes, R.T. Analysis of silver coins from colonial Brazil by hand held XRF and micro-XRF. *Appl. Radiat. Isotopes* **2020**, *166*, 109409. [CrossRef]
4. Gharib, A.; Mohamed, H.; Abdel Ghany, N. Nondestructive techniques in the study of a gilded metallic sword from the Islamic Art Museum. *Egypt. J. Archaeol. Restor. Stud.* **2018**, *8*, 15–21.

5. Giumlia-Mair, A.; Lucchini, E. Surface analyses on modern and ancient copper based fakes. *Surf. Eng.* **2005**, *21*, 406–410. [CrossRef]
6. Chiavari, C.; Martini, C.; Prandstraller, D.; Niklasson, A.; Johansson, L.-G.; Svensson, J.-E.; Åslund, A.; Bergsten, C.J. Atmospheric corrosion of historical organ pipes: The influence of environment and materials. *Corros. Sci.* **2008**, *50*, 2444–2455. [CrossRef]
7. Lehmann, E.H.; Hartmann, S.; Speidel, M.O. Investigation of the content of ancient tibetan metallic buddha statues by means of neutron imaging methods. *Archaeometry* **2009**, *52*, 416–428. [CrossRef]
8. Chilton, J.P. The Corrosion of Metals. *J. R. Soc. Arts* **1971**, *119*, 614–629.
9. Amirudin, A.; Thieny, D. Application of electrochemical impedance spectroscopy to study the degradation of polymer-coated metals. *Prog. Org. Coat.* **1995**, *26*, 1–28. [CrossRef]
10. Beech, I.B.; Sunner, J. Biocorrosion: Towards understanding interactions between biofilms and metals. *Curr. Opin. Biotechnol.* **2004**, *15*, 181–186. [CrossRef]
11. Neff, D.; Reguer, S.; Bellot-Gurlet, L.; Dillmann, P.; Bertholon, R. Structural characterization of corrosion products on archaeological iron: An integrated analytical approach to establish corrosion forms. *J. Raman Spectrosc.* **2004**, *35*, 739–745. [CrossRef]
12. Baglioni, P.; Chelazzi, D.; Giorgi, R.; Poggi, G. Colloid and Materials Science for the Conservation of Cultural Heritage: Cleaning, Consolidation, and Deacidification. *Langmuir* **2013**, *29*, 5110–5122. [CrossRef] [PubMed]
13. Turner-Walker, G. *A Practical Guide to the Care and Conservation of Metals*; Headquarters Administration of Cultural Heritage, Council for Cultural Affairs: Taichung City, Taiwan, 2008; ISBN 9789860172980.
14. Comensoli, L.; Maillard, J.; Albini, M.; Sandoz, F.; Junier, P.; Joseph, E. Use of Bacteria To Stabilize Archaeological Iron. *Appl. Environ. Microbiol.* **2017**, *83*, 1–14. [CrossRef]
15. Albini, M.; Letardi, P.; Mathys, L.; Brambilla, L.; Schröter, J.; Junier, P.; Joseph, E. Comparison of a bio-based corrosion inhibitor versus benzotriazole on corroded copper surfaces. *Corros. Sci.* **2018**, *143*, 84–92. [CrossRef]
16. Guaragnone, T.; Casini, A.; Chelazzi, D.; Giorgi, R. PVA-based peelable films loaded with tetraethylenepentamine for the removal of corrosion products from bronze. *Appl. Mater. Today* **2020**, *19*, 100549. [CrossRef]
17. Joseph, E.; Simon, A.; Mazzeo, R.; Job, D.; Wörle, M. Spectroscopic characterization of an innovative biological treatment for corroded metal artefacts. *J. Raman Spectrosc.* **2012**, *43*, 1612–1616. [CrossRef]
18. Palomar, T.; Ramírez Barat, B.; Cano, E. Evaluation of cleaning treatments for tarnished silver: The conservator's perspective. *Int. J. Conserv. Sci.* **2018**, *9*, 81–90. [CrossRef]
19. Watkinson, D. Preservation of Metallic Cultural Heritage. In *Shreir's Corrosion*; Elsevier: Amsterdam, The Netherlands, 2010; pp. 3307–3340. ISBN 9780444527882.
20. Argyropoulos, V.; Giannoulaki, M.; Michalakakos, G.P.; Siatou, A. A Survey of the Type of Corrosion Inhibitors and Protective Coatings Used for the Conservation of Metal Objects from Museum Collections in the Mediterranean Basin. In *Strategies for Saving our Cultural Heritage. Proceedings of the International Conference on Conservation Strategies for Saving Indoor Metallic Collections, Cairo (Egypt). TEI of Athens, Athens*; Academic Press: Cambridge, MA, USA, 2007; pp. 166–170.
21. Wolfe, J.; Grayburn, R. A review of the development and testing of Incralac lacquer. *J. Am. Inst. Conserv.* **2017**, *56*, 225–244. [CrossRef]
22. Perera, D.Y. Physical ageing of organic coatings. *Prog. Org. Coat.* **2003**, *47*, 61–76. [CrossRef]
23. Couture-Rigert, D.E.; Sirois, P.J.; Moffatt, E.A. An investigation into the cause of corrosion on indoor bronze sculpture. *Stud. Conserv.* **2012**, *57*, 142–163. [CrossRef]
24. Jaeger, T. Short Communication Removal of Paraffin Wax in the Re-treatment of Archaeological Iron. *J. Am. Inst. Conserv.* **2008**, *47*, 217–223. [CrossRef]
25. Giuliani, C.; Pascucci, M.; Riccucci, C.; Messina, E.; Salzano de Luna, M.; Lavorgna, M.; Ingo, G.M.; Di Carlo, G. Chitosan-based coatings for corrosion protection of copper-based alloys: A promising more sustainable approach for cultural heritage applications. *Prog. Org. Coat.* **2018**, *122*, 138–146. [CrossRef]
26. Mari, Y. Il Restauro Della Grande Croce del Pollaiolo: Un Intervento All'interno del Cantiere Organizzato per l'Altare di san Giovanni. In *E l'informe si fa forma...: studi intorno a Santa Maria del Fiore in ricordo di Patrizio Osticresi*; Mandragora: Florence, Italy, 2012; pp. 281–286. ISBN 9788874615346.
27. Cagnini, A.; Gennai, S.; Mazzoni, M.D. La Banderuola di Palazzo Vecchio: Storia, vicende conservative, restauro. *OPD Restauro* **2012**, *24*, 13–32.
28. Phenix, A. Effects of Organic Solvents on Artists Oil Paint Films: Swelling. In *New Insights into the Cleaning of Paintings: Proceedings from the Cleaning 2010 International Conference, Universidad Politecnica de Valencia and Museum Conservation Institute*; Smithsonian Institution: Washington, DC, USA, 2013. [CrossRef]
29. Varnai, V.M.; Macan, J.; Ljubicic Calusic, A.; Prester, L.; Kanceljak Macan, B. Upper respiratory impairment in restorers of cultural heritage. *Occup. Med.* **2011**, *61*, 45–52. [CrossRef]
30. Duan, W.; Meng, F.; Wang, F.; Liu, Q. Environmental behavior and eco-toxicity of xylene in aquatic environments: A review. *Ecotoxicol. Environ. Saf.* **2017**, *145*, 324–332. [CrossRef]
31. De Silva, M.; Henderson, J. Sustainability in conservation practice. *J. Inst. Conserv.* **2011**, *34*, 5–15. [CrossRef]
32. Kumar, R.; Sagar, P. Preface on Green Conservation of the Museum Objects. In *A treatise on Recent Trends and Sustainability in Crafts & Design*; Gupta, T., Mistry, B., Gupta, B.S., Eds.; Excel India: Mumbai, India, 2017; pp. 118–124. ISBN 978-93-86724-21-2.
33. Appelbaum, B. Criteria for Treatment: Reversibility. *J. Am. Inst. Conserv.* **1987**, *26*, 65–73. [CrossRef]

34. Brandi, C. *Teoria del Restauro*; Ed. di storia e letteratura: Rome, Italy, 1963.
35. Di Marino, D.; Shalaby, M.; Kriescher, S.; Wessling, M. Corrosion of metal electrodes in deep eutectic solvents. *Electrochem. Commun.* **2018**, *90*, 101–105. [CrossRef]
36. Heitz, E. Corrosion of Metals in Organic Solvents. In *Advances in Corrosion Science and Technology*; Springer: Boston, MA, USA, 1974; pp. 149–243. ISBN 9781461590590.
37. Brajer, I.; Fossé-Le Rouzic, M.; Shashoua, Y.; Taube, M.; Chelazzi, D.; Baglioni, M.; Baglioni, P. The removal of aged acrylic coatings from wall paintings using microemulsions. In *ICOM-CC 17th Triennial Conference Preprints, Melbourne, Australia, 15–19 September 2014*; Bridgland, J., Ed.; art. 1103; International Council of Museums: Paris, France, 2014; ISBN 9789290124108.
38. Chelazzi, D.; Giorgi, R.; Baglioni, P. Microemulsions, Micelles, and Functional Gels: How Colloids and Soft Matter Preserve Works of Art. *Angew. Chem. Int. Ed.* **2018**, *57*, 7296–7303. [CrossRef] [PubMed]
39. Baglioni, P.; Berti, D.; Bonini, M.; Carretti, E.; Dei, L.; Fratini, E.; Giorgi, R. Micelle, microemulsions, and gels for the conservation of cultural heritage. *Adv. Colloid Interface Sci.* **2014**, *205*, 361–371. [CrossRef]
40. Mazzoni, M.; Alisi, C.; Tasso, F.; Cecchini, A.; Marconi, P.; Sprocati, A.R. Laponite micro-packs for the selective cleaning of multiple coherent deposits on wall paintings: The case study of Casina Farnese on the Palatine Hill (Rome-Italy). *Int. Biodeterior. Biodegrad.* **2014**, *94*, 1–11. [CrossRef]
41. Malik, M.A.; Hashim, M.A.; Nabi, F.; AL-Thabaiti, S.A.; Khan, Z. Anti-corrosion ability of surfactants: A review. *Int. J. Electrochem. Sci.* **2011**, *6*, 1927–1948.
42. Fratini, E.; Carretti, E. CHAPTER 10. Cleaning IV: Gels and Polymeric Dispersions. In *Nanoscience for the Conservation of Works of Art*; Royal Society of Chemistry: London, UK, 2013; pp. 252–279. ISBN 978-1-84973-566-7.
43. Rogovina, L.Z.; Vasil'ev, V.G.; Braudo, E.E. Definition of the concept of polymer gel. *Polym. Sci. Ser. C* **2008**, *50*, 85–92. [CrossRef]
44. Baglioni, P.; Baglioni, M.; Bonelli, N.; Chelazzi, D.; Giorgi, R. Smart Soft Nanomaterials for Cleaning. In *Nanotechnologies and Nanomaterials for Diagnostic, Conservation and Restoration of Cultural Heritage*; Elsevier: Amsterdam, The Netherlands, 2019; pp. 171–204. [CrossRef]
45. Wolbers, R. Restoration'92: Conservation, training, materials and techniques: Latest developments. In Proceedings of the Preprints to the Conference Held at the RAI International Exhibition and Congress Centre, Amsterdam, The Netherlands, 20–22 October 1992; pp. 74–75, ISBN 9781871656183.
46. Wolbers, R. *Cleaning Painted Surfaces: Aqueous Methods*; Archetype Publications: London, UK, 2000; ISBN 1873132360.
47. Van Loon, A.; Hartman, L.E.; van den Burg, J.; Haswell, R.; Pottasch, C. *The Development of an Aqueous Gel Testing Procedure for the Removal of Lead-Rich Salt Crusts on the Surface of Paintings by Giovanni Antonio Pellegrini (1675–1741) in the "Golden Room" of the Mauritshuis*; Springer: Berlin/Heidelberg, Germany, 2019; pp. 283–296. [CrossRef]
48. Toreno, G.; Isola, D.; Meloni, P.; Carcangiu, G.; Selbmann, L.; Onofri, S.; Caneva, G.; Zucconi, L. Biological colonization on stone monuments: A new low impact cleaning method. *J. Cult. Herit.* **2018**, *30*, 100–109. [CrossRef]
49. Cushman, M.; Wolbers, R. A new approach to cleaning iron stained marble surfaces. *WAAC Newsl.* **2007**, *29*, 23–28.
50. Mazzuca, C.; Micheli, L.; Carbone, M.; Basoli, F.; Cervelli, E.; Iannuccelli, S.; Sotgiu, S.; Palleschi, A. Gellan hydrogel as a powerful tool in paper cleaning process: A detailed study. *J. Colloid Interface Sci.* **2014**, *416*, 205–211. [CrossRef] [PubMed]
51. Delattre, C.; Bouvet, S.; Le Bourg, E. Gellan gum and agar compared to aqueous immersion for cleaning paper. In *Gels in the Conservation of Art*; Angelova, L.V., Ormsby, B., Townsend, J.H., Wolbers, R., Eds.; Archetype Publications: London, UK, 2017; pp. 57–61. ISBN 9781909492509.
52. Guilminot, E.; Leroux, M.; Raimon, A.; Chalvidal, C. Projet collaboratif sur l'utilisation des gels pour le traitement des métaux: Démarche et fonctionnement. In Proceedings of the Journées des Restaurateurs en Archéologie, Journée d'Étude, de Recherche et d'Innovation Lyon, France, 29–30 November 2018; p. 14.
53. Baij, L.; Hermans, J.; Ormsby, B.; Noble, P.; Iedema, P.; Keune, K. A review of solvent action on oil paint. *Herit. Sci.* **2020**, *8*, 43. [CrossRef]
54. Duncan, T.T.; Berrie, B.H.; Weiss, R.G. A Comparison between Gel and Swab Cleaning: Physical Changes to Delicate Surfaces. In *Gels in the Conservation of Art*; Angelova, L.V., Ormsby, B., Townsend, J.H., Wolbers, R., Eds.; Archetype Publications: London, UK, 2017; pp. 250–256. ISBN 9781909492509.
55. Guilminot, E.; Gomez, A.; Raimon, A.; Leroux, M. Use of Gels for the treatment of Metals. In *Proceedings of the Metal 2019 Proceedings of the Interim Meeting of the ICOM-CC Metals Working Group*; Chemello, C., Brambilla, L., Joseph, E., Eds.; International Councils of Museums—Committee for Conservation: Paris, France, 2019; p. 473. ISBN 9789290124573.
56. Smith, S.S. Layer by layer: The removal of complex soiling on a collection of modern art bronzes using buffered pH-adjusted aquous gels. In *Gels in the Conservation of Art*; Angelova, L.V., Ormsby, B., Townsend, J.H., Wolbers, R., Eds.; Archetype Publications: London, UK, 2017; pp. 349–355. ISBN 9781909492509.
57. Parisi, E.I.; Bonelli, N.; Carretti, E.; Giorgi, R.; Ingo, G.M.; Baglioni, P. Film forming PVA-based cleaning systems for the removal of corrosion products from historical bronzes. *Pure Appl. Chem.* **2018**, *90*, 507–522. [CrossRef]
58. Feldman, D. Poly(Vinyl Alcohol) Recent Contributions to Engineering and Medicine. *J. Compos. Sci.* **2020**, *4*, 175. [CrossRef]
59. Chiellini, E.; Corti, A.; D'Antone, S.; Solaro, R. Biodegradation of poly (vinyl alcohol) based materials. *Prog. Polym. Sci.* **2003**, *28*, 963–1014. [CrossRef]
60. Duplat, V.; Rouchon, V.; Desloges, I.; Papillon, M.C. Steel versus Paper: The Conservation of a Piece of Modern Art Consisting of a Rust Print on Paper. *J. Pap. Conserv.* **2009**, *10*, 26–34.

61. Varghese, S.A.; Rangappa, S.M.; Siengchin, S.; Parameswaranpillai, J. Natural polymers and the hydrogels prepared from them. In *Hydrogels Based on Natural Polymers*; Elsevier: Amsterdam, The Netherlands, 2020; pp. 17–47. [CrossRef]
62. Dalvi-Isfahan, M.; Hamdami, N.; Le-Bail, A. Effect of freezing under electrostatic field on selected properties of an agar gel. *Innov. Food Sci. Emerg. Technol.* **2017**, *42*, 151–156. [CrossRef]
63. Cremonesi, P. Rigid Gels and Enzyme Cleaning. *Smithson. Contrib. Museum Conserv.* **2012**, *3*, 179–183.
64. Tamura, M.; Takagi, K. Towards the sustainable use of agar/agarose in conservation: A case study of the Izu peninsula, Japan. In *Gels in the Conservation of Art*; Angelova, L.V., Ormsby, B., Townsend, J.H., Wolbers, R., Eds.; Archetype Publications: London, UK, 2017; pp. 152–154. ISBN 9781909492509.
65. Bertasa, M.; Chiantore, O.; Poli, T.; Riedo, C.; Di Tullio, V.; Canevali, C.; Sansonetti, A.; Scalarone, D. A study of commercial agar gels as cleaning materials. In *Gels in the Conservation of Art*; Angelova, L.V., Ormsby, B., Townsend, J.H., Wolbers, R., Eds.; Archetype Publications: London, UK, 2017; pp. 11–18. ISBN 9781909492509.
66. Wolbers, R. Terminology and properties of selected gels. In *Gels in the Conservation of Art*; Angelova, L.V., Ormsby, B., Townsend, J.H., Wolbers, R., Eds.; Archetype Publications: London, UK, 2017; pp. 381–394. ISBN 9781909492509.
67. Marchand, G.; Chevallier, R.; Guilminot, E.; Rossetti, L.; Lemoine, S. Study of the conservation treatment applied to the archaeological horn silver artifacts. In *Proceedings of the Interim Meeting for the International Council of Museums Committee for Conservation Metal Working Group, Metal 2013*; International Councils of Museums—Committee for Conseravtion and Historic Scotland: Edinburgh, Scotland, 2013; pp. 245–250. ISBN 9781849171427.
68. Fays, M. *«D'or, D'argent et de Pate Noire: Incrustations Révélées» Étude et Conservation-Restauration de Cinq Objets Islamiques en Alliage Cuivreux Incrustés*; Institut National du Patrimoine: Paris, France, 2018.
69. Wolbers, R.; Rivers, S.; Yamashita, Y. Corroded applied lead-based decoration (hyomon) on Japanese lacquer: Principles and case studies. *Stud. Conserv.* **2014**, *59*, S191–S194. [CrossRef]
70. Létrange, A.; Hourdet, D.; Guerrier, J.; Pons, E. Comparison of three hydrogels for cleaning tarnished silver threads using electrochemical treatment. In *Gels in the Conservation of Art*; Angelova, L.V., Ormsby, B., Townsend, J.H., Wolbers, R., Eds.; Archetype Publications: London, UK, 2017; pp. 369–371. ISBN 9781909492509.
71. São João, J.; Branco, L.C.; Leite Fragoso, S. Trials fo agar gels and task-specific salts for the electrochemical reduction of silver sulphide on silver leaf. In *Gels in the Conservation of Art*; Angelova, L.V., Ormsby, B., Townsend, J.H., Wolbers, R., Eds.; Archetype Publications: London, UK, 2017; pp. 287–291. ISBN 9781909492509.
72. Domon Beuret, E.; Mathys, L.; Brambilla, L.; Albini, M.; Cevey, C.; Bertholon, R.; Junier, P.; Joseph, E. Biopatines: Des champignons au service des alliages cuivreux. In Proceedings of the Cahier n°22—XXVIIIe Journées des restaurateurs en archéologie, Arles, 16–17 October 2014; ARAAFU: Paris, France, October 2015.
73. Barberà Giné, A.; Marín Ortega, S. The removal of Paraloid B-72 coatings with aqueous gelled systems: Roman frescoesfrom Empúries, Catalonia. In *Gels in the Conservation of Art*; Angelova, L.V., Ormsby, B., Townsend, J.H., Wolbers, R., Eds.; Archetype Publications: London, UK, 2017; pp. 363–365. ISBN 9781909492509.
74. Nie, J.; Wang, Z.; Hu, Q. Difference between Chitosan Hydrogels via Alkaline and Acidic Solvent Systems. *Sci. Rep.* **2016**, *6*, 1–8. [CrossRef] [PubMed]
75. Carneiro, J.; Tedim, J.; Fernandes, S.C.M.; Freire, C.S.R.; Silvestre, A.J.D.; Gandini, A.; Ferreira, M.G.S.; Zheludkevich, M.L. Chitosan-based self-healing protective coatings doped with cerium nitrate for corrosion protection of aluminum alloy 2024. *Prog. Org. Coat.* **2012**, *75*, 8–13. [CrossRef]
76. Carneiro, J.; Tedim, J.; Fernandes, S.C.M.; Freire, C.S.R.; Gandini, A.; Ferreira, M.G.S.; Zheludkevich, M.L. Chitosan as a smart coating for controlled release of corrosion inhibitor 2-mercaptobenzothiazole. *ECS Electrochem. Lett.* **2013**, *2*. [CrossRef]
77. Brunet, M.; Cochard, A.; Deshayes, C.; Brouca-Cabarrecq, C.; Robbiola, L.; Olivier, J.-M.; Sciau, P. Study of Post-World War II French Aeronautical Aluminium Alloy and Coatings: Historical and Materials Science Approach. *Stud. Conserv.* **2020**, *65*, 103–117. [CrossRef]
78. Wang, Y.N.; Dong, C.F.; Zhang, D.W.; Ren, P.P.; Li, L.; Li, X.G. Preparation and characterization of a chitosan-based low-pH-sensitive intelligent corrosion inhibitor. *Int. J. Miner. Metall. Mater.* **2015**, *22*, 998–1004. [CrossRef]
79. Lai, H.; Liu, S.; Yan, J.; Xing, F.; Xiao, P. Facile Fabrication of Biobased Hydrogel from Natural Resources: L-Cysteine, Itaconic Anhydride, and Chitosan. *ACS Sustain. Chem. Eng.* **2020**. [CrossRef]
80. Murakami, S.; Aoki, N.; Matsumura, S. Bio-based biodegradable hydrogels prepared by crosslinking of microbial poly(γ-glutamic acid) with L-lysine in aqueous solution. *Polym. J.* **2011**, *43*, 414–420. [CrossRef]
81. Prati, S.; Volpi, F.; Fontana, R.; Galletti, P.; Giorgini, L.; Mazzeo, R.; Mazzocchetti, L.; Samorì, C.; Sciutto, G.; Tagliavini, E. Sustainability in art conservation: A novel bio-based organogel for the cleaning of water sensitive works of art. *Pure Appl. Chem.* **2018**, *90*, 239–251. [CrossRef]
82. Baglioni, P.; Bonelli, N.; Chelazzi, D.; Chevalier, A.; Dei, L.; Domingues, J.; Fratini, E.; Giorgi, R.; Martin, M. Organogel formulations for the cleaning of easel paintings. *Appl. Phys. A* **2015**, *121*, 857–868. [CrossRef]
83. Pianorsi, M.D.; Raudino, M.; Bonelli, N.; Chelazzi, D.; Giorgi, R.; Fratini, E.; Baglioni, P. Organogels for the cleaning of artifacts. *Pure Appl. Chem.* **2017**, *89*, 3–17. [CrossRef]
84. Yiming, J.; Sciutto, G.; Prati, S.; Catelli, E.; Galeotti, M.; Porcinai, S.; Mazzocchetti, L.; Samorì, C.; Galletti, P.; Giorgini, L.; et al. A new bio-based organogel for the removal of wax coating from indoor bronze surfaces. *Herit. Sci.* **2019**, *7*, 1–12. [CrossRef]

85. Duncan, T.T.; Berrie, B.H.; Weiss, R.G. Soft, Peelable Organogels from Partially Hydrolyzed Poly(vinyl acetate) and Benzene-1,4-diboronic Acid: Applications to Clean Works of Art. *ACS Appl. Mater. Interfaces* **2017**, *9*, 28069–28078. [CrossRef] [PubMed]
86. Chelazzi, D.; Fratini, E.; Giorgi, R.; Mastrangelo, R.; Rossi, M.; Baglioni, P. Gels for the Cleaning of Works of Art. In *Gels and Other Soft Amorphous Solids*; ACS Publications: Washington, DC, USA, 2018; pp. 291–314. [CrossRef]
87. Yates, M.R.; Barlow, C.Y. Life cycle assessments of biodegradable, commercial biopolymers—A critical review. *Resour. Conserv. Recycl.* **2013**, *78*, 54–66. [CrossRef]
88. Tabone, M.D.; Cregg, J.J.; Beckman, E.J.; Landis, A.E. Sustainability Metrics: Life Cycle Assessment and Green Design in Polymers. *Environ. Sci. Technol.* **2010**, *44*, 8264–8269. [CrossRef]
89. Jiménez-González, C.; Curzons, A.D.; Constable, D.J.C.; Cunningham, V.L. Expanding GSK's Solvent Selection Guide?application of life cycle assessment to enhance solvent selections. *Clean Technol. Environ. Policy* **2004**, *7*, 42–50. [CrossRef]
90. Calvo-Flores, F.G.; Monteagudo-Arrebola, M.J.; Dobado, J.A.; Isac-García, J. Green and Bio-Based Solvents. *Top. Curr. Chem.* **2018**, *376*, 18. [CrossRef]
91. Anastas, P.; Eghbali, N. Green Chemistry: Principles and Practice. *Chem. Soc. Rev.* **2010**, *39*, 301–312. [CrossRef] [PubMed]
92. Prat, D.; Hayler, J.; Wells, A. A survey of solvent selection guides. *Green Chem.* **2014**, *16*, 4546–4551. [CrossRef]
93. van der Wal, H.; Sperber, B.L.H.M.; Houweling-Tan, B.; Bakker, R.R.C.; Brandenburg, W.; López-Contreras, A.M. Production of acetone, butanol, and ethanol from biomass of the green seaweed Ulva lactuca. *Bioresour. Technol.* **2013**, *128*, 431–437. [CrossRef] [PubMed]
94. Capello, C.; Fischer, U.; Hungerbühler, K. What is a green solvent? A comprehensive framework for the environmental assessment of solvents. *Green Chem.* **2007**, *9*, 927. [CrossRef]
95. Lens, C.; Malet, G.; Cupferman, S. Antimicrobial activity of Butyl acetate, Ethyl acetate and Isopropyl alcohol on undesirable microorganisms in cosmetic products. *Int. J. Cosmet. Sci.* **2016**, *38*, 476–480. [CrossRef] [PubMed]
96. Gettens, R.J.; Stout, G.L. *Painting Materials. A Short Encyclopaedia*; Dover Publications, Inc.: New York, NY, USA, 1966; ISBN 9780486215976.
97. La Nasa, J.; Orsini, S.; Degano, I.; Rava, A.; Modugno, F.; Colombini, M.P. A chemical study of organic materials in three murals by Keith Haring: A comparison of painting techniques. *Microchem. J.* **2016**, *124*, 940–948. [CrossRef]
98. Chércoles Asensio, R.; San Andrés Moya, M.; de la Roja, J.M.; Gómez, M. Analytical characterization of polymers used in conservation and restoration by ATR-FTIR spectroscopy. *Anal. Bioanal. Chem.* **2009**, *395*, 2081–2096. [CrossRef]
99. Gu, Y.; Jérôme, F. Bio-based solvents: An emerging generation of fluids for the design of eco-efficient processes in catalysis and organic chemistry. *Chem. Soc. Rev.* **2013**, *42*, 9550–9570. [CrossRef]
100. Ghaffar, T.; Irshad, M.; Anwar, Z.; Aqil, T.; Zulifqar, Z.; Tariq, A.; Kamran, M.; Ehsan, N.; Mehmood, S. Recent trends in lactic acid biotechnology: A brief review on production to purification. *J. Radiat. Res. Appl. Sci.* **2014**, *7*, 222–229. [CrossRef]
101. Kerton, F.; Marriott, R. *Alternative Solvents for Green Chemistry*; Royal Society of Chemistry: London, UK, 2013; ISBN 9781849735957.
102. Nikles, S.M.; Piao, M.; Lane, A.M.; Nikles, D.E. Ethyl lactate: A green solvent for magnetic tape coating. *Green Chem.* **2001**, *3*, 109–113. [CrossRef]
103. Pereira, C.S.M.; Silva, V.M.T.M.; Rodrigues, A.E. Ethyl lactate as a solvent: Properties, applications and production processes—A review. *Green Chem.* **2011**, *13*, 2658. [CrossRef]
104. Kua, Y.L.; Gan, S.; Morris, A.; Ng, H.K. Ethyl lactate as a potential green solvent to extract hydrophilic (polar) and lipophilic (non-polar) phytonutrients simultaneously from fruit and vegetable by-products. *Sustain. Chem. Pharm.* **2016**, *4*, 21–31. [CrossRef]
105. Vinçotte, A.; Beauvoit, E.; Boyard, N.; Guilminot, E. Effect of solvent on PARALOID® B72 and B44 acrylic resins used as adhesives in conservation. *Herit. Sci.* **2019**, *7*, 42. [CrossRef]
106. Esson, J.M.; Scott, R.; Hayes, C.J. Chemistry and Art: Removal of Graffiti Ink from Paints Grounded in a Real-Life Scenario. *J. Chem. Educ.* **2018**, *95*, 400–402. [CrossRef]
107. Prati, S.; Sciutto, G.; Volpi, F.; Rehorn, C.; Vurro, R.; Blümich, B.; Mazzocchetti, L.; Giorgini, L.; Samorì, C.; Galletti, P.; et al. Cleaning oil paintings: NMR relaxometry and SPME to evaluate the effects of green solvents and innovative green gels. *New J. Chem.* **2019**, *43*, 8229–8238. [CrossRef]
108. Gao, F.; Bai, R.; Ferlin, F.; Vaccaro, L.; Li, M.; Gu, Y. Replacement strategies for non-green dipolar aprotic solvents. *Green Chem.* **2020**, *22*, 6240–6257. [CrossRef]
109. Jia, Y.; Sciutto, G.; Mazzeo, R.; Samorì, C.; Focarete, M.L.; Prati, S.; Gualandi, C. Organogel Coupled with Microstructured Electrospun Polymeric Nonwovens for the Effective Cleaning of Sensitive Surfaces. *ACS Appl. Mater. Interfaces* **2020**, *12*, 39620–39629. [CrossRef]
110. Kaplan, D.L. Introduction to Biopolymers from Renewable Resources. *Biopolym. Renew. Resour.* **1998**, 1–29. [CrossRef]
111. Knutson, C.M.; Schneiderman, D.K.; Yu, M.; Javner, C.H.; Distefano, M.D.; Wissinger, J.E. Polymeric Medical Sutures: An Exploration of Polymers and Green Chemistry. *J. Chem. Educ.* **2017**, *94*, 1761–1765. [CrossRef]
112. Li, Z.; Zhang, B.; Jia, S.; Ma, M.; Hao, J. Novel supramolecular organogel based on β-cyclodextrin as a green drug carrier for enhancing anticancer effects. *J. Mol. Liq.* **2018**, *250*, 19–25. [CrossRef]
113. Peelman, N.; Ragaert, P.; De Meulenaer, B.; Adons, D.; Peeters, R.; Cardon, L.; Van Impe, F.; Devlieghere, F. Application of bioplastics for food packaging. *Trends Food Sci. Technol.* **2013**, *32*, 128–141. [CrossRef]
114. Calabrò, P.S.; Grosso, M. Bioplastics and waste management. *Waste Manag.* **2018**, *78*, 800–801. [CrossRef] [PubMed]
115. Bayer, I.S.; Guzman-Puyol, S.; Heredia-Guerrero, J.A.; Ceseracciu, L.; Pignatelli, F.; Ruffilli, R.; Cingolani, R.; Athanassiou, A. Direct Transformation of Edible Vegetable Waste into Bioplastics. *Macromolecules* **2014**, *47*, 5135–5143. [CrossRef]

116. Onen Cinar, S.; Chong, Z.K.; Kucuker, M.A.; Wieczorek, N.; Cengiz, U.; Kuchta, K. Bioplastic Production from Microalgae: A Review. *Int. J. Environ. Res. Public Health* **2020**, *17*, 3842. [CrossRef] [PubMed]
117. Keshavarz, T.; Roy, I. Polyhydroxyalkanoates: Bioplastics with a green agenda. *Curr. Opin. Microbiol.* **2010**, *13*, 321–326. [CrossRef]
118. Khatami, K.; Perez-Zabaleta, M.; Owusu-Agyeman, I.; Cetecioglu, Z. Waste to bioplastics: How close are we to sustainable polyhydroxyalkanoates production? *Waste Manag.* **2021**, *119*, 374–388. [CrossRef] [PubMed]
119. Hassan, M.A.; Yee, L.-N.; Yee, P.L.; Ariffin, H.; Raha, A.R.; Shirai, Y.; Sudesh, K. Sustainable production of polyhydroxyalkanoates from renewable oil-palm biomass. *Biomass Bioenergy* **2013**, *50*, 1–9. [CrossRef]
120. Yilmaz, M.; Soran, H.; Beyatli, Y. Determination of poly-β-hydroxybutyrate (PHB) production by some *Bacillus* spp. *World J. Microbiol. Biotechnol.* **2005**, *21*, 565–566. [CrossRef]
121. Scalioni, L.V.; Gutiérrez, M.C.; Felisberti, M.I. Green composites of poly(3-hydroxybutyrate) and curaua fibers: Morphology and physical, thermal, and mechanical properties. *J. Appl. Polym. Sci.* **2017**, *134*, 1–13. [CrossRef]
122. Balaji, S.; Gopi, K.; Muthuvelan, B. A review on production of poly β hydroxybutyrates from cyanobacteria for the production of bio plastics. *Algal Res.* **2013**, *2*, 278–285. [CrossRef]
123. Walsh-Korb, Z.; Ruiz-Fourcade, S.; Avérous, L. Responsive bio-based gels for the preservation and treatment of archaeological wooden objects. In *Gels in the Conservation of Art*; Angelova, L.V., Ormsby, B., Townsend, J.H., Wolbers, R., Eds.; Archetype Publications: London, UK, 2017; pp. 294–296. ISBN 9781909492509.
124. Samorì, C.; Galletti, P.; Giorgini, L.; Mazzeo, R.; Mazzocchetti, L.; Prati, S.; Sciutto, G.; Volpi, F.; Tagliavini, E. The Green Attitude in Art Conservation: Polyhydroxybutyrate–based Gels for the Cleaning of Oil Paintings. *ChemistrySelect* **2016**, *1*, 4502–4508. [CrossRef]
125. Fawell, J. Risk assessment case study—Chloroform and related substances. *Food Chem. Toxicol.* **2000**, *38*, S91–S95. [CrossRef]
126. Baij, L.; Buijs, J.; Hermans, J.J.; Raven, L.; Iedema, P.D.; Keune, K.; Sprakel, J. Quantifying solvent action in oil paint using portable laser speckle imaging. *Sci. Rep.* **2020**, *10*, 10574. [CrossRef] [PubMed]